ELECTROMAGNETIC FIELDS IN ELECTRICAL ENGINEERING

ELECTROMAGNETIC FIELDS IN ELECTRICAL ENGINEERING

Edited by

A. Savini

University of Pavia
Pavia, Italy

and

J. Turowski

Technical University of Lodz
Lodz, Poland

PLENUM PRESS • NEW YORK AND LONDON

Library of Congress Cataloging in Publication Data

International Symposium on Electromagnetic Fields in Electrical Engineering (1987: Pavia, Italy)

Electromagnetic fields in electrical engineering / edited by A. Savini and J. Turowski.

p. cm.

"Proceedings of the International Symposium on Electromagnetic Fields in Electrical Engineering, held September 23-25, 1987, in Pavia, Italy"—T.p. verso.

Includes bibliographies and index.

ISBN 0-306-42992-6

1. Electric engineering—Congresses. 2. Electromagnetic fields—Congresses. I. Savini, A. II. Turowski, J. III. Title.

TK5.I724 1987 88-22672

621.3—dc19 CIP

Proceedings of an International Symposium on Electromagnetic Fields in Electrical Engineering, held September 23-25, 1987, in Pavia, Italy

A Division of Plenum Publishing Corporation
233 Spring Street, New York, N.Y. 10013

Printed in the United States of America

PREFACE

This book is the collection of the contributions offered at the International Symposium on Electromagnetic Fields in Electrical Engineering, ISEF '87, held in Pavia, Italy, in September 1987. The Symposium was attended by specialists engaged in both theoretical and applied research in low-frequency electromagnetism. The charming atmosphere of Pavia and its ancient university provided a very effective environment to discuss the latest results in the field and, at the same time, to enjoy the company of colleagues and friends coming from over 15 countries. The contributions have been grouped into 7 chapters devoted to fundamental problems, computer programs, transformers, rotating electrical machines, mechanical and thermal effects, various applications and synthesis, respectively. Such a classification is merely to help the reader because a few papers could be put in several chapters.

Over the past two decades electromagnetic field computations have received a big impulse by the large availability of digital computers with better and better performances in speed and capacity. Many various methods have been developed but not all of them appear convenient enough for practical engineering use. In fact, the technical and industrial challenges set some principal attributes and criteria for good computation methods. They should be relatively easy to use, fit into moderately sized computers, yield useful design data, maintain flexibility with minimum cost in time and effort. Taking this into account, the main task of a researcher consists in reducing the still intractable problem of solving three-dimensional multiphase time-varying non-linear coupled fields to that of finding approximate solutions for simplified or partial models of the problem. In order to select the most effective computation tool for each application fundamental research as well as accurate comparison of various methods is indispensable. This is one of the most important messages which came out of the Symposium.

The editors would like to thank all the authors for their valuable contributions and, in particular, the senior authors for their patient and priceless job in organizing the various chapters. Thanks are also due to the sponsors of the Symposium for supporting it financially. Finally the editors wish to express gratitude to the Publisher for offering the opportunity of addressing an international audience much greater than that present in Pavia.

A. Savini
J. Turowski

CONTENTS

1. FUNDAMENTAL PROBLEMS

2. COMPUTER PROGRAMMES

Theoretical aspects

Practical Use

3. TRANSFORMERS

4. ELECTRICAL MACHINES

Synchronous Machines, DC and Linear Motors

Induction Machines

5. MECHANICAL AND THERMAL EFFECTS

6. VARIOUS APPLICATIONS

7. SYNTHESIS

1. FUNDAMENTAL PROBLEMS

Introductory remarks

P. Hammond

Electrical Engineering Department
University of Southampton
Southampton, Hampshire, England

Some writers regard electromagnetism as a well-understood and even as a closed field of study. Maxwell's equations can be accepted and all that is needed is to find appropriate solutions. This, however, is an onlooker's view, practitioners know that there are many fundamental problems to be investigated. Partly this is due to the vast range of electromagnetic phenomena and the even greater range of applications. Then there is the inherent complication of the interaction of electromagnetic energy with matter. Also there are geometrical features inherent in electromagnetic fields and only recently have numerical methods become available which make it possible to study the importance of local and global geometries.

It is, therefore, not surprising to find that at the Symposium a full session was devoted to fundamental problems. The papers presented in that session are reproduced in the following pages.

The paper by J. Anuszczyk deals with the important practical problem of predicting the losses in the cores of power transformers and rotating machines. Such losses are caused by eddy current and hysteresis effects, the latter being essentially non-linear because of the domain structure of the iron. Moreover the iron has directional properties and the field has rotational components. To find an engineering solution for the calculation of losses requires great ingenuity.

The paper by K. Bessho et al. describes an experimental method of producing strong magnetic fields by means of eddy currents in a very simple but effective manner. Here the phenomenon is well known but the application is novel.

The paper by I.R. Ciric addresses the fundamental problem of devising efficient computation schemes for magnetic fields in multiply-connected regions containing currents and magnetic material.

The papers by P. Graneau and J. Nasilowski deal with the very fundamental question of force distributions in current-carrying conductors. Both of them present experimental evidence for disruptive axial forces and raise the question how such forces can arise. One proposed explanation suggests that Ampere's original formula for forces on current elements deserves closer study, because it allows for an axial force. Another proposal is that the explanation may lie in considering the effect of the non-uniform current distribution. These matters were debated in the session, but they were by no means resolved.

POWER LOSSES IN ELECTROTECHNICAL SHEET STEEL

UNDER ROTATIONAL MAGNETIZATION

Jan Anuszczyk

Institute of Electrical Machines and Transformers

Technical University of Lodz, Poland

INTRODUCTION

In magnetic circuits of electric machines, as well as in nodes of three-phase transformer cores, rotational magnetization occurs. Calculations of iron losses in this kind of magnetization are based upon the knowledge of rotational field distribution in particular elements of the considered magnetic circuit. The rotational field inside the iron core was calculated by means of reluctance network method[1]. Calculations were carried out for a ring sample magnetic circuit and for a three-phase induction motor magnetic circuit. In this paper an idea of calculations and results of examinations of basic losses in a ring sample laminated core havebeen presented.

IRON LOSSES UNDER ROTATIONAL MAGNETIZATION

Let us consider any hodograph of rotational flux density $\vec{B}$, determined in (x,y) plane of core plate in an elementary zone resulting from discretization and related to the core volume V. The actual value of magnetic field power losses in an elementary volume V, limited by a surface A, is represented by a Poynting vector $\vec{S}$ in the considered zone and at a given time

$$p = -\int_A \vec{S}\, d\vec{A} \qquad (1)$$

while

$$\vec{S} = \vec{E} \times \vec{H} \qquad (2)$$

where $\vec{E}$ and $\vec{H}$ vectors of electric and magnetic strengths, respectively. Power losses due to Poynting vector is, in this case, a sum of losses resulting from eddy-currents and hysteresis

$$p = p_e + p_h = \int_V \sigma E^2 dV + \int_V \frac{\partial}{\partial t}(H\ B)\, dV \qquad (3)$$

where σ - electrical sheet conductivity.

In the considered case of two-dimensional field, determined in the plane of core plating, the vector $\vec{S}$ is defined by its component S_z, vector $\vec{E}$ by its components E_x, E_y and vector $\vec{H}$ by its components H_x, H_y - Fig. 1.

Eddy-curent losses of power

Writing the Maxwell equation, concerning ferromagnetic material

$$\text{rot}\ \vec{H} = \vec{J} = \sigma\vec{E} \tag{4}$$

where $\vec{J}$ - eddy-current density vector, and taking into account the components of field, we obtain

$$-\frac{\partial H_y}{\partial z} = \sigma E_x, \quad \frac{\partial H_x}{\partial z} = \sigma E_y \tag{5}$$

Hence the actual power of eddy-current losses

$$p_e = \frac{1}{\sigma}\left[\left(\frac{\partial H_y}{\partial z}\right)^2 + \left(\frac{\partial H_x}{\partial z}\right)^2\right] \tag{6}$$

The actual power defined by (6) is the power $p_e(\varphi)$, where φ stands for an angle, determining the position of field strengh $\vec{H}$ at the given time. The mean loss of power during one rotation cycle of vector $\vec{H}$ (or $\vec{B}$) by 2π rd

$$P_e = \frac{1}{2\pi}\int_0^{2\pi} p_e(\varphi)\,d\varphi \tag{7}$$

The above expression concerns losses in an elementary volume of a ferromagnetic material.

Power losses resulting from rotational hysteresis

In case rotational magnetization, the magnetic field strengh and flux density have different direction

$$\vec{H}(\varphi) = H(\varphi)\{\cos\varphi,\ \sin\varphi\} \tag{8}$$

$$\vec{B}(\varphi) = \mu(H)\,H(\varphi)\{\cos(\varphi - \theta_B),\ \sin(\varphi - \theta_B)\} \tag{9}$$

where θ_B - angle between $\vec{B}$ and $\vec{H}$ in a given element and at a given moment of time. The energy of hysteresis losses in volume V, regardless of the kind of magnetization is, according to (3), given by an integral

$$w_h = \int_V H\ B\ dV \tag{10}$$

Substituting the dependencies[3] (8),(9) into (10), we obtain energy losses per unit volume

$$w_h(\varphi) = H\cos\theta_B\frac{\partial(\mu H)}{\partial\varphi} + \mu H^2\left(1 - \frac{\partial\theta_B}{\partial\varphi}\right)\sin\theta_B \tag{11}$$

Power of hysteresis losses per one cycle is described by the expression

$$P_h = \frac{1}{T}\int_0^{2\pi} w_h(\varphi)\, d\varphi \tag{12}$$

where T - duration of a single magnetization cycle.

The determination the hysteresis angle Θ_B is one of the most important problems at the stage of preparing initial parameters for losses calculations. In order to determine this angle of a ferromagnetic material, the following assumptions have been made:

- given actual values of field strength and flux density in the cycle of rotational magnetization correspond to the same displacement of vectors $\bar{H}$ and $\bar{B}$, as under alternating magnetization (for the same frequency of both magnetizations).
- each pair of values B_m, H_m of the averaging magnetization characteristic corresponds to an equivalent elliptical hysteresis loop, which in equivalent to the factual dynamic hysteresis loop of the material.

Having assumed the above, we obtain

$$\sin \Theta_B = \frac{P_h T}{\pi B_m H_m} \tag{13}$$

In this equation power P_h and time T refer to one cycle alternating magnetization.

RESULTS

Results of calculations of rotational field and losses are presented, concerning a ring sample made of two kinds of electrotechnical sheet:

- steel siliconless sheet Fe64.50 (Si less than 0.3 %),
- steel transformer sheet M4.

The examined sample is ring-shaped and consists of individual sheets, the conformity is of rolling directions of adjacent sheets being preserved.

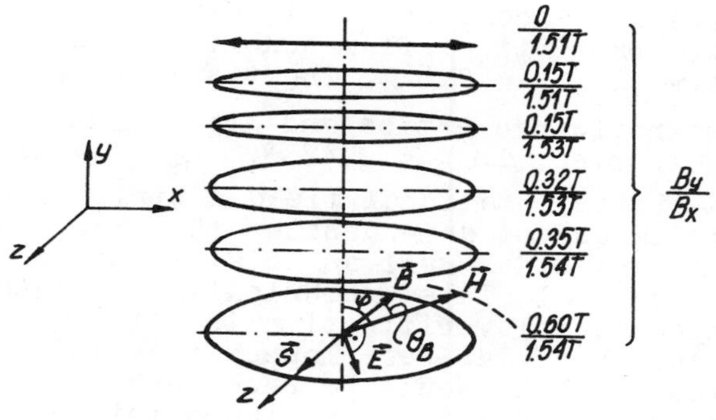

Fig. 1. Rotational field in a ring yoke made of siliconless steel (average flux density along the yoke height B = 1.53 T)

The sample was of the following dimensions: outside diameter 0.208 m, inside diameter 0.151 m. The hodographs of rotational flux density inside the sample core, in the plane of sheets, are shown in Figs. 1 and 2.

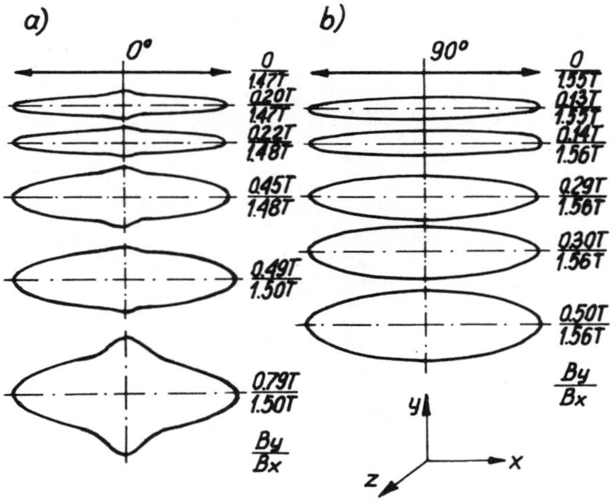

Fig. 2. Rotational field in a ring yoke made of transformer steel. (a) in the elements along direction of the rolling sheet, B = 1.48 T; (b) in the elements with the angle 90° in relation to the rolling direction, B =1.56 T

Presented results concern the case of rotational magnetization, with sinusoidal distribution of magnetic flux along the inside circumference of the sample. Similar magnetization occurs in a yoke of a three-phase induction motor. In a stator yoke of dimensions close to those of the examined sample, similar proportions between the components B_x, B_y of flux density were found.

The examined siliconless steel possesses practically isotropic magnetic properties: flux density anisotropy ΔB_{25} =0.06 T, loss anisotropy $A_{p1.5}$ = 2.6 %. The transformer steel has distinct anisotropic properties and high values of flux density and loss anisotropies: ΔB_{25} = 0.47 T, $A_{p1.5}$ = 48.7 %.

On the basis of rotational field distribution (Figs. 1 and 2) and making use of dependencies (5 ÷ 12), eddy-current and hysteresis losses, as well as total losses in the core of a ring sample were calculated numerically. In case of a ring sample the experimental verification is relatively simple, for there is no need to divide core losses into those in stator yoke and in teeth, which would be necessary if a magnetic circuit of an induction motor were taken into consideration. The calculations were verified experimentally by means of a special device, measuring losses in a rotationally magnetized ring sample[2]. A practical identity of results of calculations and measurements of total iron losses was obtained - Fig. 3.

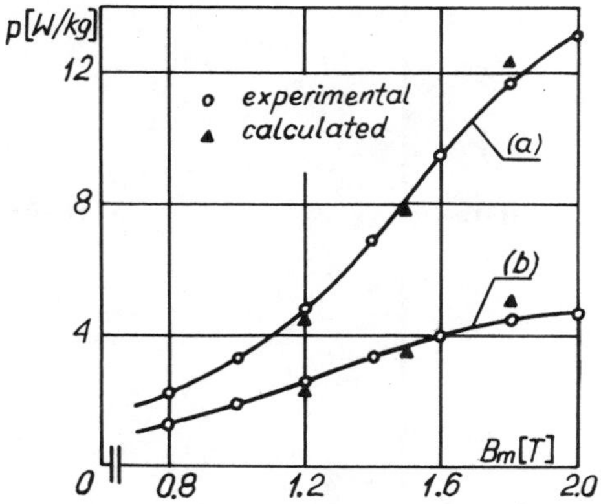

Fig. 3. Power losses inside the core of a ring sample under rotational magnetization. (a) steel siliconless sheet; (b) steel transformer sheet.

CONCLUSIONS

The method of calculation of iron losses, presented above, takes into consideration $B_m = f(H_m)$ magnetization curve and hysteresis-angle $\Theta_B = f(B_m)$ characteristic of a ferromagnetic, as well as of its basic physical parameters, such as the electrical conductivity and mass density. The calculations are based upon the hodographs of rotational flux density inside the core. This method can be applied to both isotropic and anisotropic magnetic materials. The increase of losses under rotational magnetization in relation to those under alternating magnetization depends on the material and on the dimensions of the ring sample, which is connected with the value of flux density ratio B_y/B_x. As regards losses in a mixed sample, determined by Epstein's apparatus, the losses under rotational magnetization are higher by about 30 %, and in a steel transformer sheet by about 45 %. Percentage values of loss increase refer to flux density range of $1.4 \div 1.7$ T and to the examined ring sample of diameters 0.208/0.151 m.

REFERENCES

1. J. Anuszczyk, Evaluation of rotary magnetization in stator yoke of induction motor by equivalent reluctance network, Rozpr. Elektrotech., 1:83 (1985), in Polish.
2. J. Anuszczyk, A method and a device for measurements of power losses in ferromagnetics under rotational magnetization, Polish Patent No.127489 (1984).
3. W. Wolff, "Drehmagnetisierung in Dynamoblechen", Darmstadt (1979).

AC HIGH MAGNETIC FIELD GENERATION
DUE TO EDDY-CURRENT CONCENTRATION EFFECT

K. Besso, S. Yamada, M. Kooto, and T. Minamitani*

Electrical Energy Conversion Laboratory
Faculty of Technology, Kanazawa University

*Kanazawa Murata Electric Company

ABSTRACT

The paper describes the high magnetic field generation by the use of eddy-current effects. The main features of the new type generator as well as some of results obtained both numerically and experimentally are presented. A short account of this research has also been presented.

INTRODUCTION

The main idea of the high field generation is based on the shielding effect of eddy currents as shown in Fig.1. The eddy currents in a conducting plate can be induced by either AC magnetic field excitation or by moving a plate in a DC magnetic field. The latter was used in the first generator, which has been described in the papers[1,2]. Such a system is well-suited to the DC high magnetic field generation. It can also be used, of course, for AC field generation when the excitation field is of AC nature. As the field concentration slightly increases with frequency, it does not seem to be necessary to use a dynamic device for AC field generation. That is why, the authors suggest that for an AC field a static device with a time-dependent excitation should be used. The first attempt, based on two- or four-plates conception has already been done and the arrangement of this device is shown in Fig.2. The details of the generator as well as results obtained are given in the paper[3]. Although the results seemed to be promising, the authors worked out the new device using a special conducting cylinder. The paper is devoted to this approach.

H-SHAPED CYLINDER TYPE GENERATOR

The first model of a conducting cylinder is H-shaped and is shown in Fig.3. The expected eddy-current lines are shown in Fig.4. It can be seen that due to eddy currents excited by the coil wound around the cylinder, the AC magnetic flux is concentrated in the hole. To decrease leakage flux the device is placed in the ferromagnetic laminated yoke.

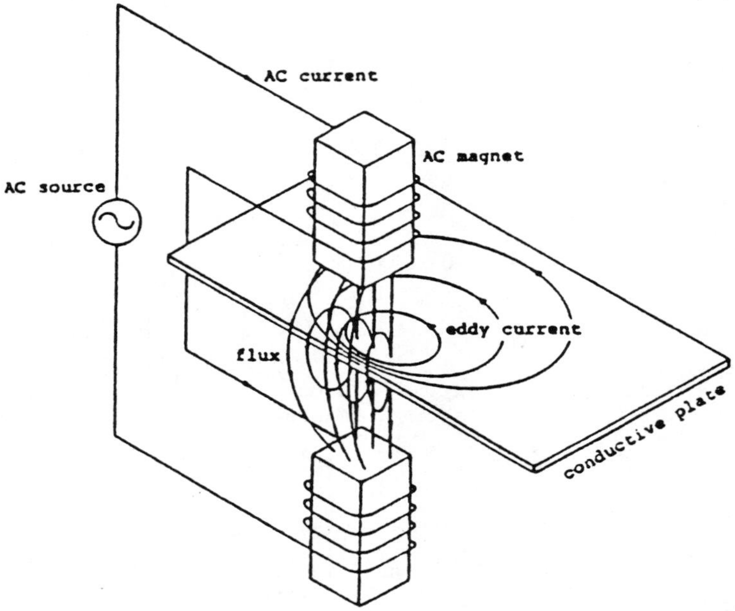

Fig. 1 Magnetic shield effect of conductive plate

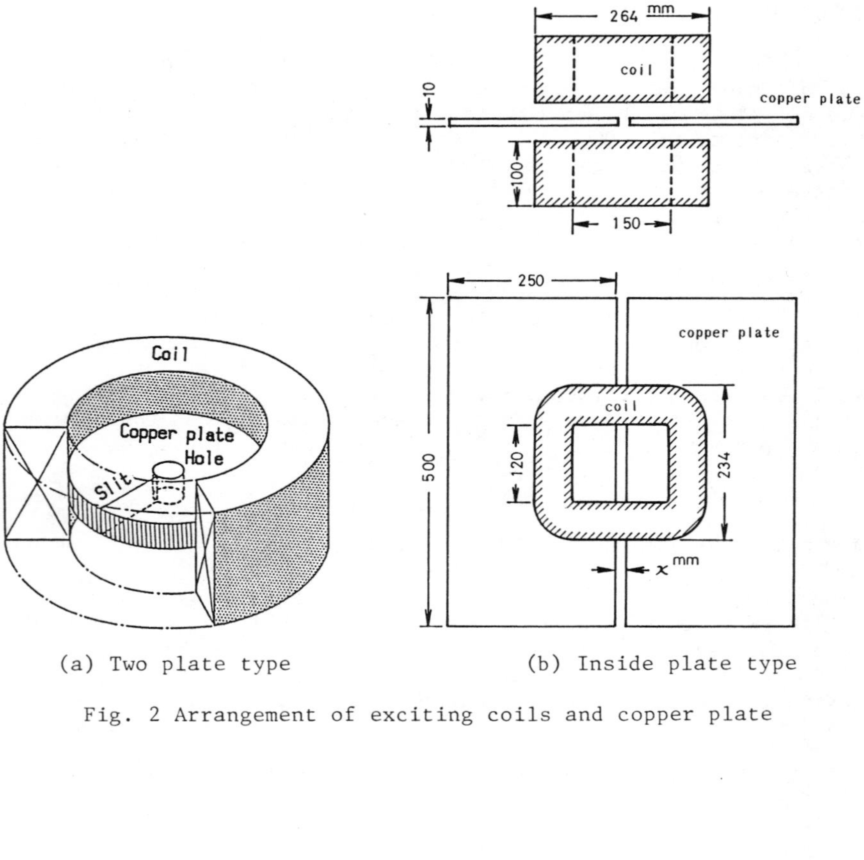

(a) Two plate type

(b) Inside plate type

Fig. 2 Arrangement of exciting coils and copper plate

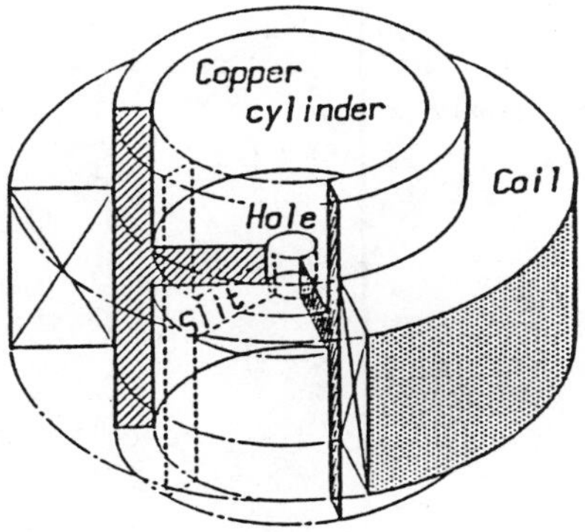

Fig. 3 H-shape cylinder type coil

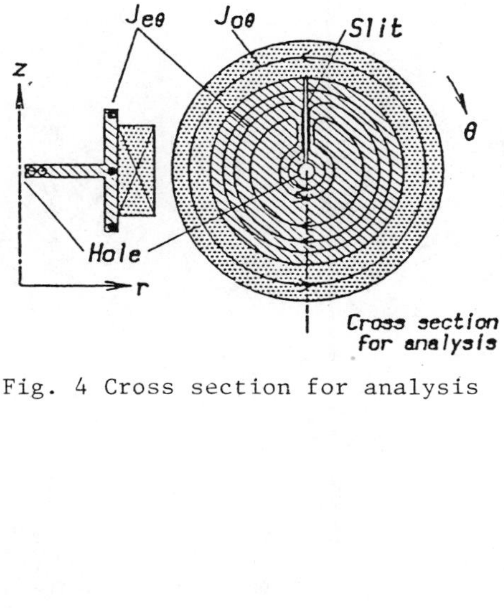

Fig. 4 Cross section for analysis

The analysis of the magnetic field

The mathematical analysis of the magnetic field has been made under some assumptions. Firstly, the presence of the yoke is neglected. Secondly, the eddy currents are assumed to flow only cylindrically. However, it is clear that the eddy currents also have the r-components near the slit. But analyzing the r-z plane away from slit this assumption is correct enough. The presence of the slit and the resulting flow of eddy currents will be taken into account by constraint condition. Thirdly, all material coefficients are constant and variables in time are purely sinusoidal.

According to the above conditions magnetic vector potential A has only one component A_θ which satisfies the following equation,

$$\frac{\partial}{\partial z}\{\frac{1}{\mu}(\frac{\partial A_\theta}{\partial z})\}+\frac{\partial}{\partial r}\{\frac{1}{\mu}(\frac{\partial rA_\theta}{\partial r})\}=-J_{o\theta}+J_{e\theta} \tag{1}$$

where

$J_{o\theta}$: applied current density, $J_{e\theta}$: eddy current density,
μ : magnetic permeability.

The eddy-current density $J_{e\theta}$ is expressed in the form of

$$J_{e\theta}=\sigma\,(j\omega A_\theta+\frac{1}{r}\frac{\partial\phi}{\partial\theta}) \tag{2}$$

where

ϕ :scalar potential, σ :conductivity,
ω :angular frequency.

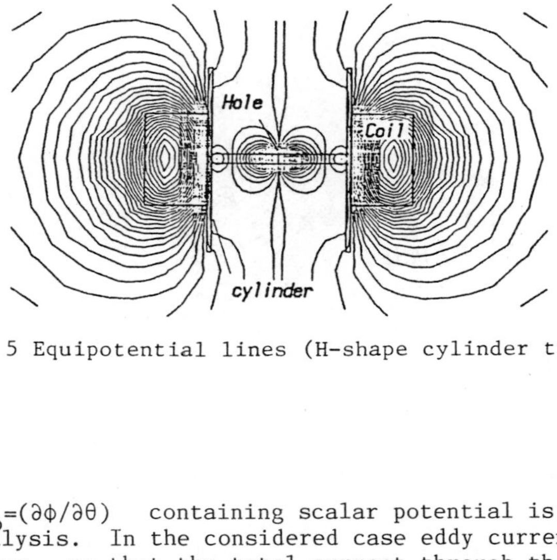

Fig. 5 Equipotential lines (H-shape cylinder type)

The term $C_o=(\partial\phi/\partial\theta)$ containing scalar potential is zero in 2-D and axisymmetric analysis. In the considered case eddy currents flow in and out the r-z plane, so that the total current through the r-z plane is equal to zero. Hence, we assumed the term Co to be constant in order to satisfy the constraint condition. The constant Co is as follows,

$$C_o=\frac{\partial\phi}{\partial\theta}=-\frac{\int j\omega\sigma A_\theta\,dS}{\sigma\int 1/r\,dS} \tag{3}$$

The introduction of constraint condition usually leads to an integro-differential formulation. Here the equations (1),(2), and (3) are solved iteratively. The above mathematical model has been solved by means of the finite element method. The magnetic flux lines for the H-shaped cylinder are shown in Fig.5. The analysis can be made for either constant voltage or constant current since the circuit equation is coupled with the field ones.

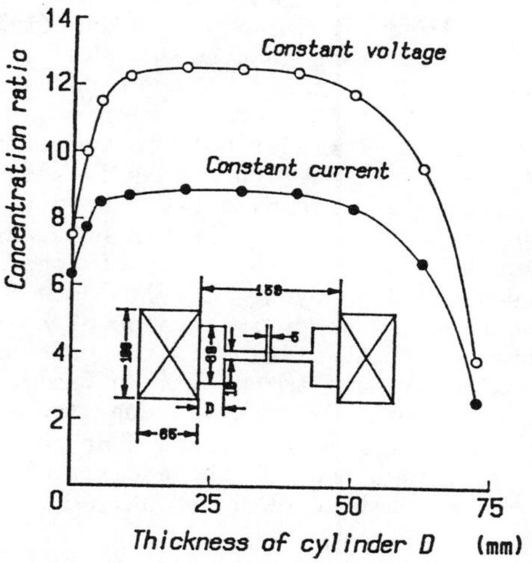

Fig. 6 Relation between the thickness of cylinder and the concentration ratio

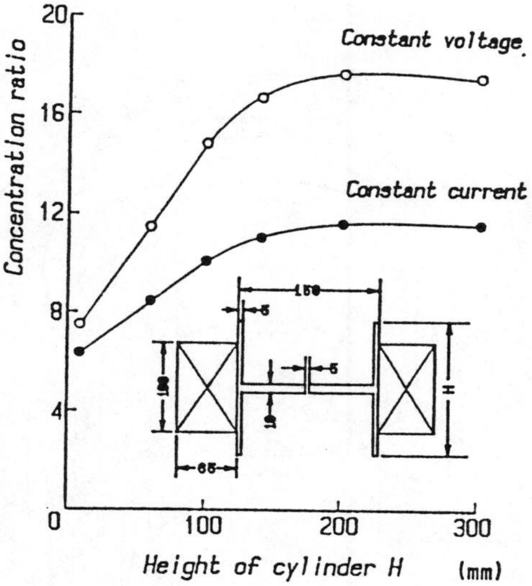

Fig. 7 Relation between the height of cylinder and the concentration ratio

The analysis of the cylinder properties

In order to choose the proper dimensions of the cylinder, some variants of various geometry have been examined. All the calculations were made for exciting frequency 60 Hz and for conductivity of the proper cylinder for copper. The relationship between ratio of field concentration and cylinder dimensions with the supply method as a parameter were established. The ratio of field concentration means here the ratio of magnetic flux density in the cylinder hole to that without the cylinder. In Fig.6 the dependence on cylinder thickness is shown. It is seen that the constant input voltage gives considerably bigger field concentration than that of constant current. Also it can be seen that there is an interval of thickness within which the ratio of concentration is almost the same and decreases rapidly beyond it. It leads, in consequence, to important conclusion that is the optimal value of thickness which gives satisfactory concentration of the field. In the case examined the cylinder thickness should be slightly more than 7 mm. In Fig.7, in turn, the height of cylinder is being considered. The curves confirm again the advantage of voltage supply. It can be also concluded that the height of cylinder should be slightly bigger than that of the exciting coil but not too much as the ratio of concentration becomes constant.

MULTILAYER CYLINDER TYPE

Recently the new model of high field generator has been introduced. The structure and the eddy-current distribution are shown in Figs.8 and 9. The exciting coils are placed between two layers. The device has been named "Ohyama coil" by the authors. As the example for calculation the two-layer device has been chosen. In Fig.10 the flux lines are shown and it can be seen that the concentration ratio considerably increases. It results from the fact that the second layer is like an external shield.

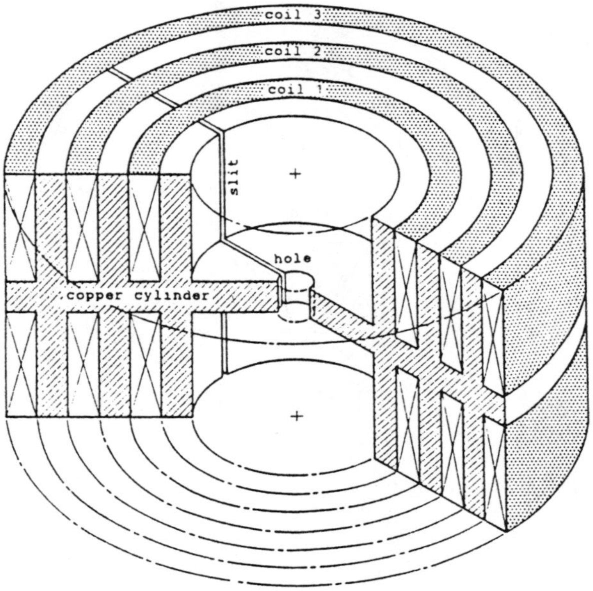

Fig. 8 Multilayer eddy-current type coil (Ohyama coil)

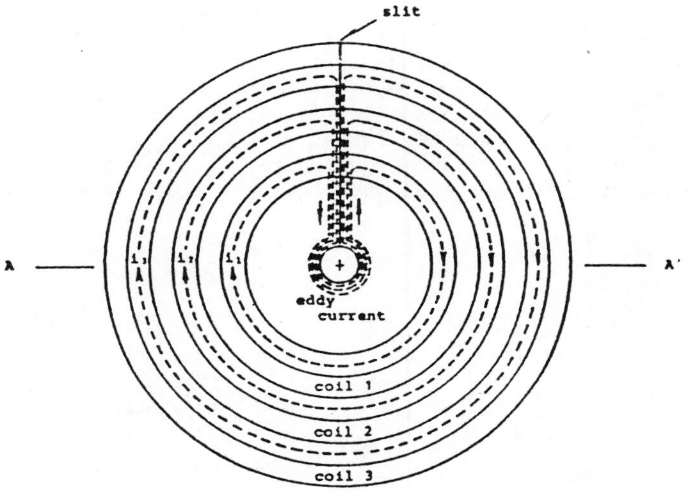

(a) Plane

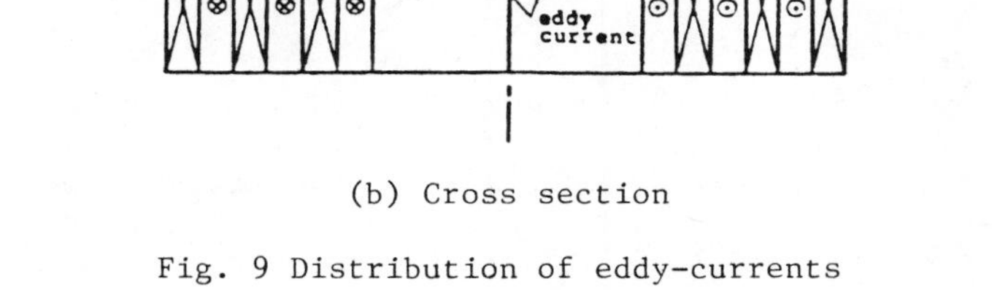

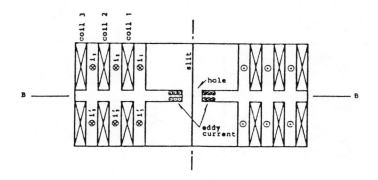

(b) Cross section

Fig. 9 Distribution of eddy-currents

COMPARISON AND THE AC MAGNETIC FIELD

Compared with various type models, the results of experiment are shown in Fig.11. There are the waveforms of concentrated flux density for each structures. The experiments are done at the same exciting condition. The very high effectiveness is especially seen when using a recently investigated multilayer device.

The effect of field concentration has been also checked experimentally. The three-layer generator was supplied with an input voltage of 1350 V and frequency 60 Hz. Then the AC flux density of 11.2 T was obtained with input electric power of 156 kVA. In order to compare the results the plate-type device was tested and the results were far worse, applying the input voltage of 3400 V and the power of 1697 kVA, the flux density of 7.4 T was obtained. In Fig.12 the concentrated flux density for these two types of generator as a function of input voltage is presented. This numerical and experimental analysis shows very clearly that the multilayer device gives much better results than former ones.

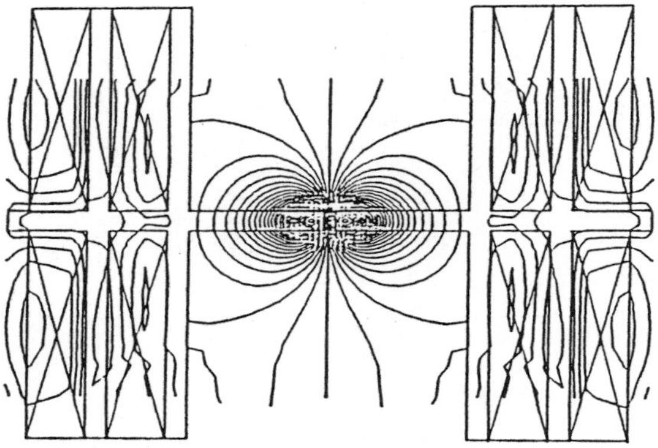

Fig. 10 Equipotential line (two layer type)

CONCLUSIONS

The paper was to show the method of AC high magnetic field generation. The use of this aim is that the effect of eddy currents appeared to be very effective. The recently investigated multilayer device is most suited for the AC generator. That is why the authors suppose that this way of high magnetic field generation has a very promising future.

ACKNOWLEDGEMENT

This work is constantly supported by the Ministry of Education of Japan, especially by Mr. Cho Ohyama. To express their gratitude the authors named the multilayer cylinder type device after him.

Moreover, the authors wish to thank Toshiba Corporation for manufacturing the equipment and also Hokuriku Electric Power Corporation for supplying the high voltage power.

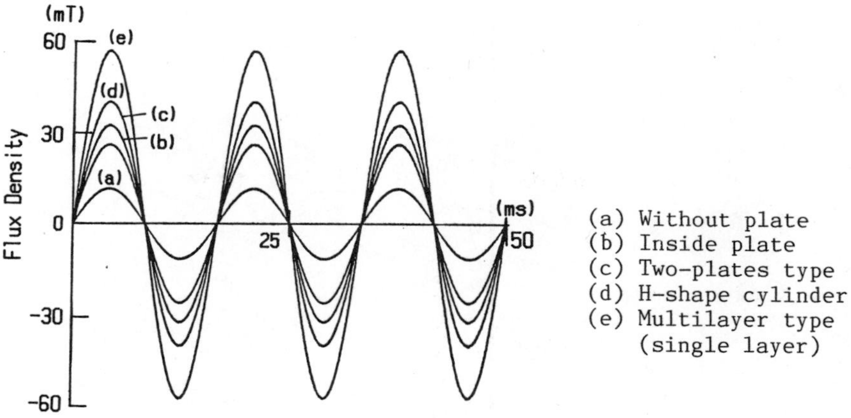

Fig. 11 Waveforms of flux density on the hole

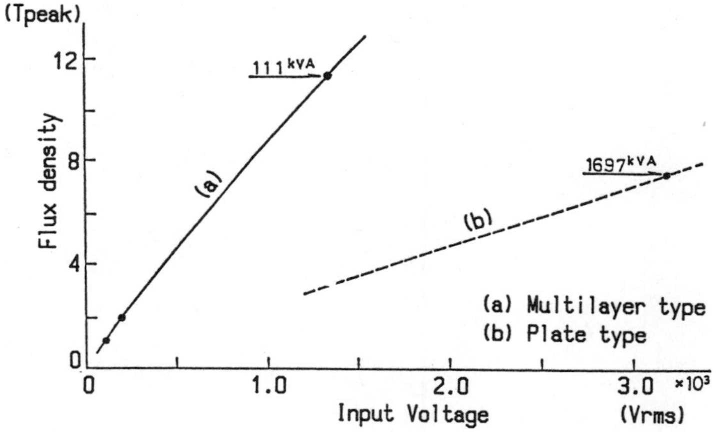

Fig. 12 Relation between the AC voltage and the maximum flux density

REFERENCES

1. K.Bessho, et.al. : " High-speed Rotating Disc Type Generator for High Magnetic Field", IEEE Transactions on Magnetics, Vol.MAG-19, No.5, 207-209, 1983.
2. K.Bessho, et.al. : " Asymmetrical Eddy Currents and Concentration Effect of Magnetic Flux in a High-Speed Rotating Discs", IEEE Transactions on Magnetics, Vol.MAG-21, No.6, 1747-1749, 1985.
3. K.Bessho, et.al. : "AC High Magnetic Field Generator Based on the Eddy Current Effect", IEEE Transactions on Magnetics, Vol.MAG-22, No.5, 970-972, 1986.

SCALAR POTENTIAL MODELS FOR MAGNETIC FIELDS OF VOLUME CURRENT DISTRIBUTIONS

I.R. Ciric

Department of Electrical Engineering
University of Manitoba
Winnipeg, Canada

ABSTRACT

A scalar potential function for magnetic fields in the presence of given volume current distributions is defined, by using an equivalent distribution of fictitious magnetization. The general formulation of the magnetic field problems, based on the equations and boundary conditions satisfied by this scalar potential, is presented. For practical configurations, the magnetic field due to given volume current distributions can be determined from that due to surface distributions of fictitious magnetic charge. Two examples illustrate the modelling procedure and the efficiency of the proposed method with respect to methods developed so far.

SCALAR POTENTIAL FOR MAGNETIC FIELD PROBLEMS

In the presence of given volume current distributions, the stationary or quasistationary magnetic field is described by the equations:

$$\nabla \times \mathbf{H} = \mathbf{J}, \tag{1}$$

$$\nabla \cdot \mathbf{B} = 0, \tag{2}$$

$$\mathbf{B} = \mu_o(\mathbf{H} + \mathbf{M}), \tag{3}$$

where $\mathbf{H}$, $\mathbf{B}$, $\mathbf{J}$, and $\mathbf{M}$ are the field intensity, the magnetic induction, the volume current density, and the magnetization vector, respectively, and μ_o is the permeability of free space. Across the surfaces of discontinuity of the field quantities, Eqs. (1) and (2) are replaced, correspondingly, with

$$\hat{\mathbf{n}}_{12} \times (\mathbf{H}_2 - \mathbf{H}_1) = \mathbf{J}_s, \tag{4}$$

$$\hat{\mathbf{n}}_{12} \cdot (\mathbf{B}_2 - \mathbf{B}_1) = 0, \tag{5}$$

$\hat{\mathbf{n}}_{12}$ being the normal unit vector from side 1 to side 2 of each surface, and $\mathbf{J}_s$ the surface current density eventually present.

Consider a distribution of fictitious magnetization $\mathbf{M}_c$ such that its equivalent Ampèrian volume current density[1] is just the given current density $\mathbf{J}$,

$$\nabla \times \mathbf{M}_c = \mathbf{J}. \tag{6}$$

$\mathbf{M}_c$ may be different from zero outside the regions with $\mathbf{J}\neq 0$ but its curl must be zero wherever $\mathbf{J}=0$. Since we do not impose any a priori conditions for its divergence, $\mathbf{M}_c$ is not uniquely determined by Eq. (6), and simple expressions for $\mathbf{M}_c$ can be found extremely easily for practical current distributions (as shown in the next Sections). Equations (1) and (6) yield

$$\nabla\times(\mathbf{H} - \mathbf{M}_c) = 0 \qquad (7)$$

and hence $\mathbf{H} - \mathbf{M}_c$ can be derived from a scalar potential Φ_c,

$$\mathbf{H} - \mathbf{M}_c = -\nabla\Phi_c. \qquad (8)$$

Once the fictitious magnetization is chosen, the field intensity is given by

$$\mathbf{H} = \mathbf{M}_c - \nabla\Phi_c. \qquad (9)$$

The equation satisfied by Φ_c for an isotropic medium, without true permanent magnetization, for instance, can be derived from (2), (3) (written as $\mathbf{B} = \mu\mathbf{H}$), and (9) in the form

$$\nabla^2\Phi_c + \frac{\nabla\mu}{\mu}\cdot\nabla\Phi_c = -\frac{\rho_c}{\mu}, \qquad (10)$$

where μ is the permeability of the medium and ρ_c is a volume density of fictitious magnetic charge corresponding to $\mathbf{M}_c$,

$$\rho_c = -\nabla\cdot(\mu\mathbf{M}_c). \qquad (11)$$

In the case of a linear, isotropic, and homogeneous medium, Φ_c satisfies the Poisson equation

$$\nabla^2\Phi_c = -\frac{\rho_c}{\mu}, \qquad (12)$$

with $\rho_c/\mu = -\nabla\cdot\mathbf{M}_c$. If $\mathbf{M}_c$ in Eq. (6) is chosen such that its volume divergence is equal to zero, then Φ_c satisfies the Laplace equation within linear, isotropic, and homogeneous materials,

$$\nabla^2\Phi_c = 0. \qquad (13)$$

The conditions for the tangential and normal components of $\nabla\Phi_c$ at the surfaces of discontinuity of the quantities $\mathbf{H}$, $\mathbf{B}$, and $\mathbf{M}_c$ are obtained from Eqs. (4), (5), and (9):

$$\hat{\mathbf{n}}_{12}\times(\nabla\Phi_{c_1} - \nabla\Phi_{c_2}) = \mathbf{J}_s + \mathbf{J}_{sc}, \qquad (14)$$

$$\mu_1\frac{\partial\Phi_{c_1}}{\partial n_{12}} - \mu_2\frac{\partial\Phi_{c_2}}{\partial n_{12}} = \rho_{sc}, \qquad (15)$$

where

$$\mathbf{J}_{sc} = -\hat{\mathbf{n}}_{12}\times(\mathbf{M}_{c_2} - \mathbf{M}_{c_1}), \qquad (16)$$

$$\rho_{sc} = -\hat{\mathbf{n}}_{12}\cdot(\mu_2\mathbf{M}_{c_2} - \mu_1\mathbf{M}_{c_1}). \qquad (17)$$

We remark that, according to the Ampèrian model of the magnetized media, from the point of view of the macroscopic magnetic field produced in free space, the distributions of volume current $\mathbf{J}$ (Eq. (6)) and of fictitious surface current $-\mathbf{J}_{sc}$ (Eq. (16)) are equivalent to the distribution of magnetization $\mathbf{M}_c$. On the other hand, we see that $\mathbf{M}_c$ is equivalent to the fictitious distributions of volume charge ρ_c (Eq.(11)) and of surface charge ρ_{sc} (Eq. (17)). The result of the above analysis is that the field intensity (in Eq. (9)) due to a distribution of volume current density $\mathbf{J}$ can be obtained as the sum of $\mathbf{M}_c$ and the field intensity due to the distributions ρ_c, ρ_{sc}, and $\mathbf{J}_{sc}$.

Equation (14) shows that wherever $\mathbf{J}_s + \mathbf{J}_{sc} = 0$, the tangential component of $\nabla\Phi_c$ is continuous, which is equivalent to the continuity of the potential itself,

$$\Phi_{c_1} = \Phi_{c_2}. \tag{18}$$

When $\mu_1 = \mu_2 \equiv \mu$, Eq. (15) becomes

$$\frac{\partial\Phi_{c_1}}{\partial n_{12}} - \frac{\partial\Phi_{c_2}}{\partial n_{12}} = \frac{\rho_{sc}}{\mu}, \tag{19}$$

with $\rho_{sc}/\mu = -\hat{\mathbf{n}}_{12}\cdot(\mathbf{M}_{c_2} - \mathbf{M}_{c_1})$.

The modelling procedure presented is flexible in the sense that $\mathbf{M}_c$ in Eq. (6) can be chosen in an optimum way in order to reduce the necessary amount of computation. For magnetic field problems related to a large class of practical electromagnetic systems, models can always be constructed to contain only fictitious surface charge distributions within the region considered. By using this type of models, the calculation of the field due to volume currents is reduced to that of the field due to surface charges. At the same time, one can see that, if the given distribution of stationary or quasistationary current inside a region is entirely modelled in terms of a fictitious magnetization and only charge distributions, then the scalar potential Φ_c is a single-valued function of position in that region, as for an electrostatic field.

A rigorous analysis of the modelling method for the case of generalized distributions of given volume, surface, and line currents was presented recently by the author.[2]

MODELLING OF LONG, STRAIGHT CONDUCTORS

As an illustrative example, let us consider an infinitely long, straight conductor of rectangular cross section, with a constant volume current density along the positive z axis, $\mathbf{J}=\hat{\mathbf{z}}J$, as shown in Fig. 1. $\mathbf{M}_c$ in Eq. (6) can be chosen, for instance, as

$$\mathbf{M}_c(x,y) = \left| \begin{array}{ll} \hat{\mathbf{y}}Jx & \text{inside the conductor} \\ \hat{\mathbf{y}}Ja & \text{for } x \in (a,a'),\ y \in (-\frac{b}{2}, \frac{b}{2}) \\ 0 & \text{elsewhere,} \end{array} \right. \tag{20}$$

where $\hat{\mathbf{y}}$ is the unit vector along the positive y axis. The volume charge

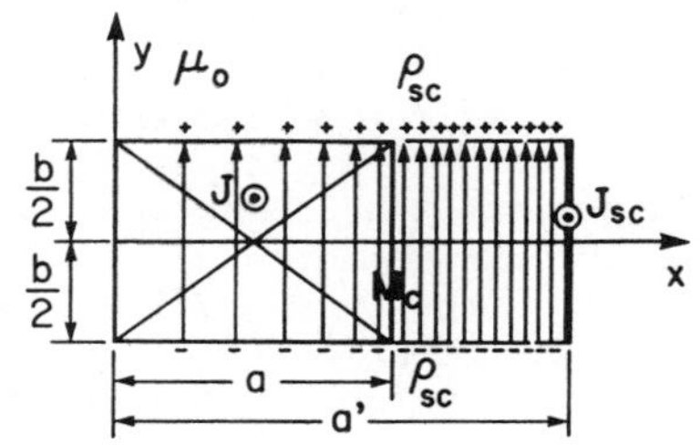

Fig. 1. Cross section of rectangular conductor carrying current.

density in Eq. (11) is equal to zero and Eqs. (17) and (16) yield

$$\rho_{sc} = \begin{cases} \pm\,\mu_o J x & \text{for } x\,\epsilon(0,a),\ y=\pm\frac{b}{2} \\ \pm\,\mu_o J a & \text{for } x\,\epsilon(a,a'),\ y=\pm\frac{b}{2}\,, \end{cases}$$

$$\mathbf{J}_{sc} = \hat{\mathbf{z}}Ja \qquad \text{for } x=a',\ y\,\epsilon(-\frac{b}{2},\frac{b}{2}), \tag{21}$$

where μ_o is the permeability of the medium inside and outside the conductor. The field intensity produced by the conductor is given by the sum of $\mathbf{M}_c$ and the field intensity due to ρ_{sc} and $\mathbf{J}_{sc}$. For an unbounded, homogeneous space this field is expressed in terms of single integrals instead of the double integral over the conductor cross section in the Biot-Savart formula. Another advantage of the model presented is that the current sheet in Eq. (21) can be placed at any $x=a'\epsilon(a,\infty)$, which allows a much simpler formulation of related boundary-value problems (see the next Section). The same type of simple models can be developed in the direction of the negative x axis, and also with the magnetization $\mathbf{M}_c$ along x axis.

This modelling technique can be applied to conductors of an arbitrary cross section , as well as to toroidal conductors,[2] thus reducing the field computation to the evaluation of line integrals. It should be noted that the same procedure can be used when the current density is not constant over the conductor cross section.

FORMULATION OF BOUNDARY-VALUE PROBLEMS

To illustrate the usefulness of the modelling technique presented, consider the rectangular opening of an electromagnetic device with two conductors of rectangular cross section, as shown in Fig. 2. The magnetic core is assumed to be of an ideal ferromagnetic material ($\mu\to\infty$) and the field in the opening is approximated to have a two-dimensional structure. The currents carried by the two conductors are uniformly distributed over their cross sections, and are equal in magnitude and opposite in direction, of intensity I. For the magnetic field problem in $x\,\epsilon(0,d)$, $y\,\epsilon(0,h)$, the boundary condition is that of a zero tangential component of the field intensity.

The current distribution of each of the two conductors can be modelled as shown in the previous Section, by ending the distributions of fictitious

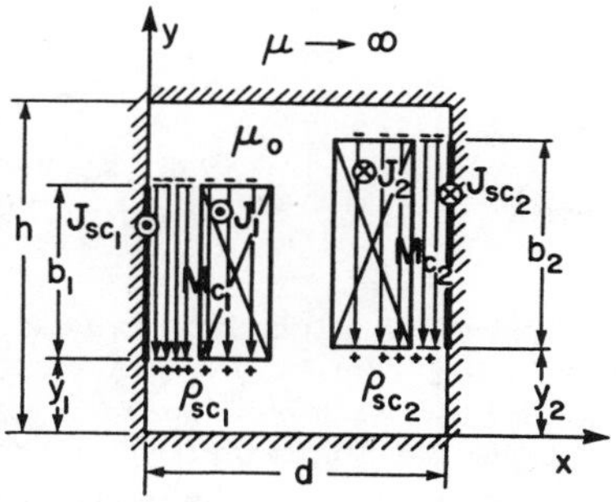

Fig. 2. Cross section of a rectangular cavity with two conductors.

magnetization $\mathbf{M}_{c_1}$ and $\mathbf{M}_{c_2}$ either at x=0 or at x=d, such that the model contains a distribution of magnetization and only surface charge inside the cavity (see Eqs. (20) and (21)). The corresponding current sheets are therefore placed on the boundary and have the densities

$$\mathbf{J}_{sc_1} = \hat{\mathbf{z}}I/b_1 \quad \text{for } x=0,\ y \in (y_1, y_1+b_1),$$

$$\mathbf{J}_{sc_2} = -\hat{\mathbf{z}}I/b_2 \quad \text{for } x=d,\ y \in (y_2, y_2+b_2). \tag{22}$$

Inside the two-dimensional region the scalar potential Φ_c (in Eq. (8)) is a single-valued function of position and satisfies the Laplace equation (13) everywhere, except the points on the lines C where the charge density $\rho_{sc} \neq 0$. Using Eqs. (8), (14), and (22), the conditions on the region boundary Γ can be expressed, for instance, as follows:

$$\Phi_c\Big|_{\substack{x=0\\ y\in(o,y_1)}} = \Phi_c\Big|_{\substack{x\in(0,d)\\ y=0}} = \Phi_c\Big|_{\substack{x=d\\ y\in(0,y_2)}} = 0,$$

$$\Phi_c\Big|_{\substack{x=d\\ y\in(y_2,y_2+b_2)}} = -\frac{I}{b_2}(y-y_2),$$

$$\Phi_c\Big|_{\substack{x=d\\ y\in(y_2+b_2,h)}} = \Phi_c\Big|_{\substack{x\in(0,d)\\ y=h}} = \Phi_c\Big|_{\substack{x=0\\ y\in(y_1+b_1,h)}} = -I, \tag{23}$$

$$\Phi_c\Big|_{\substack{x=0\\ y\in(y_1,y_1+b_1)}} = -\frac{I}{b_1}(y-y_1).$$

The boundary-value problem for Φ_c is, therefore, an interior Dirichlet problem. Its solution can be expressed in terms of the corresponding Green function $G(x,y;x',y')$,[3] as

$$\Phi_c(x,y) = \frac{1}{4\pi\mu_o}\int_C \rho_{sc}(x',y')G\,d\ell' - \frac{1}{4\pi}\oint_\Gamma \Phi_c(x',y')\frac{\partial G}{\partial n'}d\ell', \tag{24}$$

where $\partial/\partial n'$ denotes the derivative along the outward normal. The component of Φ_c given by the integral over the region boundary Γ is identical to the classical magnetic scalar potential corresponding to the field intensity which would be produced in the region only by the two current sheets in Eq. (22). This component is independent of the position and dimensions of the two conductors along the x direction. The first term in the right-hand side of Eq. (24) is identical to the electrostatic potential which would be produced in the region considered by a charge distribution of density ρ_{sc}, if the entire boundary were kept at zero potential. This type of solution is much simpler and more useful for calculating local field quantities, as well as global quantities (inductances, forces), than the solution corresponding to vector potential or multivalued scalar potential formulations. Equivalent models can be constructed with the distributions of magnetization $\mathbf{M}_{c_1}$ and/or $\mathbf{M}_{c_2}$ along the x axis. The same type of modelling and solution is applicable when the conductor cross-sectional sides are not parallel to the cavity walls. In such a case, the models present on the boundary not only a surface current distribution, but also a surface charge distribution.

Similar models can be used in the case of axisymmetric configurations or even more general systems.[2] For regions of an arbitrary geometry, when the Dirichlet Green function is not available in an analytic form, the scalar potential Φ_c can be computed by using boundary scalar integral equations formulated on the basis of the modelling technique presented.

CONCLUSION

The new modelling method for given current distributions allows for a field problem to construct the most appropriate models in order to reduce substantially the necessary amount of computation, by using an associated scalar potential which is a single-valued function of position. At the same time, these models allow an easier physical interpretation of the results, since the field due to volume current distributions is determined from that due to surface charge distributions. The simple examples considered illustrate the method and its efficiency. The method presented can be applied to all practical current distributions which can be decomposed in straight current tubes of finite length and current tubes in the form of a portion of toroid. It can be readily extended to systems with anisotropic or nonlinear materials.

REFERENCES

1. E. Durand, "Electrostatique et Magnétostatique," Masson, Paris (1953).
2. I.R. Ciric, New models for current distributions and scalar potential formulations of magnetic field problems, J. Appl. Phys. 61:2709 (1987).
3. P.M. Morse and H. Feshbach, "Methods of Theoretical Physics," McGraw-Hill, New York (1953).

THE ELECTRODYNAMICS OF AMPERE AND NEUMANN

Peter Graneau

Center for Electromagnetics Research
Northeastern University
Boston, MA 02115, U.S.A.

ABSTRACT

A brief outline of the history of the Ampere-Neumann electrodynamics of metals is provided. It was developed in France and in Germany during the 19th century. The old theory is based on instantaneous action-at-a-distance. The paper points out to what extent it agrees with presently taught relativistic electromagnetic field theory and then delineates an area where the old and new theories disagree. Four groups of experiments with metallic conductors are cited which, in the area of disagreement, favor the Ampere-Neumann electrodynamics. A fifth group of experiments is mentioned which reveals a new electrodynamic force in dense arc plasma which is orders of magnitude stronger than the Lorentz force. Precisely such a force has emerged from the modern extension of the old theory. This seems to broaden the empirical basis of the Ampere-Neumann electrodynamics to include dense plasma conductors.

1. INTRODUCTION

The Ampere-Neumann electrodynamics is the electromagnetic theory which was developed, taught, and practiced in Europe during the 19th century. Its philosophical basis is instantaneous action-at-a-distance, or more concisely 'simultaneous far-action' between two particles or bodies. This was the action principle used by Newton in his theory of gravitation.

Ampere derived his electrodynamics from a series of experiments with metallic conductors. Strictly speaking, hisempirical force law applies only to currents flowing in metals. Recent experiments have indicated that it may also apply to dense plasma. Electron beams in vacuum do not obey the Ampere-Neumann electrodynamics. This is equally true for charges drifting in vacuum or electrolytes. Virtually all nuclear physics experiments, therefore, lie outside the scope of the A-N electrodynamics.

The present paper is based on the book of reference [1]. All mathematical analyses, references to original work, and the diagramatic description of many experiments are contained in the book. The review paper of reference [2] is an abbreviated treatment of the same subject area.

2. HISTORICAL FACTS

In 1820 Oersted demonstrated the influence of an electric current on a magnetic needle. Immediately thereafter Ampere established the existence of forces between current-carrying wires. Within two years of intense experimentation Ampere had found the empirical law governing the mechanical interaction of two metallic current elements. The elements were matter elements and not elements of the 'electric fluid'. Like Newton and Coulomb, Ampere treated his fundamental force law as representing simultaneous far-actions. Unlike the scalar mass and charge of Newton and Coulomb, Ampere's law involved vectorial elements of matter which could produce repulsions or attractions, depending on certain angles [1].

Coulomb forces remain the foundation of modern electromagnetism, but Ampere forces have disappeared from textbooks. The Lorentz force formula for two current elements was first proposed by Grassmann who, in 1845, considered it to be a simultaneous far-action law. Lorentz retained Grassmann's vector mathematics but interpreted it by contact-actions with the magnetic field which, in the meantime, had been postulated by Maxwell. With regard to the correct magnetic force law Maxwell [6] himself was ambiguous. He claimed both Ampere's and Grassmann's formula gave results which were compatible with all experiments. On this point he later turned out to be mistaken.

Lorentz equated charges travelling in vacuum to current elements, but realized the new elements did not obey Ampere's force law. He suggested what has become known as the magnetic Lorentz force component to make up for the deficiency. Subsequently he suggested Ampere's metallic current element could be replaced by the drifting electron in the metal lattice. The positive ions of the lattice were supposed to make no contribution to the magnetic force. He proved this to be in accord with many--but not all--experimental results.

Hertz and Lorentz found that Maxwell's equations had a problem with induction by relative motion. This was repaired with the Lorentz transformations. With them Einstein wrote the special theory of relativity. The aggregate of Maxwell's equations, the Lorentz force law, and special relativity will be referred to as relativistic electromagnetism. It assumes contact actions between charges and the electromagnetic field. The 'substance' of the field is free energy which, according to relativity, possesses inertial mass and mechanical momentum. This energy is either stationary or it travels with the velocity of light. Throughout most of the 20th century, relativistic electromagnetism has been considered a closed book.

3. AREA OF DISAGREEMENT

The ponderomotive force between two current elements, when one produces a magnetic field at the position of the other, in modern theory is given by the magnetic component of the Lorentz force. It is mathematically identical to the Grassmann force. According to Grassmann's formula the force exerted by element 1 on element 2 is not equal and opposite to the force exerted by element 2 on 1. This violates Newton's third law of motion.

The Grassmann force is usually written in the form of a triple vector product. This may be split into two simple vectors [1]. One is found to be an attraction or repulsion vector complying with the third law. It is also contained in the Ampere force for the same two current elements. It is the second vector of the Grassmann formula which violates the third law. The offender may be termed a relativistic vector which is not present in the Ampere force. When calculating the force on one current element due to all the elements in a separate closed circuit, the relativistic vector integrates to zero, leaving only the newtonian vector of the Grassmann (Lorentz) force. When performing the same integration with Ampere's force law, an identical result is obtained. Hence a single closed loop integration removes the relativistic aspect of the Grassmann and Lorentz forces, and reduces relativistic electromagnetism to a simultaneous far-action theory. Since the validity of the A-N electrodynamics is restricted to metallic conduction, both theories are in complete numerical agreement on the operation of ordinary motors and generators where the far-actions are between complete circuits.

The reconciliation of the theories, via a closed loop integration, is absent when calculating the reaction forces between two parts of the same circuit. The two theories still agree on the net reaction forces between the two parts, but they disagree on the distribution of the reaction forces. This disagreement can be tested by experiment. One manifestation of the disagreement are longitudinal Ampere forces which act along the streamlines of current flow.

In the modern extension of the old electrodynamics [1] it became clear that a new kind of ponderomotive electrodynamic forcemust exist which was unknown to Ampere and Neumann. In the case of two parallel, coplanar amperian current elements, it tries to swing one around the other at constant radius. The turning moment has been called alpha-torque. It tries to expand the conductor material laterally, that is perpendicular to the flow of the current. The alpha-torque directly opposes pinch forces. It is orders of magnitude stronger than Lorentz and Ampere pinch forces. The metal lattice prevents significant lateral expansion, but the alpha-torque effect should be easily observable in liquid metals and dense plasmas, if the latter contain amperian current elements.

4. EXPERIMENTAL RESOLUTION

The disagreement between relativistic electromagnetism and the A-N electrodynamics has been repeatedly put to experimental tests. In all cases nature was found to be on the side of Ampere and Neumann [1].

The prediction of longitudinal forces prompted Ampere and de LaRive to stage a successful demonstration [3]. This was sometimes called the hairpin experiment. It has been repeated by a number of investigators during the past 160 years. In the present century it has been interpreted as being the result of the Lorentz force. Since the experiment does not distinguish between the two force distributions, it remains inconclusive. Neumann had its own, very different, classroom demonstration of longitudinal forces which cannot be explained by the Lorentz force. Still later Hering [8] performed a series of longitudinal force experiments, some more convincing than others. In all the early experiments, wire conductors were seen to move in the direction of their axes while floating on liquid mercury.

Nine years ago the author repeated Ampere's hairpin experiment and noticed that the longitudinal forces on the hairpin legs had their reaction in visible jets in the liquid mercury. This gave rise to several other experiments exhibiting electromagnetic jets in liquid metal which were explainable only with Ampere's longitudinal forces [1].

A third group of experiments, conforming with Ampere's law and conflicting with Lorentz's, was initiated by Nasilowski in Poland. He was the first to produce, in 1960, wire fragmentation in the solid state by a large current pulse. According to Ampere's law, the pulse will induce considerable tension in the wire at a time when its temperature rises and the tensile strength decreases. The Ampere tension is then capable of rupturing the wire in many places successively while the current continues to flow through arcs across the fracture gaps. Nasilowski's finding was confirmed at MIT with straight and curved wires.

Five years ago Pappas [10] opened the door to yet another family of experiments which have been confirming the A-N electrodynamics. He converted part of a metallic circuit to a pendulum hanging from the ceiling. Mercury cups connected it to the stationary part of the circuit. When a current pulse was passed through the pendulum, it would swing away from the stationary part of the circuit. If relativistic electromagnetism were correct, the mechanical momentum aquired by the pendulum should have been balanced by the destruction of an equal amount of field energy-momentum. Pappas found the field energy required for momentum conservation was orders of magnitude greater than the energy that could have been supplied by the current source. Pappas' experimental result was confirmed in the author's MIT laboratory.

The experiments by Pappas and Graneau proved conclusively that conduction currents in metals were not subject to local field energy-momentum action. In addition it was shown by both investigators that the pendulum was pushed by longitudinal forces, as required by Ampere's law, and not pulled by transverse forces, as predicted by Lorentz's force law. These results have a direct bearing on the operation of railguns. The railgun recoil should, by relativistic electromagnetism, be magnetic pressure on the field which is not felt by the rails. Ampere's law, on the other hand, requires the existence of longitudinal recoil forces in the rails [1] which should distort the rails and have a detrimental effect on the performance of the accelerator. Experiments at MIT have unambiguously confirmed rail distortion as a result of the Ampere recoil mechanism.

One can find reports in the literature of relatively small currents expelling liquid mercury from the butt joint between two solid conductors. In all cases the heat generated in the film was much too small to furnish a thermodynamic explanation. The author found that liquid mercury filling a 1/8th-inch gap between horizontal ½-inch square copper conductors laid in a groove in a dielectric board would be expelled vertically upward when an increasing DC current through the mercury joint reached about 1000 A. The mercury was pushed upward against the pinch force and had to pull a vacuum underneath it in the groove. The temperature in the mercury did not reach 100°C. Hydrostatic pinch pressure could not explain this effect, even if it were much larger than it is, because it cannot exert an outward force which is greater than the inward pinch. Hence relativistic electromagnetism offers no hint of an explanation.

It has now become clear that experiments performed in the past four years with electric arcs in saltwater reveal hitherto unknown electrodynamic forces. The explosion of water arcs was shown not to be due to the generation and superheating of steam. The water temperature did not even reach the boiling point. Liquid water was expelled from arc cavities and punched holes through metal plates some distance away from the arc. At first it seemed natural to believe the cold water arc explosions were driven by longitudinal Ampere forces. Then Aspden [7] pointed out that the longitudinal forces, as well as Lorentz forces, were orders of magnitude too small to account for the powerful explosions.

Photographic evidence gathered with short atmospheric arcs, carrying as much current as lightning strokes, has proved that the resulting shockwave in air is not caused by an omnidirectional, adiabatic gas expansion process [11]. Therefore thunder also appears to be the result of a cold arc plasma explosion.

The alpha-torque forces of the A-N electrodynamics [1] have been found to act in the right direction for cold arc plasma explosions. They are also much larger than the longitudinal Ampere forces. This is encouraging, but it must not be taken as proof that dense plasma conductors obey the A-N electrodynamics.

REFERENCES

[1] Graneau, P., Ampere-Neumann electrodynamics of metals, Hadronic Press, Nonantum MA 02195, USA (1985).

[2] Graneau, P., "Ampere-Neumann electrodynamics of metallic conductors", Fortschritte der Physik, Vol.34, p.457 (1986).

[3] Ampere, A.M., Theorie mathematique des phenomenes electro-dynamiques, Albert Blanchard, Paris (1958).

[4] Grassmann, H.G., "A new theory of electrodynamics", Annalen der Physik und Chemie, Vol.64, p.1 (1845).

[5] Neumann, F.E., "Die mathematischen Gesetze der inducirten elektrischen Stroeme", Akademie der Wissenschaften, Berlin (1845).

Vorlesungen ueber elektrische Stroeme, Teubner, Leipzig (1884).

[6] Maxwell, J.C., A treatise on electricity and magnetism, Oxford University Press, Oxford (1873).

[7] Aspden, H., "Anomalous electrodynamic explosions in liquids", IEEE Transactions on Plasma Science, Vol. PS-14, p.282 (1986).

[8] Hering, C., "Electromagnetic forces: A search for more rational fundamentals and a proposed revision of the laws", Journal AIEE, Vol.42, p.139 (1923).

[9] Nasilowski, J., [Unduloids and striated disintegration of wires], Exploding wires, Vol.3, Plenum, New York (1964).

[10] Pappas, P.T., "The original Ampere force and Biot-Savart and Lorentz forces", Nuovo Cimento, Vol.76B, p.189 (1983).

[11] Graneau, P., "Cold explosions of short atmospheric arcs", Proceedings of the 1986 International Tesla Symposium, Colorado College, Colorado Springs CO 80905-1095, USA.

A CONTRIBUTION TO THE THEORY OF SKIN-EFFECT

Jan Nasiłowski

Instytut Elektrotechniki

04-703 Warsaw, Poland

INTRODUCTION

We can easily measure transient phenomena connected with the operation of a circuit but the knowledge of how the current increases or decreases in the circuit does not mean that we know the distribution of the current density over the cross-section of the conductors. There are no convincing experiments to prove the existing theory because we cannot measure the current density distribution in the conductors.

The existing theory[1,2,3] maintains that the sine wave alternating current in a conductor shows "the tendency to concentrate in the outer part, or skin, rather than be distributed uniformly over the cross-section of the conductor" and that 50Hz current flows uniformly over the cross-section of the conductors.

According to me this theory is open to criticism.

CRITICAL REMARKS

1. The theory of skin-effect connects this phenomenon with the frequency f of the alternating current in the conductors.

Let us note that this frequency is a parameter which the circuit "recognizes" only after at least one period has elapsed. It seems incorrect to assume that the current distribution in the conductor from the very beginning of the current flow should be dependent on a parameter which will not be known until later (Fig. 1).

When switch S in the circuit shown in Fig. 1a is closed, in the circuit may in extreme cases be either exponential current rise (d.c. source) or a sine wave current (a.c. source). If we consider the current rise starting from zero, the beginning can be the same (Fig. 1b).

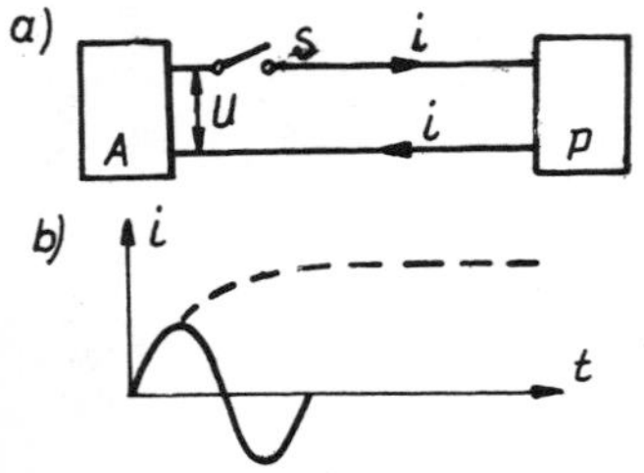

Fig. 1a. Arrangement of the circuit
A - source of the current
P - power receiver
Fig. 1b. Common current increase

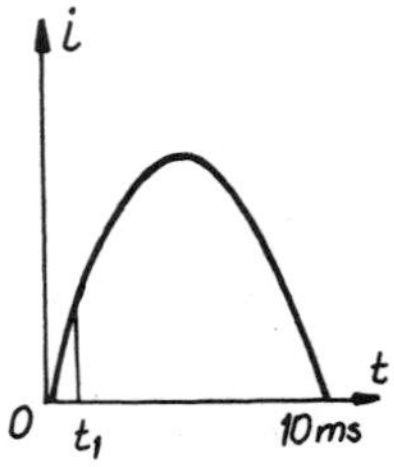

Fig. 2. Half-wave of the 50 Hz current

At the moment of closing switch S the conductors in the circuit receive two signals:
i) A voltage signal U which determines the direction of the movement of the electrons,
ii) A current signal di/dt which is independent of the form of the current.

Some current distribution over the cross-section of the conductors must happen immediately from the beginning of the current flow, therefore it cannot be governed by the frequency but rather by the rate of the current rise di/dt, which is immediately apparent.

2. It seems to be a fault of the existing theory of skin-effect that it does not take into consideration the important fact that for I = constant we have di/dt = 0 and j = constant over the cross-section of the conductors.

In the case of sine wave a.c. currents we have twice in each period a situation where di/dt = 0, nevertheless the existing theory neglects the fact that at those instants the condition of uniform current density is fulfilled[4].

3. The increase in the resistance of conductors for a.c. $(R_{\sim}/R_{=} > 1)$ cannot be considered as the proof of the skin-effect, because a greater current density in the skin layer, as well as a greater current density in the axial area of the wire will cause increased resistance of the conductors due to non-uniformity of the current distribution,

4. McLachlan[1] suggests that the 50 Hz sine wave currents do not cause non-uniformity in the current density distribution of fine wires.

Let us accept this opinion and consider the half-wave of the 50Hz current at the instant t_1 during the rising of the current (Fig. 2). If current density distribution is uniform then the distribution of the magnetic field intesity along of the radius of the wire for the instant t_1 should be as shown in Fig. 3. The same intensity for the instant t_1 + dt will be represented by more steeper straight line. From Fig. 3 it follows that:

$$(dH_r/dt) > (dH_x/dt) > (dH_o/dt) = 0 \qquad (1)$$

therefore the induced emfs in the filaments with cross-section $l \cdot dr$ should be

$$l dr(dH_r/dt) > l dr(dH_x/dt) > l dr(dH_o/dt) = 0 \qquad (2)$$

These induced emfs obey Lenz law and will rise with the distance from the axis. They will in different degrees counteract the current rise. Therefore our initial assumption that j = constant and $H = kx$ can no longer be accepted.

From the above we can draw the following conclusion: For any $di/dt \neq 0$ the $dj/dx \neq 0$. The change of the current in the circuit disturbs the uniform current density in the conductors.

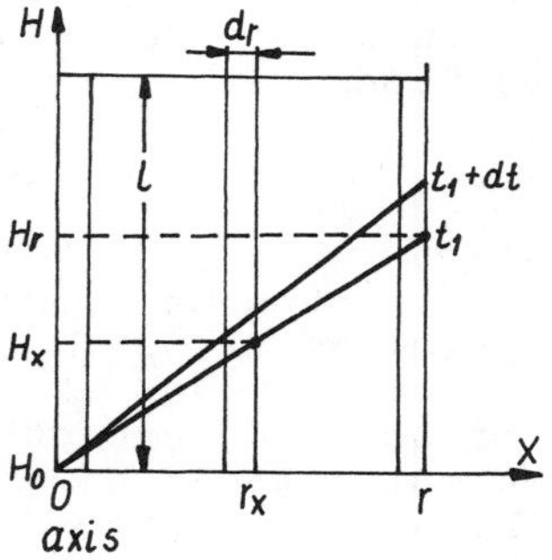

Fig. 3. Magnetic field intensity in the wire in case of uniform current density

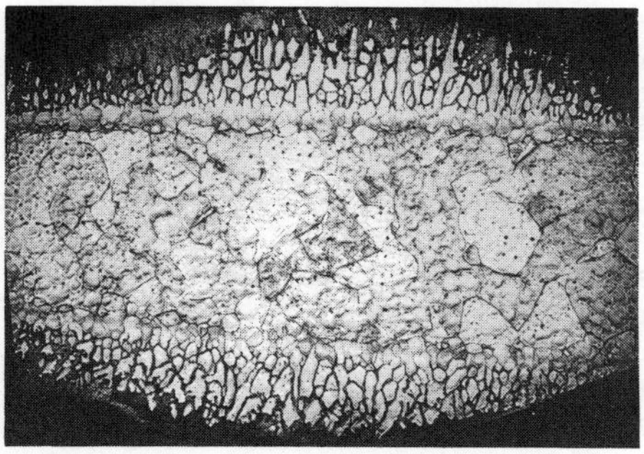

Fig. 4. Internal structure of the wire after single pulse current flow. Melted outer layer, recrystallized core

5. The next McLachlans conclusions[1] deduced from the mathematical theory of the skin-effect is that in the case of a high enough frequency of sine wave current, practically the whole current flows in the skin of the wire, and only a small part of the current flows in the axial zone, but in the reverse direction. The partially reverse current flow against the direction of applied voltage enables rejection of that theory.

EXPERIMENTAL EVIDENCE OF SKIN-EFFECT

Experimental data presented below have been known[5] since 1961 but the explanation of these experiments was dubious[6].

When transforming copper wires into unduloids the skin melting was observed[7], confirmed by the examination of the internal structure of the wire. Fig. 4 shows that the wire of the initial diameter of 0.75 mm. was not completely melted. The core of the wire was not melted but it was recrystallized, while the outer layer crystallized from the liquid state (dendritic structure is seen).

It was believed that skin-melting was not caused by skin-distribition of the current density because the rate of the current changes were slow (Fig. 5). The d.c. single pulse can be considered as a half wave of about 8 Hz sine wave, and for such a low frequency the current density j was considered as constant. The temperature rise calculated for uniform current density distribution:

$$\Delta\vartheta = f\left(\int_0^t j^2\, dt\right)$$

for the instant of the current crest was about 200 °C, and for the end of the current flow about 930 °C.

This temperature is not high enough for melting Cu-wire, but we can see that the outer layer of the wire was nevertheless melted. There is a paradox: Skin melting without skin distribution of the current density.

Let us attempt to solve this paradox. Let us consider one centimeter length of a straight, long Cu-wire suspended horizontally in the air, isolated from magnetic or thermic influences of other conductors. We will analyze two filaments of the wire: axial and superficial each of cross-section dq (Fig. 6). The real current will be replaced by the trapezoidal current i' (Fig. 5).

The voltage drop across the section of axial filament at the initial instant of the current increase, when the resistivity can be assumed constant and uniform across the whole wire and equal to ρ_1 is given by:

$$u_a = di_o \rho_1 \frac{1}{dq} = j_o \rho_1 \tag{3}$$

However for the surface filament the voltage drop is given by:

$$u_s = di_r \rho_1 \frac{1}{dq} + L_r \frac{di}{dt} = j_r \rho_1 + L_r \frac{di}{dt} \tag{4}$$

In the above formula the inductive voltage drop caused by the whole current i of the conductor is shown.
Because $u_a = u_s$ we obtain:

$$j_o \rho_1 = j_r \rho_1 + L_r \frac{di}{dt} \tag{5}$$

Equation 5 can only be true for $j_o > j_r$. During the current rise, the density of the current in the axial area of the wire should be greater than that at the circumference. The current increases more rapidly in the area around the axis of the wire.

During the time when i' is constant (Fig. 5) the current density should also be uniform and during this time the resistivity of the material of the wire will tend to equalize over the cross-section. We assume that at the instant of decrease of the current ($di/dt < 0$) we will have resistivity ρ_2 uniform for the whole wire.

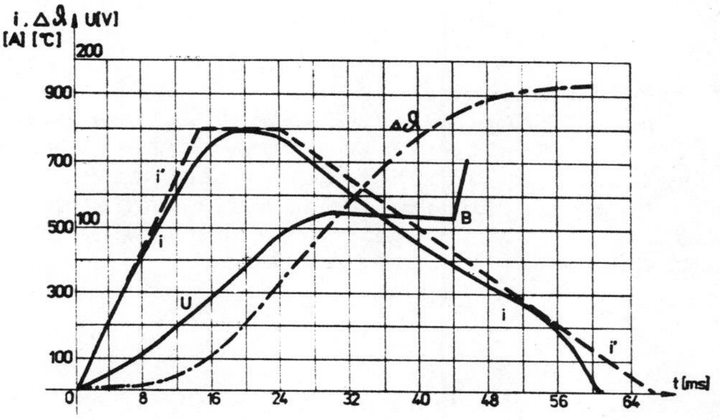

Fig. 5. The single pulse current in d.c. circuit which caused changes of the internal structure as shown in Fig. 4

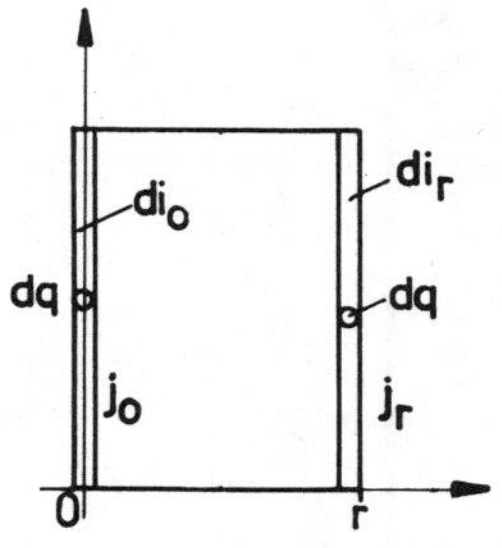

Fig. 6. The two examined filaments of the wire

Therefore for the axial filament we can express the voltage drop as:

$$u_a = j_o \rho_2 \qquad (6)$$

while for the surface filament the voltage drop will be given by:

$$u_s = j_r \rho_2 - L_r \left|\frac{di}{dt}\right| \qquad (7)$$

Comparing equations 6 and 7 we obtain:

$$j_o \rho_2 = j_r \rho_2 - L_r \left|\frac{di}{dt}\right| \qquad (8)$$

This equation is true when $j_o < j_r$. During the decrease of the current, the current density in the axial area is smaller than at the circumference of the wire. The current decays more rapidly in the axial area.

This course of events explains melting of the surface layer of the wire (Fig. 4), when the mean temperature rise of the wire over the room temperature is only about 930°C. Skin distribution of the current during decrease of the current can produce additional heat for observed skin melting.

DISCUSSION

Considering in the similar manner one period of the sine wave alternating current we find pulsating changes of the current desities in the conductor. The current in the conductor rises like a liquid column with convex meniscus and decays like a liquid column with concave meniscus. The hight of the "meniscus" depends on di/dt. For di/dt = 0 there is no meniscus. Current density is uniform.

REFERENCES

1. N.W.McLachlan: Bessel functions for engineers, Oxford, 1955
2. A.E.Knowlton (ed.): Standard handbook for electrical engineers, Chapters: 2-98, 4-27, New York, 1941
3. J.Thewlis (ed.): Enc. dictionary of physik vol. 6, Oxford, 1962
4. M.G.Haines: The inverse skin-effect, Proc. Phys. Soc. 1959 p. 576
5. J.Nasilowski: Phenomena connected with the disintegration of conductors overloaded by short-circuit current (Polish), Przegl. Elektrotechniczny 1961, No 10.
6. J.Nasilowski: Skin-effect at the single pulse d.c. current, (Polish), Prace Inst. Elektrotechniki, 1984, No 131.
7. J.Nasilowski: Unduloids and striated disintegration of wires, Exploding Wires vol.3, Plenum, New York 1964.

2. **COMPUTER PROGRAMMES – Theoretical aspects**

Introductory remarks

R. Sikora

Technical University of Szczecin
Al. Piastow 19
Szczecin, Poland

During the last twenty years engineers and scientists in electromagnetic fields calculation have used computer programs to solve an increasingly large number of their problems. The construction of efficient and economical numerical programs depends on a great variety of factors. This subsection consists of five papers that discuss numerically efficient algorithms for electromagnetic fields calculation in various electrodynamical systems.

In the paper by B. Beland and D. Gamache Maxwell's equations are solved for a sinusoidal current in a cylindrical conductor, assuming a rectangular magnetization curve. The problem is solved numerically because the position of the flux wave, the voltage drop are the losses cannot be expressed as simple functions. Computer results are compared with experimental data.

K. Komeza and S. Wiak consider the method of transient electromagnetic field calculation in devices used to magnetic forming of metals. The axisymmetric problem is described by the nonlinear differential equation in the cylindrical coordinate system. Using the difference method in variational and classical approaches, the spatial - temporal distribution of the electromagnetic field is calculated. The condition of stability of the DuFort-Frankel diagram is given. The authors also describe the reasons of existance and the way of elimination of local nonstabilities.

The paper by W. Krajewski is devoted to the analysis of electromagnetic shielding problems using the boundary element method. The author considers the problems with given boundary conditions and the one with open boundary. Two numerical examples are presented. The numerical difficulties associated with the distribution of electromagnetic fields over unbounded regions have been the focus of considerable research during the last ten years and many new methods have been devised, e.g. the infinite element method. Up to now we have no clear answer which method is the best in electromagnetic field calculations.

A. Krawczyk considers the boundary element method in the analysis of the transient open boundary problems. Theoretical considerations are illustrated by numerical examples.

The paper by J.K. Sykulski and P. Hammond is devoted to the interactive graphical method of field calculation, which the authors called a method of Tubes and Slices. The method is both economical and accurate and can be considered as an alternative to other well established methods but unfortunately it can not be used to eddy current problems.

IMPEDANCE AND LOSSES IN MAGNETIC CYLINDRICAL CONDUCTORS

Bernard Béland

Department of Electrical Engineering
University of Sherbrooke
Sherbrooke (Québec)
Canada J1K 2R1

Daniel Gamache

CEGEP of Sherbrooke
Sherbrooke (Québec)
Canada J1E 2J8

ABSTRACT

Maxwell's equations are solved for a sinusoidal current in a cylindrical conductor. The material is assumed to have a rectangular magnetization curve of the form $B = B_s$ sign H. The magnetization process is described. Initially, the conduction, because of the shielding effect, takes place only on the outside of the conductor. As time progresses, the current circulates in an increasing section of the conductor. The problem is solved mathematically. Tables are presented to simplify the computation process. The results give the voltage drop, the voltage waveform, and the losses. Experimental results confirm the validity of the hypothesis.

INTRODUCTION

In a previous paper [1], flat magnetic conductors were analyzed using a rectangular magnetisation curve and a sinusoidal current. The analysis is repeated in this paper under the same conditions except that the conductor has a circular cross section. This problem is much more complex than the previous one. The position of the flux wave, the voltage drop, and the losses cannot be expressed as simple functions and one must use numerical analysis. However, the equations can be normalized and the computations done once for all. Tables and curves are then made that allow one to solve any particular problem.

Following an electrical fault, there are sometimes currents that flow in magnetic conductors. These are non-linear because of the magnetic characteristic. A method that has shown some success in predicting the behaviour of magnetic materials is that of a rectangular B-H curve of the form [2-6]

$$B = B_s \text{ sign } H \tag{1}$$

in which B_s is the induction at saturation.

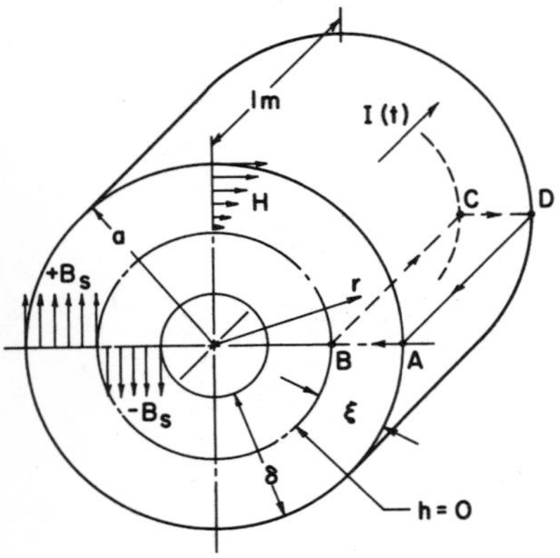

Fig. 1 - Cylinder in steady-state conditions

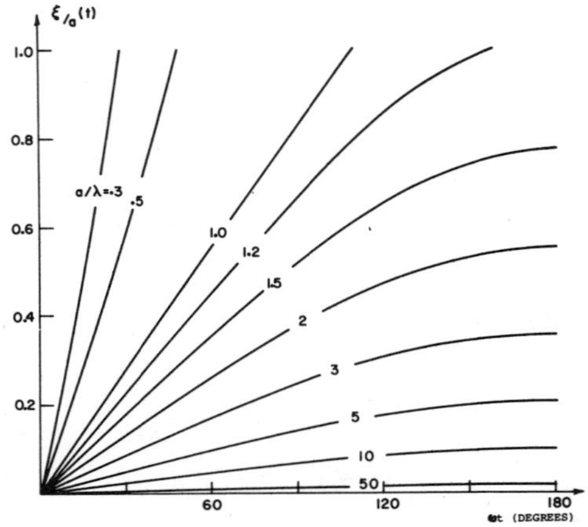

Fig. 2 - Penetration of the flux wave

PROCESS OF MAGNETIZATION

Because of a previous cycle, at time t = 0, under steady-state conditions, let us assume that the material is saturated to $-B_s$ up to a depth δ measured from the surface of the cylinder. The value of δ could be anywhere between 0 and a, the radius of the cylinder. That situation is shown in Figure 1. As time progresses, the current circulates in the outside surface of the cylinder. A magnetic field is thus created around the cylinder which gradually changes the induction from $-B_s$ to $+B_s$. The induction change can not occur instantaneously over the whole cross section; it progresses from the outside surface of the cylinder toward the center. A collapsing surface is thus defined at which surface the magnetic field (H) is zero and the induction changes sign. The position of that flux wave is measured by ξ from the outside surface.

The flux wave penetrates the material to a depth δ at the end of the first half-cycle and the process repeats itself at every other half-cycle. If $a > \delta$, the cylinder of radius $(a-\delta)$ is never influenced by the current and the magnetic field. If the flux wave reaches the center of the cylinder before the end of a half-cycle, the whole material is then saturated and the current is distributed uniformly over the cross section for the rest of the half-cycle.

MATHEMATICAL ANALYSIS

Let A(t) be the conducting cross section of the conducteur which is defined by the radius a and $(a-\xi)$. The current density is

$$J(t) = \frac{I_m \sin \omega t}{\pi\ \xi\ (2a - \xi)} \qquad (2)$$

Maxwell's equation applied to path IJKL of figure 1 gives

$$J(t) = 2\ B_s\ \sigma \frac{d\xi}{dt} \qquad (3)$$

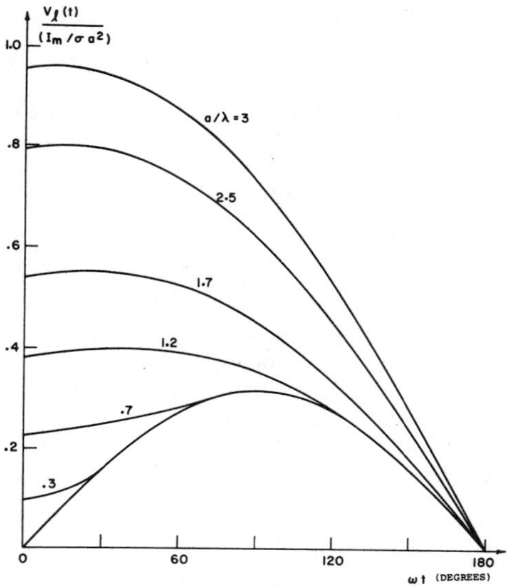

Fig. 3 - Voltage drop in a cylinder

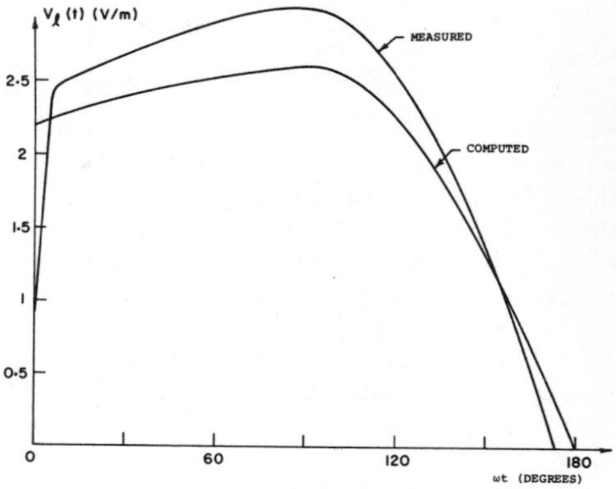

Fig. 4 - Experimental and computed voltage waveform

after integration and using the initial conditions $\xi = 0$ at $t = 0$, leads to

$$\left(\frac{\xi}{a}\right)^2 \left(1 - \frac{\xi}{3a}\right) = \frac{\lambda^2}{a^2} \sin^2 \frac{\omega t}{2} \tag{4}$$

in which $I_m = 2\ \pi a\ H_m$ and

$$\lambda = \sqrt{\frac{2\ H_m}{\omega\ \sigma\ B_s}} \tag{5}$$

Equation (4) gives the position of the flux wave as a function of time although it is not in the form of $\xi = f(t)$. It can, however, be expressed in the form of $t = f(\xi)$. Figure 2 gives the position of the flux wave at any time t for a few values of the parameter a/λ. In equation (4), if $\xi/a = 1$ and $t = \pi/\omega$, then one finds the value of a/λ for which the flux wave reaches the center of the cylinder exactly at the end of a half-cycle. This occurs for $a/\lambda = \sqrt{3/2}$. If a $a/\lambda > \sqrt{3/2}$, the flux wave never reaches the center. The maximum depth of penetration is given by

$$\frac{\delta}{a} \sqrt{1 - \frac{\delta}{3a}} = \frac{\lambda}{a} \tag{6}$$

If $a/\lambda \to \infty$, then one has the case of thick plates that has been analyzed previously [1].

If $a/\lambda < \sqrt{3/2}$, the flux wave reaches the centre of the cylinder before the first half-cycle. The whole cylinder is then saturated and no more flux changes can take place. The current is therefore uniform over the whole section for the rest of the half-cycle. The time at which this happens is

$$t_c = \frac{2}{\omega} \sin^{-1} \left(\sqrt{\frac{2}{3}}\ \frac{a}{\lambda}\right) \tag{7}$$

The voltage drop per unit length in the cylinder is then given by

$$v(t) = \frac{I_m \sin \omega t}{\pi\sigma a^2 \frac{\xi}{a}(2 - \frac{\xi}{a})} , \quad 0 < t < t_c \tag{8a}$$

$$v(t) = \frac{I_m \sin \omega t}{\pi\sigma a^2} , \quad t_c < t < \frac{\pi}{\omega} \tag{8b}$$

This is obtained by the product of the current times the resistance of the cylinder that participates to the conduction. Obviously, if $a/\lambda > \sqrt{3/2}$, only part (8a) of equation (8) is retained with $t_c = \pi/\omega$. Figure 3 gives the voltage waveform for a few values of a/λ.

The power is obtained by the time average of the product of the voltage times the current. One obtains

$$W_\ell = \frac{I_m^2}{\pi^2 \sigma a^2} \left\{ \begin{array}{l} 4(\frac{a}{\lambda})^3 \int_0^{\delta/a} \frac{\xi}{a} \sqrt{(1-\frac{\xi}{a})[1-(\frac{a}{\lambda})^2 (\frac{\xi}{a})^2 (1-\frac{\xi}{3a})} \, d(\frac{\xi}{a}) \\ +\frac{\pi}{2} - \sin^{-1}(\sqrt{\frac{2}{3}} \frac{a}{\lambda}) \sqrt{\frac{2}{3}} \frac{a}{\lambda} \sqrt{1-\frac{2}{3}(\frac{a}{\lambda})^2} [1-\frac{4}{3}(\frac{a}{\lambda})^2] \end{array} \right\} \tag{9}$$

If $a/\lambda \geqslant 3/2$, only the integral term needs be considered; the rest is equal to zero. If $a/\lambda < \sqrt{3/2}$, the upper limit of integration δ/a is unity. Equation (9) can be computed once for all. Table I gives the value of the term in the brackets of equation (9) as a function of the parameter a/λ. Once the losses are evaluated, it is simple to find the apparent resistance per unit length (R_ℓ)

$$R_\ell = \frac{2 W_\ell}{I_m^2} \tag{10}$$

While this is not done here, it is easy, but lengthy to find the RMS value of the voltage, the apparent power, and the power factor.

EXPERIMENTAL WORK

To verify the validity of the theory many rods of different diameters were tested at numerous values of the current. Only a few tests will be reported. A sinusoidal current was forced in rods about 650-mm long using a tuned L-C series filter. The rods were installed in the low-voltage side of a transformer, while the filtering was done in the high-voltage side. The power was measured with an electrodynamic wattmeter, the voltage coil of which was driven by an amplifier. The voltage drop in the rod was taken at two points, 0,5 m distant. The pick-up wires were run in close proximity to the rod and then twisted to feed the amplifier.

Figure 4 shows the experimental and computed voltage waveform for an 6,1-mm (¼-inch) diameter steel rod that was heat-treated to 900°C for half an hour and cooled down slowly in the furnace. The conductivity of the material is $6{,}8 \times 10^6$ Siemen per meter. The peak exciting current is 590 A and the value of a/λ is 0,72. One notes the characteristic voltage

Table I - Losses as a function of a/λ

a/λ	$\frac{\pi\sigma a^2 W_\ell}{I_m^2}$	a/λ	$\frac{\pi\sigma a^2 W_\ell}{I_m^2}$
0,01	0,159	1,5	0,241
0,1	0,159	2	0,305
0,4	0,160	5	0,705
0,7	0,169	10	1,38
1,0	0,188	20	2,73
1,2	0,206	50	6,78

Table II - Computed and measured powers and resistances

I_m (A)	a/λ	W_ℓ (ω/m)		R_ℓ(Ω/m) ×10^{-2}	
		comp.	mea.	comp.	mea.
14,4	3,15	1,89	1,83	1,83	1,77
72	1,83	29	28,3	1,13	1,11
280	1,01	300	313	0,75	0,78
590	0,72	1190	1460	0,68	0,8

discontinuity at t = 0. That jump is noted even for relatively low current although it is not as abrupt. The non-linear nature of the problem is readily apparent.

Table II shows some experimental results from the same rod as above. These results are compared with those obtained from computations. In addition to the power loss per unit length, the table also gives the apparent resistance. Although the exciting current varies by a ratio of 40, the predicted values are usually better than 5%. At the higher currents, the rod was heated at a very rapid rate and it was difficult to take a reliable reading. Within a few seconds, the rod was already hot and its characteristic had changed appreciably. In the times it took for the wattmeter needle to reach its position, the rod had already heated by some 50°C and the conductivity had changed by 30%. Tests were also ran on many other rods with similar results but they are not reported here.

For these comparisons, the choice of B_s is that of the induction that is obtained on the actual magnetization curve at an excitation of one quarter the maximum magnetic field. This was suggested in reference 6.

CONCLUSION

The behaviour of circular magnetic conductors under a sinusoidal current has been predicted with the hypothesis of a square magnetization curve. Maxwell's equations were solved under that hypothesis. The theory predicts the losses and the apparent resistance within 5% in most cases as was verified experimentally.

REFERENCES

1. B. Béland and D. Gamache, "The Impedance of Flat Plate Steel Conductors", Proceedings of the International Conference on Electrical Machines, Part I, Lausanne, Suisse, 18-21, septembre 1984, pp. 5-8.
2. W. MacLean, "Theory of Strong Electromagnetic Waves in Massive Iron", Journal of Applied Physics, Tome 25, 1954, pp. 1267-1270.
3. H.M. McConnell, "Eddy-Current Phenomena in Ferromagnetic Materials", AIEE Trans., Jul. 1954, pp. 226-235.
4. P.D. Agarwal, "Eddy-Current Losses in Solid and Laminated Iron", AIEE Trans., May 1959, pp. 169-181.
5. B. Béland and J. Robert,"Eddy-Current Losses in Saturated Cylindres", AIEE, Trans., 1964, pp. 99-111.
6. B. Béland, "Eddy-Currents in Circular, Square and Rectangular Rods", IEE Proc., Vol. 130, Pt. A, No. 3, May 1983, pp. 112-120.

ERRORS OF SOLUTION OF CLASSICAL AND VARIATIONAL FINITE DIFFERENCE METHOD APPLIED TO TRANSIENT ELECTROMAGNETIC FIELD ANALYSIS

K. Komęza and S. Wiak

Institute of Electrical Machines and Transformers
Technical University of Łódź, Poland

INTRODUCTION

The process of entering the electromagnetic field into ferromagnetics has a very strong influence on the dynamics of electrical machines, especially in the case of field forcing and decaying of exciting current. The other, especially important when applying pulse magnetic field, is magnetic forming of ferromagnetics [1,2]. The process of magnetic forming is based on applying the Maxwell pressure interacting on conductors and ferromagnetics placed within the transient magnetic field. The transient magnetic field is generated in cylindrical winding, while the formed tube is placed inside the winding. The amplitude of flux density in the workpiece wall is 20 ÷ 50 T, while the pressure is 10^9 N/m^2. This problem has been widely solved in world literature, but magnetic permeability of the workpiece has been assumed to be constant [1,2,5]. The searched model system of magnetic forming (Figure 1) has a ferromagnetic tubular core defined by electric conductivity and static magnetic permeability $\mu_{st} = f(r,z,t)$. This problem has led to a two-dimensional problem, this transient electromagnetic field is described by the following field equation:

$$\nu \frac{\partial^2 A_\theta}{\partial z^2} + \frac{\partial \nu}{\partial z} \cdot \frac{\partial A_\theta}{\partial z} + \frac{\nu}{r} \cdot \frac{\partial A_\theta}{\partial r} - \frac{\nu A_\theta}{r^2} + \frac{A_\theta}{r} \cdot \frac{\partial \nu}{\partial r} + \frac{\partial^2 A_\theta}{\partial r^2} + \frac{\partial A_\theta}{\partial r} \cdot \frac{\partial \nu}{\partial r} =$$

$$= - J_w(t) + \frac{\partial A_\theta}{\partial t} \qquad (1)$$

where: $\nu = \frac{1}{\mu_{st}}$, $\quad A_\theta = A_\theta(r,z,t)$.

Having assumed infinite length of the core and sufficient length of exciting winding to neglect the end effects, this problem can be reduced for technical purposes to a one-dimensional problem; thus the equation describing transient electromagnetic field in this case has the following form:

$$\frac{\nu}{r} \cdot \frac{\partial A_\theta}{\partial r} + \frac{A_\theta}{r} \cdot \frac{\partial \nu}{\partial r} - \nu \frac{A_\theta}{r^2} + \frac{\partial \nu}{\partial r} \cdot \frac{\partial A_\theta}{\partial r} + \nu \frac{\partial^2 A_\theta}{\partial r^2} = - J_w(t) + \frac{\partial A_\theta}{\partial t} \qquad (2)$$

where: $A_\Theta = A_\Theta(r,t)$.

Additionally, thickness of winding is assumed to be negligibly small.

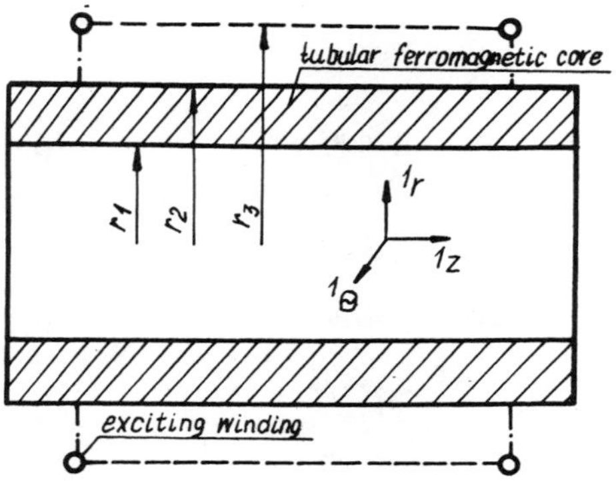

Fig.1. Axial cross section of coil-tubular core system for magnetic forming

THE DIFFERENCE METHOD IN THE VARIATIONAL APPROACH APPLIED TO THE APPROXIMATION OF THE POSED PROBLEM

Equation (1) describing transient electromagnetic field has been transformed to this form:

$$1_\Theta \operatorname{curl}_\Theta \left(1_z\left(\nu \frac{A_\Theta}{r} + \nu \frac{\partial A_\Theta}{\partial r}\right) - 1_r \nu \frac{\partial A_\Theta}{\partial z}\right) = J_w(t) - \sigma \frac{\partial A_\Theta}{\partial t} . \tag{3}$$

The considered region is divided into subregions of arbitrary dimensions. Each considered region is represented by static magnetic permeability μ_{st} and conductivity σ . Figure 2 presents an exemplary division. For an arbitrary chosen subregion closed with line "abcdefgha", surrounding point (i,j) equation (3) (after applying Stoke´s theorem) is transformed to this form:

$$\oint \left(1_z\left(\nu \frac{\partial A_\Theta}{\partial r} + \nu \frac{A_\Theta}{r}\right) - 1_r \nu \frac{\partial A_\Theta}{\partial z}\right) drdz = \int_E^W \int_S^N \left(J_w(t) - \frac{\partial A_\Theta}{\partial t}\right) drdz. \tag{4}$$

After applying the DuFort and Frankel difference diagram (Figure 3) to approximate equation (4) and reducing this problem to a one-dimensional task, as has been mentioned above, the following difference equation will be obtained:

$$A_i^{k+1} \left[\frac{\nu_{NE}\, r_i}{r_i(r_i - 1/2\Delta r_i)} + \frac{\nu_{SW}\, r_i}{r_{i+1}(r_i + 1/2\Delta r_{i+1})} + \frac{\sigma_{NE}\, r_i + \sigma_{NW}\, r_{i+1}}{2\,\Delta t} \right] =$$

$$A_{i-1}^{k} \frac{2\nu_{NE}(r_i - \Delta r_i)}{r_i(r_i - 1/2\Delta r_i)} + A_{i+1}^{k} \frac{2\nu_{NW}(r_i + \Delta r_{i+1})}{r_{i+1}(r_i + 1/2\; r_{i+1})} - A_i^{k-1} \left[\frac{\nu_{NE}\, r_i}{r_i(r_i - 1/2\Delta r_i)} + \right.$$

$$+ \frac{\nu_{NW} \cdot r_i}{r_{i+1}(r_i + 1/2\Delta r_{i+1})} - \frac{\sigma_{NE}\Delta r_i + \sigma_{NW}\Delta r_{i+1}}{2 \quad t}\Bigg] + J^k_{NE}\Delta r_i + J^k_{NW}\Delta r_{i+1}. \quad (5)$$

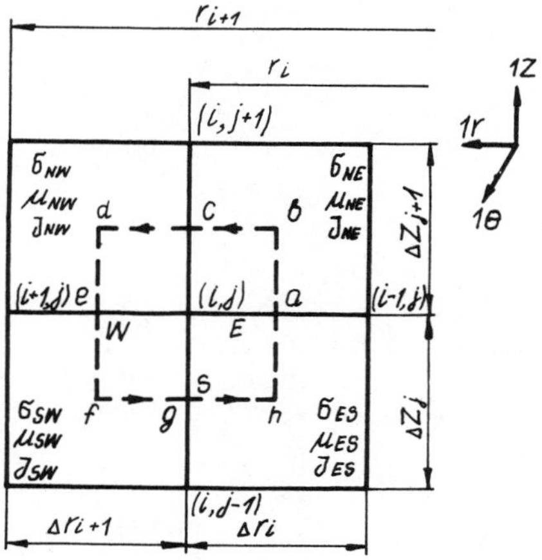

Fig.2. An (i,j) node in the mesh with its four surroundings cells

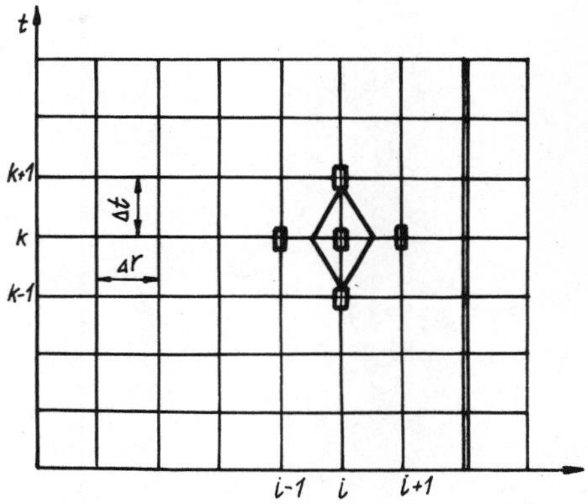

Fig.3. DuFort and Frankel difference diagram in (r,t) coordinates assumed for numerical calculation

STABILITY OF DUFORT AND FRANKEL DIFFERENCE DIAGRAM

It is proved that the DuFort and Frankel difference diagram is absolutely stable if it approximates the problem described with the linear parabolic equations (3). Let us assume $\Delta r_i = \Delta r_{i+1}$, $\nu_{NW} = \nu_{NE}$ and $\sigma_{NW} = \sigma_{NE}$.
The abocF assumption does not limit the generality of the performed proof but simplifies the notation of the equation. The difference equation (5) describing the transient electromagnetic field will be transformed, for ferromagnetics, to this form:

$$(A_i^{k+1} + A_i^{k-1}) : b_i^k - c_i^k \cdot A_{i+1}^k - d_i^k \cdot A_{i-1}^k = A_i^{k+1} - A_i^{k-1} \quad (6)$$

where:

$$b_i^k = \frac{\Delta t \ r_i}{2\sigma\Delta r^2}\left[\frac{\nu_{NE}}{r_i + 1/2\Delta r} + \frac{\nu_{SE}}{r_i - 1/2\Delta r}\right] \quad (7)$$

$$c_i^k = \frac{2\Delta t \ \nu_{NE}(r_i + \Delta r)}{r^2(r_i + 1/2\,\Delta r)} \quad (8)$$

$$d_i^k = \frac{2\Delta t \ \nu_{SE}(r_i - \Delta r)}{r^2(r_i - 1/2\,\Delta r)} \quad . \quad (9)$$

Stability of the difference diagram approximating nonlinear field equation can be proved by means of "Energy Inequalities Method" based on Richtmyer and Morton´s lemma:

Lemma (3)

A three-layer difference diagram is strongly stable if the sequence of real numbers S^k exists where $S^k = S^k(A^k)$ and two positive constants K_1 and K_2 such that

$$K_1^{-1} (\|A^{k+1}\|^2 + \|A^k\|^2) \leqslant S^k \leqslant (\|A^k\|^2 + \|A^{k+1}\|^2)K_1 \quad (10)$$

and

$$S^{k+1} - S^k \leqslant K_2 \Delta t (\|A^{k+1}\|^2 + \|A^k\|^2) \quad (11)$$

$K = 0,1 \ldots.,$ then

$$\|A^k\|^2 \leqslant \text{constant} \quad (\|A^0\|^2 + \|A^1\|^2) \quad (12)$$

which denotes the stability of the diagram. Equation (6) is multiplied by $\Delta r(A_i{}^{k+1}\ A_i{}^{k-1})$ and summing is performed for all internal points of the grid. After some necessary algebra, the following equation will be obtained:

$$(b^k,(A^{k+1})^2) + 2(b^k A^{k+1}, A^{k-1}) + (b^k,(A^{k-1})^2) - (c^k T_+ A^k, A^{k+1}) +$$
$$-(c^k T_+ A^k, A^{k-1}) - (d^k T_- A^k, A^{k+1}) - (d^k T_- A^k, A^{k-1}) = \|A^{k+1}\|^2 - \|A^{k-1}\|^2 \quad (13)$$

where:

$$T_+ A^k = A_{i+1}^k \quad (14)$$

$$T_- A^k = A_{i-1}^k . \quad (15)$$

On the basis of equation (13) the expression for S^k is formulated of the form:

$$S^k = \|A^{k+1}\|^2 + \|A^k\|^2 + (c^k T_+ A^k, A^{k+1}) - (b^k,(A^{k+1})^2) - 2(b^k A^{k+1}, A^k) +$$
$$+ (d^k T_- A^k, A^{k+1}). \quad (16)$$

After performing all necessary transformations the following inequality will be obtained:

$$S^k \geqslant \|A^k\|^2, (1 + \tfrac{1}{2}\delta + \tfrac{1}{2}\gamma - B) + A^{k+1}\ {}^2(1 + \tfrac{1}{2}\delta + \tfrac{1}{2}\gamma - 2B) \quad (17)$$

where:

$$\delta = \min\{c^k\}$$
$$\gamma = \min\{d^k\}$$
$$B = \max\{b^k\} .$$

The form S^k is positively determined if the following inequality is fulfilled:

$$1 + \tfrac{1}{2}\delta + \tfrac{1}{2}\gamma - 2 \cdot B \geqslant 0 . \quad (18)$$

Inequality (18) after necessary transformations is as follows

$$\frac{\Delta t}{\Delta r^2} \left(\frac{1}{\mu_{min}} - \frac{1}{\mu_{max}}\right) \leqslant \frac{\sigma}{2} . \quad (19)$$

For the media with strong magnetic nonlinearity it can be assumed that $\frac{1}{\mu_{min}} \gg \frac{1}{\mu_{max}}$, thus the final inequality is

$$\frac{\Delta t}{\Delta r^2} \leqslant \frac{\sigma}{2} \cdot \mu_{min} . \quad (20)$$

The difference between the forms S^{k+1} and S^k on (k+1) and k time layer has its upper boundary, which means that inequality (11) of Richtmyer and Morton´s lemma is fulfilled. On the basis of the above analysis the following corollary can be formulated:

<u>Corollary</u>

The difference diagram (6) approximating differential equation (2) is stable if inequality (20) is fulfilled.

DIFFERENCE METHOD IN CLASSICAL APPROACH APPLIED TO APPROXIMATION OF THE POSED PROBLEM

The differential equation (2) has been transformed to the difference form. DuFort and Frankel´s difference diagram has been applied to the approximation of this equation, thus the equation is as follows:

$$A^k_{i+1}(w^k_i + z^k_i + v^k_i) + A^k_{i-1}(w^k_i - z^k_i - v^k_i) - w^k_i(A^{k+1}_i + A^{k-1}_i) + A^k_i l^k_i =$$

$$= A^{k+1}_i - A^{k-1}_i \tag{21}$$

where:

$$w^k_i = \frac{2\,\Delta t}{\sigma\,\Delta r^2 \left[\mu_{st}\right]^k_i} \tag{22}$$

$$z^k_i = \frac{\Delta t}{\Delta r \cdot r_i \left[\mu_{st}\right]^k_i} \tag{23}$$

$$v^k_i = \frac{\Delta t \frac{\partial}{\partial r}\left[\mu_{st}\right]^k_i}{\Delta r \left[\mu^2_{st}\right]^k_i \sigma} \tag{24}$$

$$l^k_i = -\left(\frac{\frac{\partial}{\partial r}\left[\mu_{st}\right]^k_i}{r_i \left[\mu_{st}\right]^k_i} + \frac{1}{r^2_i}\right) \cdot \frac{2\,\Delta t}{\sigma\left[\mu_{st}\right]^k_i} \tag{25}$$

$$d^k_i = w^k_i + z^k_i - v^k_i \tag{26}$$

$$f^k_i = w^k_i - z^k_i + v^k_i \;. \tag{27}$$

After some necessary algebra, expression for S^k is as follows:

$$S^k = \|A^{k+1}\|^2 + \|A^k\|^2 - (d^k A^{k+1}, T_+ A^k) + (w^k, (A^{k+1})^2) - (f^k A^{k+1} T_- A^k) +$$

$$+ (l^k A^k, A^{k+1}) + 2(w^k A^{k+1}, A^k). \tag{28}$$

After performing all necessary transformations the following inequality is:

$$S^k \|A^{k+1}\|^2(1 + \tfrac{1}{2}l + 2 \cdot w_m - \tfrac{1}{2}D - \tfrac{1}{2} \cdot F) + \|A^k\|^2(1 + w_m - \tfrac{1}{2}D - \tfrac{1}{2}F + \tfrac{1}{2}l). \tag{29}$$

The form S^k is positively determined if the following inequality is fulfilled:

$$1 + w_m - \tfrac{1}{2}D - \tfrac{1}{2}F + \tfrac{1}{2}l \geqslant 0 \tag{30}$$

where:

$$F = \max \left\{ f_i^k \right\}$$

$$l = \min \left\{ l_i^k \right\}$$

$$w_m = \min \left\{ w_i^k \right\} \quad .$$

Inequality (30) after some algebra is as follows:

$$\frac{\Delta t}{\Delta r^2} \leqslant \frac{\sigma}{2} \cdot \mu_{min} \; , \tag{31}$$

which is equivalent to the condition of choice of time step Δt obtained from expression (20) formulated for DuFort and Frankel's difference diagram in variational from.

CONCLUSION

It is widely known that oscillations on vector potential curves, which appear during the calculation process, mean nonstability of the difference diagram. These oscillations have mainly appeared on the air-ferromagnetic intermediate surface. It seems to be very interesting choosing greater values Δt and Δr, to shorten the time of computing, not fulfilling criterion of stability. The vector potential curves with oscillations, from calculations made in such cases, can be smoothed by applying average square approximation formulas for chosen subregions and chosen time-level. The conducted analysis of stability condition of difference diagrams in variational and classical purchase gives the same expressions (20) and (31) for stability, but algebraic system equations is much more complicated for the classical difference method. It is also widely known that in this case changing the space step during the calculating process can generate additional local nonstabilities.

REFERENCES

1. Dobrogowski J., Kołaczkowski Z., Tychowski F., Electromagnetic metal forming. Polish Academy of Sciences. Poznań, Poland, 1979.
2. Knoepfel H., Pulsed high magnetic field. North-Holland, Amsterdam-London, Publ. Comp. 1970.
3. Richtmyer R.D., Morton K.W., Difference methods for initial value problems. Interscience Publishers, a Division of John Wiley and Sons, New York-London-Sydney, 1967.
4. Wiak S., Stability of Difference Diagrams in Classical and Variational Purpose Approximating a Class of Nonlinear One-Dimensional Parabolic Field Equations. Archiv für Elektr. 71 : 889 (1988).

ANALYSIS OF ELECTROMAGNETIC SHIELDING PROBLEMS USING THE BEM

Wojciech Krajewski

Department of Fundamental Research
Instytut Elektrotechniki, Warsaw, Poland

INTRODUCTION

In general, the analysis of shielding problems requires calculating of electromagnetic field distribution in the regions which can be divided into the subregions with different physical properties and different material parameters.

In the case of the magnetic shielding, non-conducting subregions with different magnetic permeability have to be considered. In the problems connected with electromagnetic shielding the conducting and non-conducting subregions have to be considered. From the mathematical point of **view** in the first of the above-mentioned cases discontinuity of the normal derivative of the magnetic vector potential on the interface between subregions appears, however, a type of governing partial differential equation is maintained. In the second case a type of governing partial differential equation on crossing the interface changes.

In the present paper, the Boundary Element Method (BEM)[1,2] has been adapted to the analysis of electromagnetic shielding problems. This method is widely applied to solving the elliptic[1,2,3], parabolic[2] and hyperbolic[6] partial differential equations but in the hitherto existing literature there is a relatively small number of papers in which the BEM is employed to the field calculations in piece-wise homogeneous regions[4,5].

In this paper the two-dimensional sinusoidal steady-state problem comprising the z-component of the current density vector only, has been considered. The problem with given boundary conditions and the one with an open boundary have been discussed.

PROBLEM WITH GIVEN BOUNDARY CONDITIONS

Let us consider the following problem shown in Fig.1. In the subregions the z-component of the magnetic vector potential $A(x,y)$ satisfies the following partial differential equations

$$\nabla^2 A - \alpha^2 A = 0 \qquad \text{for} \quad (x,y) \in \Omega_1 \tag{1a}$$

$$\nabla^2 A = 0 \qquad \text{for} \quad (x,y) \in \Omega_2 \tag{1b}$$

$$\nabla^2 A = -\mu J_e \qquad \text{for} \quad (x,y) \in \Omega_3 \tag{1c}$$

where

$\alpha^2 = j\omega\gamma\mu$
ω - angular frequency
μ - magnetic permeability
γ - conductivity
$j = \sqrt{-1}$
J_e - excitation current density

Applying the Green's second theorem and the appropriate Green's free space functions one can obtain the following boundary integral equations

$$c_i A(P_i) = \int_{\Gamma_1 \cup \Gamma_I} \frac{\partial A(P)}{\partial n} G_1^*(P,P_i)\, d\Gamma - \int_{\Gamma_1 \cup \Gamma_I} \frac{\partial G_1^*(P,P_i)}{\partial n} A(P)\, d\Gamma \quad \text{for} \quad P_i \in \Omega_1 \tag{2a}$$

$$c_i A(P_i) = \int_{\Gamma_I \cup \Gamma_2} \frac{\partial A(P)}{\partial n} G_2^*(P,P_i)\, d\Gamma - \int_{\Gamma_I \cup \Gamma_2} \frac{\partial G_2^*(P,P_i)}{\partial n} A(P)\, d\Gamma - \int_{\Omega_3} \mu J_e(P) G_2^*(P,P_i)\, d\Omega$$

$$\text{for} \quad P_i \in \Omega_2 \cup \Omega_3 \tag{2b}$$

where

$$c_i = \begin{cases} 1 & \text{for internal points} \\ 1/2 & \text{for points on a smooth boundary} \end{cases}$$

$$G_1^*(P,P_i) = \frac{1}{2\pi} K_0\left(\alpha\, r_{PP_i}\right) \tag{3}$$

$$G_2^*(P,P_i) = \frac{1}{2\pi} \ln \frac{1}{r_{PP_i}} \tag{4}$$

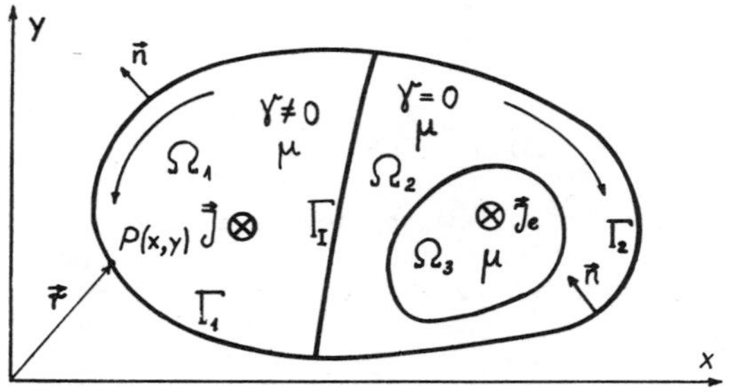

Fig. 1. Geometry of the problem

$G_1^*(P,P_i)$ and $G_2^*(P,P_i)$ are the Green's free space functions for Helmholtz's and Laplace's equations, respectively. $G_1^*(P,P_i)$ is represented by the modified Bessel function of the second kind and zero order.

The application of the BEM transforms the set of two boundary integral equations (2a) and (2b) into two sets of the following algebraic equations

$$\begin{bmatrix} H^1 & H_I^1 \end{bmatrix} \begin{bmatrix} u^1 \\ u_I^1 \end{bmatrix} = \begin{bmatrix} G^1 & G_I^1 \end{bmatrix} \begin{bmatrix} q^1 \\ q_I^1 \end{bmatrix} \tag{5a}$$

$$\begin{bmatrix} H_I^2 & H^2 \end{bmatrix} \begin{bmatrix} u_I^2 \\ u^2 \end{bmatrix} = \begin{bmatrix} G_I^2 & G^2 \end{bmatrix} \begin{bmatrix} q_I^2 \\ q^2 \end{bmatrix} + \mathbf{e} \tag{5b}$$

The components of **u** and **q** are the nodal values of the magnetic vector potential and of its normal derivative, respectively. The superscripts 1 and 2 denote the nodes belonging to the boundary of Ω_1 and of Ω_2. The traditional BEM matrices **H** and **G** are divided here into the submatrices corresponding to the boundary of $\Omega_1 \cup \Omega_2$ and to the interface Γ_I. **e** is the forcing vector.

To obtain the set of algebraic equations with the unique solution it is necessary to introduce the compatibility conditions on the interface Γ_I

For the uniform μ in the whole region the equality of the left hand and right hand normal derivative of the magnetic vector potential occurs as a result of the continuity of the tangential component of the magnetic field strength. On the other hand, the discontinuity of the magnetic vector potential appears and its jump on crossing the interface is equal to an unknown constant value A_c. This discontinuity can be expressed as follows

$$\lim_{P \to P_i} A(P) = A_I^2 \qquad \text{for} \quad P \in \Omega_2 \text{ and } P_i \in \Gamma_I \tag{6a}$$

$$\lim_{P \to P_i} A(P) = A_I^2 + A_c \qquad \text{for} \quad P \in \Omega_1 \text{ and } P_i \in \Gamma_I \tag{6b}$$

Now, we have to introduce one additional equation. This is so-called equation of constraints resulting from the Ampere's law. It is the relation between the total current through Ω_1 and the normal derivative of the magnetic vector potential on the boundary

$$\int_{\Gamma_1 \cup \Gamma_I} \frac{\partial A}{\partial n}\, d\Gamma = -\mu I_t \tag{7}$$

Now, let us combine equations (5a) and (5b) eliminating u_I^1 and taking into account that $q_I^1 \equiv q_I^2 \equiv q_I$. Finally, using equation (7) the following set of algebraic equations is obtained

$$\left[\begin{array}{ccccc|c} H^1 & H_I^1 & 0 & h_o & & 0 \\ 0 & H_I^2 & H^2 & 0 & & 0 \\ \hline & & 0 & & & 1 \end{array}\right] \left[\begin{array}{c} u^1 \\ u_I^2 \\ u^2 \\ A_c \\ \hline -\mu I_t \end{array}\right] = \left[\begin{array}{cc|c} G & G_I^1 & 0 \\ 0 & G_I^2 & G^2 \\ \hline & d^T & 0 \end{array}\right] \left[\begin{array}{c} q^1 \\ q_I \\ \hline q^2 \end{array}\right] + \left[\begin{array}{c} 0 \\ e \\ \hline 0 \end{array}\right] \qquad (8)$$

Each component of the vector $\mathbf{h_o}$ is the sum of the entries of the row of $\mathbf{H_I^1}$. The components of the vector d (for zero-order approximation of both magnetic vector potential and its normal derivative) are the lengths of the elements of Γ_1 and Γ_I.

Obviously, to obtain the vector of unknown quantities on the left-hand side, the matrix equation (8) has to be rearranged according to the kind of assumed boundary conditions.

OPEN BOUNDARY PROBLEM

In some eddy current problems there are no given boundary conditions on the conductor boundary. Here, the integral condition for the total current in the conductor exists only. These problems are named "the open boundary problems" and analysis of them using domain methods (e.g. Finite Element Method) meets some difficulties, however, the BEM is very suitable for this purpose.

It can be noticed that the problem shown in Fig.2 is the particular case of the one discussed in Section 2. Taking into account that we have $\Gamma_1 \cup \Gamma_2 = \emptyset$ here and in sequel H^1, H^2, G^1, G^2 have dimensions 0×0, it is easy to transform the matrix equation (8) into that as given by Rucker and Richter[4].

NUMERICAL EXAMPLES

Two numerical examples are presented in this section. The first one concerns the problem with given boundary conditions. Its geometry is shown in Fig.3a. The assumed dimensions and parameters of the system are as follows

$$a=0{,}1\text{m},\quad b=0{,}01\text{m},\quad c=0{,}01\text{m},\quad d=0{,}005\text{m},\quad \gamma=5{,}6\circ 10^7\,\text{S/m},\quad \mu=\mu_o,\quad f=50\text{Hz},$$

$$J_e=(1+j0)\circ 10^6\,\text{A/m}^2,\ I_t=(0+j0)\,\text{A}$$

The boundary has been divided into 27 elements and the interface has been divided into 10 elements.

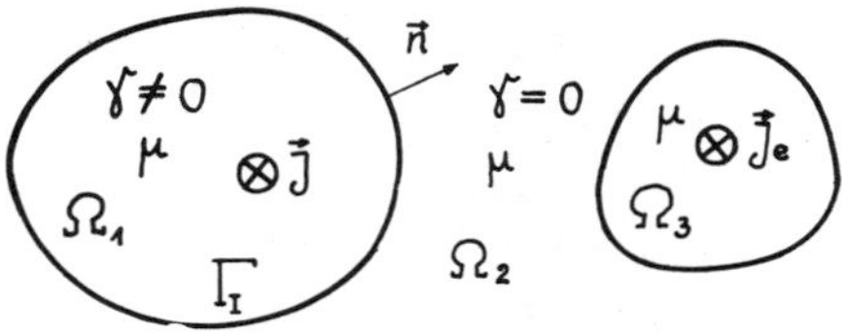

Fig. 2. Open boundary problem

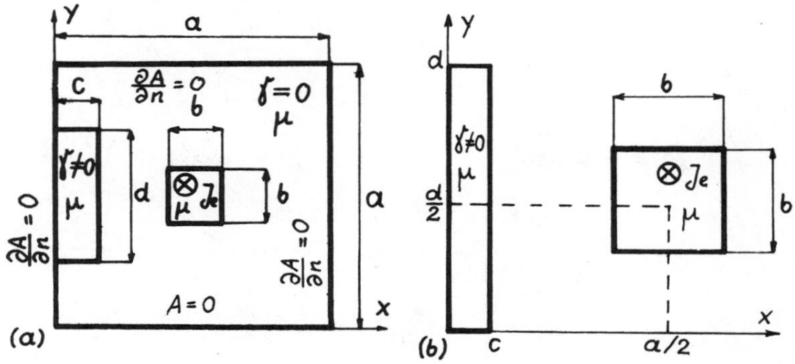

Fig. 3. Geometry of the examples. (a) Problem with given boundary conditions; (b) Open boundary problem.

The real and imaginary parts of the current density vector at the chosen points of the shield are shown in Figs. 4a and 4b.

The second example concerns the open boundary problem (Fig.3b). The assumed dimensions a, b, c, d are identical as in the first example. The interface has been divided into 22 elements. The results of numerical calculations are shown in Figs. 5a and 5b.

The relation between the magnetic vector potential and the current density vector is as follows

$$J(x,y) = -j\omega\gamma A(x,y) \tag{9}$$

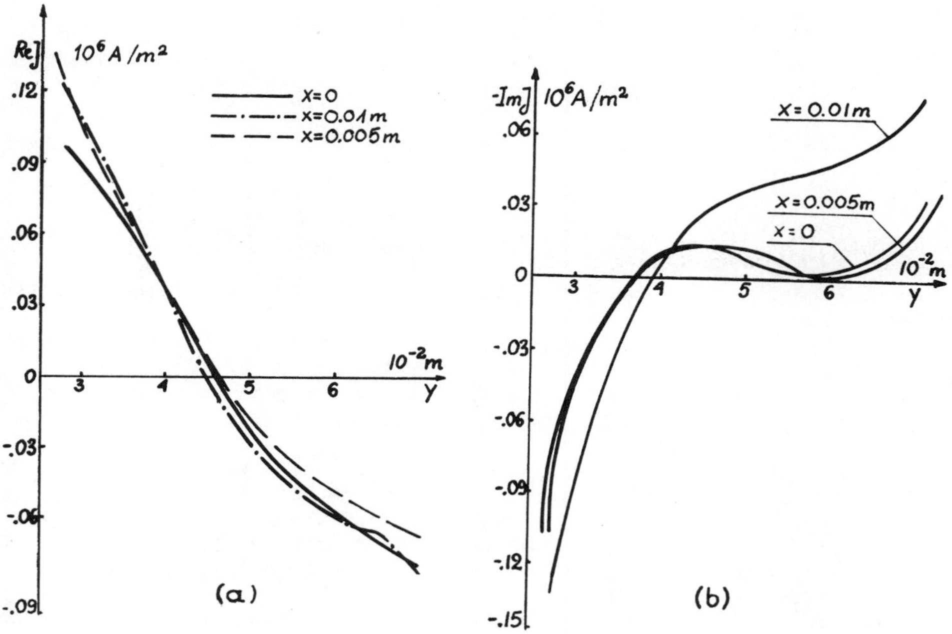

Fig. 4. Current density at the chosen points of the shield (the problem with given boundary conditions). (a) Real part; (b) Imaginary part

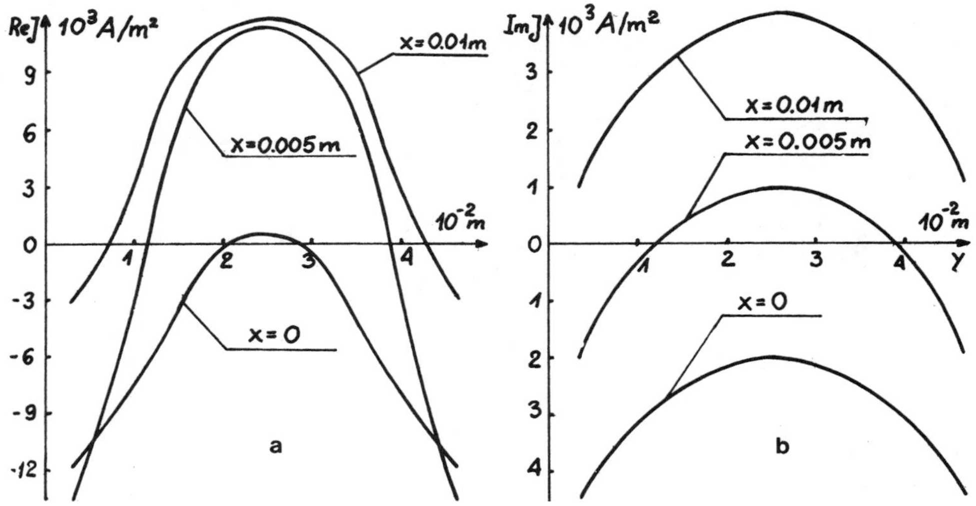

Fig. 5. Current density vector at the chosen points of the shield (open boundary problem). (a) Real part; (b) Imaginary part.

CONCLUSIONS

The application of the BEM to the analysis of 2D sinusoidal electromagnetic field in piece-wise homogeneous regions which can be divided into conducting and non-conducting subregions has been considered. The matrix equation corresponding to problem in question has been derived.

It has been shown that an open boundary problem is the particular case of the general one considered here and can be effectively solved using the same computer program.

Two examples of numerical calculations have been presented. The good agreement with physical nature of the phenomenon was obtained.

The described technique can be useful for the electromagnetic field analysis of the problems related to the electromagnetic shielding.

REFERENCES

1. C. A. Brebbia , "The Boundary Element Method for Engineers" Pentech Press, London (1980).
2. C. A. Brebbia, J. C. F. Telles and L. C. Wrobel, "Boundary Element Techniques" Springer Verlag (1984).
3. J. M. Schneider and S. J. Salon, A boundary integral formulation of the eddy current problem, IEEE Transactions on Magnetics, Vol. MAG-16. No.5, (1980), pp 1086-1088.
4. M. W. Rucker and K. R. Richter , Calculation of two-dimensional eddy current problems with the boundary element method, IEEE Transactions on Magnetics, Vol. MAG-19, No.6 (1983) pp. 2429-2432.
5. J. Poltz, E. Kuffel, Application of boundary element techniques for 2D eddy current problems, IEEE Transactions on Magnetics, Vol. MAG-21, No.6 (1985) pp. 2254-2256.
6. W. Krajewski, The method of computation of coupled electromechanical fields in metal tubes formed by a magnetic impulse, Prace Instytutu Elektrotechniki, No. 132, Warszawa (1984) pp. 71-84.

THE USE OF THE BOUNDARY ELEMENT METHOD IN TRANSIENT OPEN BOUNDARY PROBLEMS

Andrzej Krawczyk

Department of Fundamental Research
Instytut Elektrotechniki
Warsaw, Poland

INTRODUCTION

The so called "open boundary problems"(OBP) have been solved by means of various methods. Mainly, the combination of two or more methods were employed and they were usually the finite element method (FEM) and some others, like the boundary integral methods or analytical methods. Also, the OBP have been analysed by means of the FEM only, but then the artificial boundary has been put far enough from the object in question. All the approaches lack the methodological clarity and, what is more important, are cumbersome in a practical use. Moreover, the FEM, being the basis for all the above mentioned solving procedures, is not suited to the eddy current problems, especially while dealing with transient phenomena[1]. That is why, for the last few years there have been attemps to solve the OBP by the use of the BEM only. The harmonic (sinusoidal) problems have already been solved[4] and this paper is to show how to solve transient problems. These problems are very attractive for the BEM treatment as the BEM is well-suited to both transient and open boundary problems[2].

MODEL AND ASSUMPTIONS

All the problems to be considered are of 2-D nature, i.e. the object analysed is long enough to neglect the variation of field quantities along its length. Consider a set of conductors, cross-sections of which are placed in a plane (x-y)(Fig.1).The number of conductorsis Nand their cross-sections are of surfaces S_k and boundaries Γ_k (k=1,...,N).The transient phenomenon is due to either current flowing in the objects in question (externally supplied) or currents flowing in N_L independent conductors, in which eddy currents do not exist. The analysed conductors can be active (the case mentioned above) or passive and then only the eddy currents occur . The way the eddy currents flow depends generally on the connection of the conductors. This problem will be discussed later on.

For simplicity it is assumed that all the conductors are of the same permeability and equal to that of surrounding space, while conductivities are obviously different. In the k-th conductor the parabolic equation

$$\nabla^2 A_k - \mu\gamma \frac{\partial A_k}{\partial t} = -\mu j_k(t) \qquad (1)$$

is held and in the non-conducting region the Poisson equation

$$\nabla^2 A' = -\mu \sum_{i=1}^{N_L} j_i(t) \qquad (2)$$

is fulfilled. The right-hand side of (1) is given by the total current I_k (if any) and the current densities occuring in (2) are prescribed. The right-hand side of (1) represents self-excitation and can be eliminated by simple transformation. The system has to satisfy additionally the interface conditions, which on each boundary Γ_k are as follows

$$B_t' \big|_{\Gamma_k} = B_{tk} \qquad (3)$$

$$B_n' \big|_{\Gamma_k} = B_{nk} \qquad (4)$$

where B_t' and B_n' are the tangential and normal components of magnetic flux density existing in non-conducting region and B_{tk} and B_{nk} are the same components but occuring in k-th conductor. The condition (3) can be written in magnetic vector potential terms and then it becomes

$$A' \big|_{\Gamma_k} = A_k + \text{constant} \qquad (5)$$

as the magnetic vector potential is not necessarily continuous

BOUNDARY ELEMENT ANALYSIS

The BEM enables us to consider the problem in question as the problems coupled on the boundary Γ_k . The boundary Γ_k can be used as a basis both in an internal parabolic problem and in an external problem of Laplace. For the parabolic equation (1), when using the boundary element technique, one obtains the following matrix equation[2]

$$[H_k(t_m)]\ \{A_k(t_m)\} - [G_k(t_m)]\ \{B_{tk}(t_m)\} = \{F_k(t_m)\} \qquad (6)$$

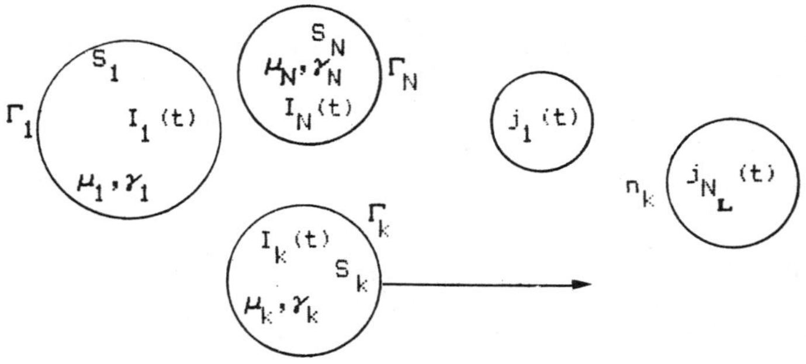

Fig.1 Set of conductors

Making use of the same technique for equation (2) one obtains[3]

$$[H_L]\ \{A'(t_m)\}-[G_L]\ \{B_t(t_m)\}=\{F_L(t_m)\} \tag{7}$$

The denotations above used are of common use in the BEM area[2,3], so they do not require to be described here. It should be mentioned only that the right-hand side vector of (6) results partially from the history of the process considered. The matrices $[H_k]$ and $[G_k]$ are generally dependent on the time being, but for the uniform time step they are constant for each instance of time and depend on the time step. It should also be noticed that the matrix equation (7) describes the external Poisson's problem. Hence the properties of the matrix $[H_L]$ occuring in the internal problem are no longer valid. The proof of the correctness of the formula (7) is based on the consideration of double-connected region with external boundary tending to infinity[3]. The right-hand side vector of (7) represents the Newtonian potential of exciting currents, or in other terms, the magnetic vector potential due to these currents. There is also a possibility of considering pure flux excitation. In such a case one should know the exciting magnetic vector potential and put it on the right-hand side of (7). The mostly used flux excitation is of homogenous nature, which allows one to write the simple relation for magnetic vector potential A:

$$A = B\ \xi \tag{8}$$

where B is the exciting flux density and ξ is the coordinate perpendicular to the flux lines. This approach seems to be more suitable in such problems than that of the FEM.

There are N matrix equations (6) each of order n, if all the boundaries are divided into n elements and one matrix equation (7) of Nxn order. Joining all the eqůations and introducing the interface conditions one finally obtains the matrix equation of 2xNxn order with a lot of zero blocks. But, as it has already been mentioned the distribution of eddy currents is determined by the way of connection of the conductors. Hence one needs the additional conditions resulting from the circuit analysis. This problem is known as the "constraint problem" and has been developed to a great extent in the FEM analysis.The conductors can be connected at infinity in series or in parallel or disconnected. Consider, at first, the latter case. Then all that is known is the total current flowing in the conductor. The relation joining the total current and boundary values is Ampere's law:

$$\int_{\Gamma_k} B_{tk}\, d\Gamma=\mu I_k \tag{9}$$

If the conductors are independent of each other or connected in series the equation (9) must be written for each of them. If they are connected in parallel, only one equation is to be written as only the total current is known, but not its distribution. Finally, the number of equations is $2xNxn+N_1$, where N_1 is the number of groups in which the parallel connection occurs. On the other hand the number of unknowns is equal to 2xNxn and in order to get quadratic matrix againone should add N_1 unknowns. They result from the fact that the vector potential is given within accuracy of some constant (see formula(5)). Thus in all independent conductors or/and parallel groups one should add the unknown constant C_k to the potential A_k . Hence, the equation (6) takes the following form

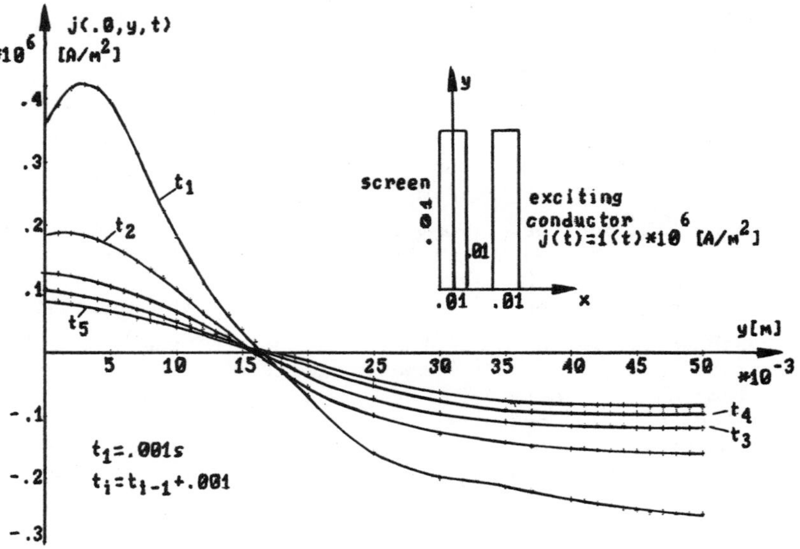

Fig.2 Eddy current distributions (current excitation)

$$[H_k]\{A_k\}-[G_k]\{B_{tk}\}+\{H_{ok}\}C_k = \{F_k\} \tag{10}$$

where elements of the vector $\{H_{ok}\}$ are calculated from

$$H_{oki} = \sum_{j=1}^{n} H_{kij} \tag{11}$$

This way the open boundary problem is entirely described. It should be stressed that the BEM allows one to compute eddy current density directly[2]. That is why one avoids errors arising due to numerical differentiation of the vector $\{A\}$.

EXEMPLARY CALCULATIONS

To illustrate the approach described above some numerical examples have been considered. All the examples are to simulate the screen effect of suddenly excited magnetic field. The screen parameters are as follows: dimension of cross-section (.01m x .1m), γ=56x10^{6} S/m μ= .4πx10^{-6}. The exciting conductor is placed next to the screen, as it is shown in Fig.2. The exciting current is of unit-step time form. The current distributions along y-axis at successive time steps and the time courses at selected space points are shown in Figs 2,3, respectively. As the illustration of the flux excitation the same screen with external homogenous field has been examined. The exciting field was also of unit-step time form. The results of eddy current distributions at succesive time instances are shown in Fig.4. It has to be stressed that the eddy currents obtained at the points close to the boundary cannot be obtained by the use of the "domain methods". All the results were obtained with 14 elements only. Hence, the order of matrix is considerably smaller than that of the FEM. On the other hand, it is well-known that the BEM needs considerable

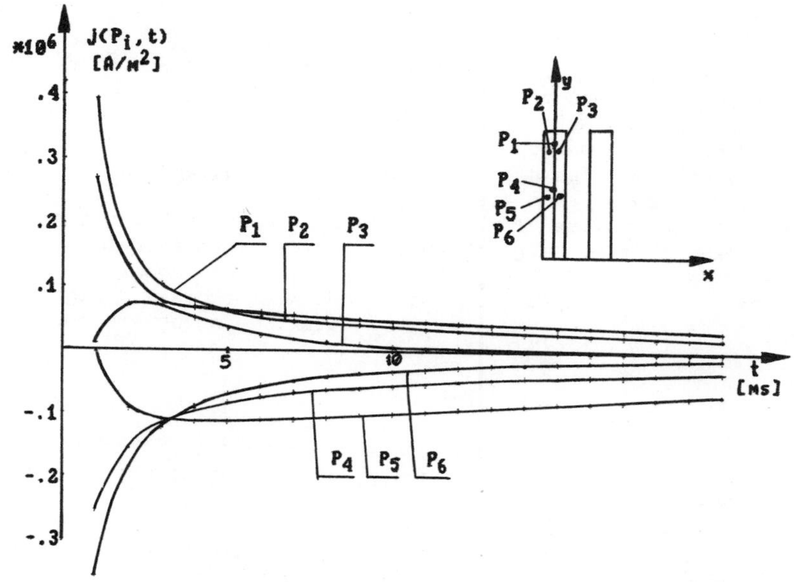

Fig.3 Eddy current courses (current excitation)

computer effort to calculate the matrix coefficients which contain the special functions. Nevertheless, using some approximation techniques one can decrease this effort in considerable way.

The obtained results, although not confirmed by any test or experiments seem to be in accordance with theoretical anticipation.

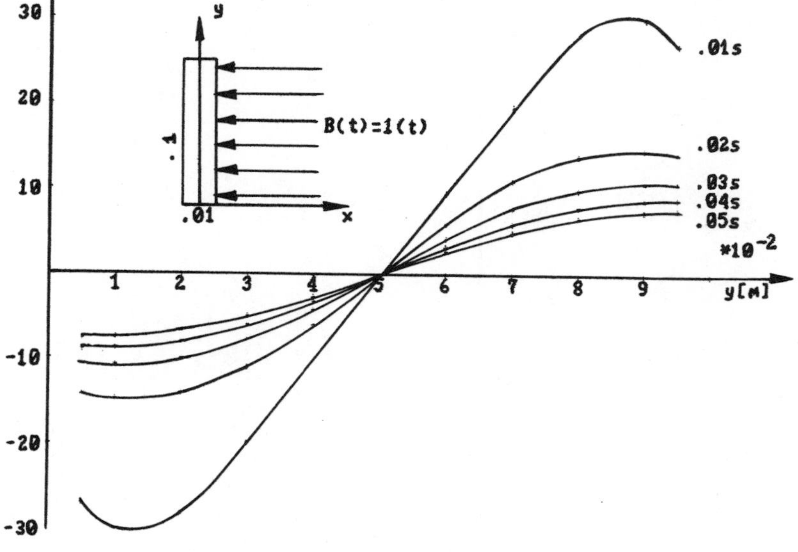

Fig. 4 Eddy current distributions (flux excitation)

FINAL REMARKS

The paper shows the possibility of the use of the BEM in the analysis of the transient open boundary problems. It is seen that the BEM enables to solve such problems with methodological clarity. Unlike the domain methods (FEM,FDM), the method in question is well-suited just to transient eddy current problems. The big advantage of the BEM is that it gives the eddy current density at time-space points an analyst wants[2]. Hence the skin-effect can be analysed far more deeper, even if exciting current varies very fast in time (e.g. the unit step). The essential disadvantage of this method is that it cannot be used in solving nonlinear problems. Nevertheless, it is expected that the method will achieve a high level of applicability in the near future as it is still being developed.

REFERENCES

1. A.Krawczyk, Application of Finite Element Method to Transient Problems of Technical Electrodynamics, Rozprawy Elektrotechniczne 2:331 (1983) (in Polish)
2. A.Krawczyk, The Calculation of Transient Eddy Currents by Means of the Boundary Element Method, in: "Boundary Element VIII", M.Tanaka and C.A. Brebbia, eds, Springer Verlag (1986)
3. C.A.Brebbia et al, "Boundary Element Techniques", Springer Verlag(1984)
4. W.M.Rucker and K.R.Richter, Calculation of Two-Dimensional Eddy Current Problems with the Boundary Element Method, IEEE Trans. on Mag. 6:2429 (1983)

CALCULATION OF ELECTRIC AND MAGNETIC FIELD

BY MEANS OF THE METHOD OF TUBES AND SLICES

J. K. Sykulski and P. Hammond

Department of Electrical Engineering
University of Southampton
United Kingdom

INTRODUCTION

Variational methods are often preferred in the calculation of electromagnetic field problems. They rely on the fact that the field as a system can be described by global parameters which have a stationary value when the system is varied in a prescribed manner. The parameters are generally formulated in terms of energy and the variational method explores the behaviour of the energy near its equilibrium value. Typical energy functionals are expressed in terms of circuit quantitites such as resistance, inductance and capacitance.

An outstanding example of a variational method is the method of finite elements. The elements span the entire system and the system energy is expressed in terms of a large number of arbitrary coefficients which describe the field. The first-order variations with respect to these coefficients are equated to zero and the values of the coefficients are obtained from the resulting large set of simultaneous equations. The method is well understood and needs no detailed discussion.

In the finite-element method the choice of the describing functions of the field and the shape of the elements is based on the desire for simplicity in the system of equations. For this reason a single global coordinate set is used. In very symmetrical configurations the global coordinates are appropriate for the local field distribution but in general there is no simple relationship between the direction of the field and the coordinate axes. Ideally of course the coordinates should follow the direction of the field. In the words of Oliver Heaviside 'space should be divided into tubes not cubes'.

This suggests that a variational method based on geometry might offer advantages over the algebraic method of finite elements of arbitrary topology. The approach of differential geometry with its emphasis on tangent spaces and congruence of sets of curves seems to be particularly well suited to vector field problems. In the differential geometry of 3-dimensional spaces, a local field vector has two properties. One of these is associated with an element of surface area and thus defines a tube of flux, while the other is associated with a line segment and thus defines a potential difference. Such vector fields, therefore, have two properties and their natural discretization is in terms of tubes of flux and slices of equipotential. This leads to the question whether these geometrical objects of tubes and slices can be embodied in a variational method.

The uniquely correct solution of a field problem produces a system of orthogonal tubes and slices. The system energy can be calculated either from the tubes or the slices and the two values be equal. In numerical work, however, unique solutions are in principle unobtainable. Clearly the enforcement of orthogonality even in an approximate manner is a costly computational process, because of the large amount of information which has to be handled. We, therefore, enquire what would happen to the system energy if the tubes and slices were not strinctly orthogonal. We should then have two ways of calculating the energy, one based on tubes and the other on slices. Very remarkably it is found that the two values provide upper and lower bounds to the energy. They, therefore, provide confidence limits and can also be used to determine the accuracy of the field representation. As this accuracy improves, so the two bounds converge on each other.

The proof of this statement has been given previously[1, 2, 3] in various contexts. The most striking form of the proof is based on Maxwell's discussion of the resistance of a conductor[4]. He pointed out that the tubes would be defined by the insertion of insulating boundaries and the slices by hingly-conducting sheets. Hence a tube solution would give an upper bound to the resistance and a slice solution would give a lower bound. From this it can be shown that tubes and slices provide dual variational principles for the system energy.

THE METHOD OF TUBES AND SLICES

We now enquire into the implementation of the method. It will be noted that each tube or slice is an independent entity which has a particular energy associated with it. Hence the calculation of the system energy involves the addition of these energies, which is a very simple numerical process easily carried out on a personal computer. There are no unknown coefficients or simultaneous equations as in the finite-element method. Instead the variation is brought about graphically by altering the tubes and slices. The process is interactive because each solution is displayed and is then modified by means of computer graphics. The analyst, therefore, has complete control and the method is particularly appropriate for use in design offices.

In CAD language, the three stages of pre-processing, solving and post-processing, have been amalgamated in the method of tubes and slices into an interactive session. Indeed, when the data is fed into the computer the distribution of tubes and slices becomes automatically known. Hence an approximate solution already exists. The 'solution' stage is replaced by calculation of energy bounds, which apart from specifying a global system parameter, provide an absolute assessment of error in the solution. Finally, the post-processing does not exist on its own but is combined with the procedure of reshaping the tube and slice distributions for the next step in this interactive scheme, a process equivalent to the mesh adaptation in finite element techniques. This is, therefore, a remarkably different concept in the CAD environment.

With so much emphasis on interactive graphics any computer implementation of the method must provide the user with a simple but fast and efficient way of data input and adaptation. The requirements in terms of hardware are minimal due to the simplicity of a numerical procedure, but a data tablet or a 'mouse' would be an advantage as a means of communication with a graphics display. There are many possible schemes which could be devised to carry out various graphics operations, with the conflicting objectives of reducing the user's effort and retaining his full control over the process. In the TAS (Tubes And Slices) program, which has been developed in Southampton, the distributions of tubes and slices are generated and subsequently reshaped with the help of 'construction points'. A set of the construction points defines a 'mesh' of quadrilaterals representing a combined system of tubes and slices. All quadrilaterals are further subdivided into 'sub-tubes' and 'sub-slices' with lines parallel to appropriate interface or boundary lines. These additional subdivisions are performed automatically but their number is user-defined.

As an illustration of the above ideals, Figure 1 shows four adjacent quadrilaterals (a) and a possible tube/slice distribution (b) generated by (a).

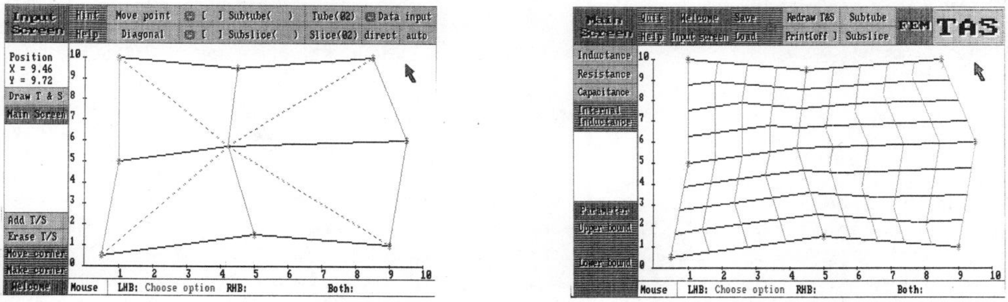

a Fig. 1 b

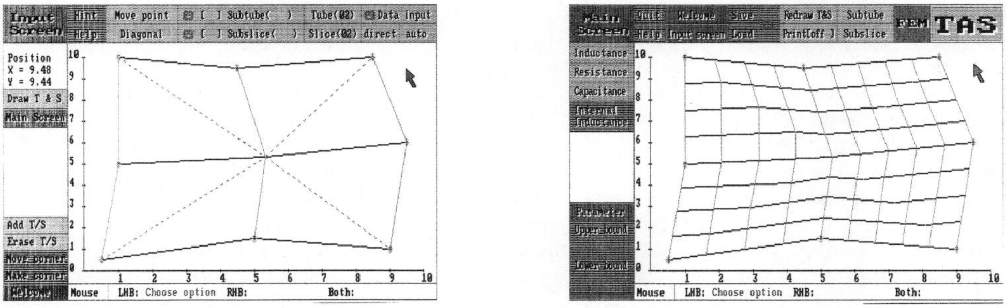

a Fig. 2 b

After the external boundary of a problem has been defined the construction points must be specified and this is done on the screen using a 'mouse'. The sequence of defining these points is important so that the program is able to distinguish between tubes and slices. In TAS a convention 'along tubes' is adopted. Introduction of contruction points completes the data preparation stage.

In order to readjust the shape of tubes/slices two techniques are available. First, the position of any construction point can be changed dynamically using a mouse, and secondly an alternative diagonal can be chosen within a quadrilateral. All points are movable including the ones lying on the boundary, in which case they slide along boundary lines. The only exceptions are the boundary corners. If the middle point of Figure 1a is moved to a new position of Figure 2a, the reshaped system of Figure 2b will be created.

The effect of changing the diagonal is illustrated in Fig. 3.

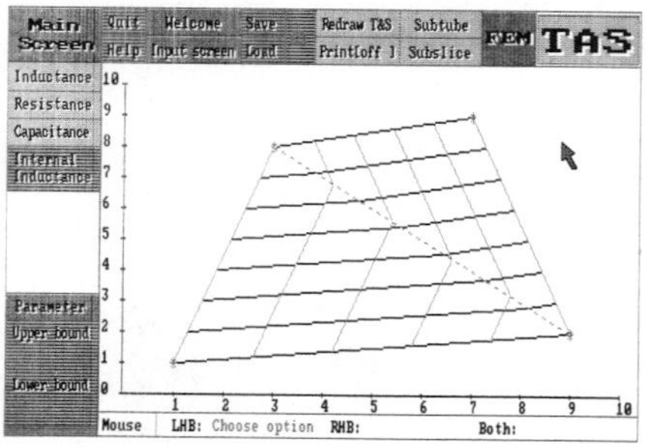

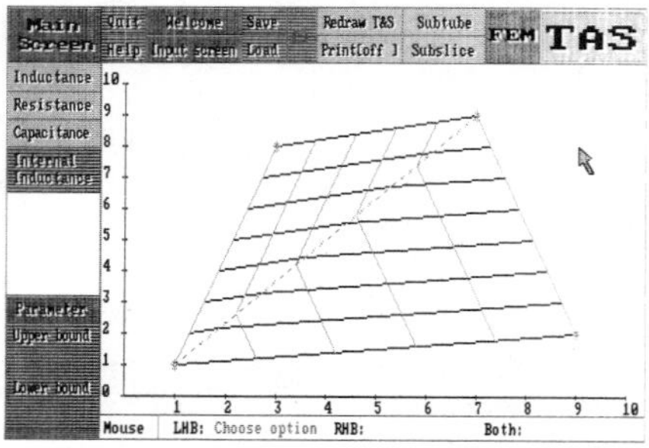

a Fig. 3 **b**

It will be noted that in TAS both systems of tubes and slices are generated simultaneously. This technique has proved convenient but is by no means necessary nor restrictive. The two distributions remain independent in a sense that the variation of either may be examined separately, and although both bounds are always calculated this can hardly be considered a disadvantage.

CONVERGENCE AND ACCURACY

An obvious way of improving the accuracy of the solution is to examine the field locally and enforce orthogonality of tubes and slices, a technique similar to the method of curvilinear squares. A relatively high number of tubes and slices is required in order to relax the constraints imposed by discretization. This is, therefore, an expensive technique and even more importantly it leaves uncertainty about the global accuracy. A perfect criterion for the global accuracy is the closeness of the upper and lower energy bounds. Each bound can be interactively improved until the difference is reduced to the acceptable value. A combination of these two techniques could also prove useful.

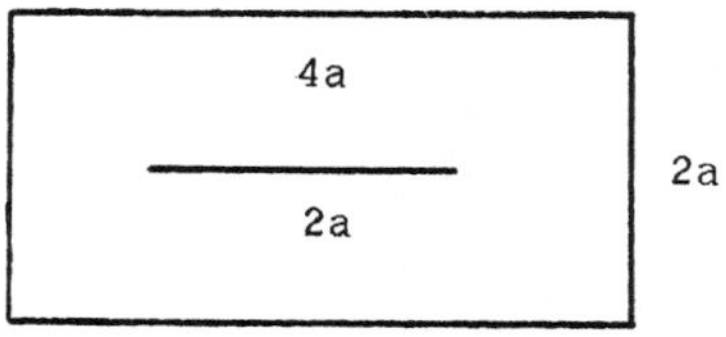

Figure 4

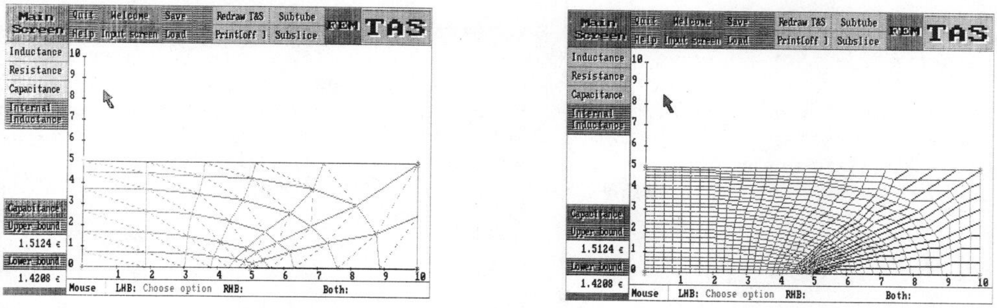

a Fig 5 b

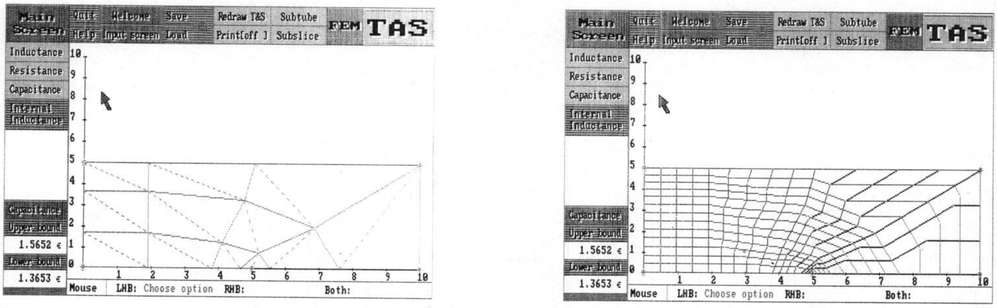

a Fig 6 b

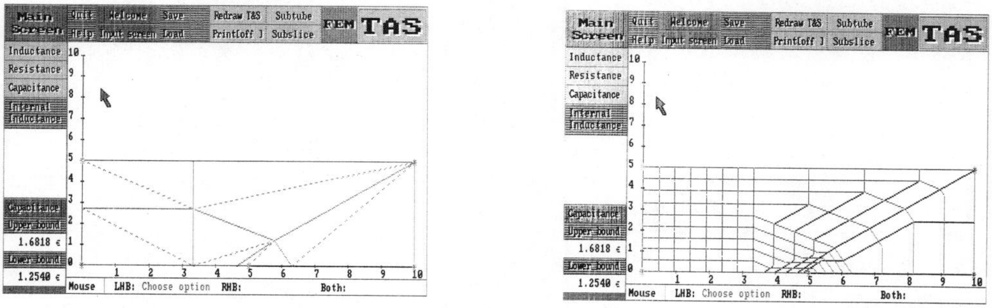

a Fig 7 b

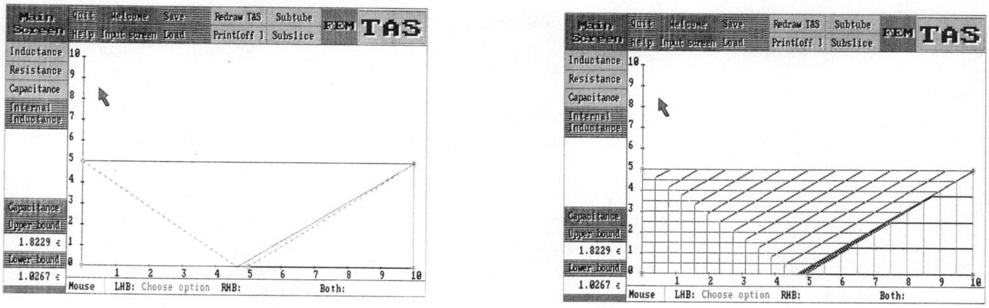

a Fig 8 b

There is, however, another interesting possibility arising from the observation that an average of the two bounds is likely to be very stable, provided we apply similar levels of approximation to the two systems of tubes and slices. If we are able to control the bounds in this manner and if we seek the global parameter then we can allow a much bigger divergence of the dual bounds, but we should no longer treat the two distributions as being independent because we want both of them to be approximated in a similar manner. To illustrate this concept let us calculate the capacitance per unit length for the coaxial system of electrodes shown in Figure 4.

Due to symmetry only a quarter of this system needs to be investigated. Figures 5 to 8 show four different solutions obtained using TAS with a decreasing number of tubes and slices. The results are summarised in Table 1. The value obtained from a finite element program is also included for comparison. As might be expected the confidence limits given by the two energy bounds deteriorate but the average value remains remarkably stable, even for the very crude field approximation of Figure 7.

TABLE 1 Capacitance per unit length

Program	all values x ϵ_0			error in C_{ave}	difference in bounds
	upper bound C_-	lower bound C_-	$C_{ave} = \frac{C_- - C_-}{2}$		
TAS 7 tubes x 6 slices	6.05	5.69	5.87	<0.5%	6%
TAS 4 tubes x 3 slices	6.26	5.46	5.86	<0.5%	14%
TAS 3 tubes x 2 slices	6.73	5.01	5.87	<0.5%	29%
TAS 2 tubes x 1 slice	7.29	4.11	5.70	2.9%	54%
finite elements (1600 elements)			5.87		

CONCLUSIONS

An interactive graphical method of field calculation, which we have called a method of Tubes and Slices, has been described. It appears that the method is both economical and accurate and that it deserves to be considered as an alternative to other well established methods.

REFERENCES

1. P Hammond: 'Energy method in electromagnetism', Clarendon Press, 1981.
2. P Hammond, M C Romero, S A Robertson: 'Fast numerical method for calculation of electric and magnetic field based on potential-flux duality', IEE Procedings, Vol 132, pt A, No 2, 1985.
3. P Hammond, Zhan Qionghua: 'Calculation of Poissonian fields by means of the method of tubes and slices', IEE Proceedings, Vol 132, Pt A, No 4, 1985.
4. J C Maxwell: 'Electrity and magnetism', Claredon Press, 1892.

2. **COMPUTER PROGRAMMES - Practical use**

Introductory remarks

J. A. Tegopoulos

National Technical University of Athens
42, October 28th Street
106 83 Athens, Greece

Computer programs may be used in different aspects of practical problems concerning electromagnetic fields. One of them, perhaps the largest, deals with the application of finite element method. A second aspect in the use of computer programmes in solving specific practical problems.

In this chapter four papers, discussed at the Symposium, are scheduled, which may be divided according to the afore mentioned categories. Thus, the first two papers fall in the first category while the other papers in the second.

In the paper by Li Biao, the perturbation FEM for electromagnetic fields is developed theoretically with reference to eddy current coupling. Equations for perturbation variation and discretization are also obtained. The analysis indicates that for non-linear problems solved by PFEM, equations become linear while the coefficient matrix remains constant. Only the column on the right hand side of equations should be recalculated. This method requires only a small number of iterations.

In the finite element simulation of electrical machines difficulties are met when charges in the position of different parts to be taken into account a very thin parts have to be modelled. The use of macro-elements, which has been dready successfully applied in the first case, is proposed by D. Shen and J.C. Sabonnadiere for the second one. They show how better accuracy in the solution can be reached, when investigating, in particular, transformer overlapping joints.

A variety of specific practical problems are tackled in the second category. Thus the paper by S. Wiak, develops a method for determining the transient current in a circuit containing non-linear elements when it is subjected to a D.C. step function excitation. The element possesses D.C. initial magnetization. Kirchhoff's law is employed in the transient state using lumped parameters as well as the one dimensional diffusion equation for the intensity of the magnetic field. From the latter the internal voltage is calculated numerically. Test cases show good agreement between the theoretical and experimental values. The method is fast but good only for simple elementsof the static type. A guide line is proposed for more complicated elements.

Finally, in the paper by K. Zakrzewski and M. Lukaniszyn, a numerical method is developed for determining the inductance of an air-core coil of rectangular crossection in the presence of magnetic screens. No eddy currents are assumed to flow in the screens. The screens are also assumed to have infinite permeability, or field impinges perpendicularly on them. The coil is assumed to be placed inside a similar but considerably larger space the boundaries of which show zero field. For the numerical solution, Finite Differences are used. For the solution of 3D Poisson's equation the direct method was used by means of FFT. The inductance is computed from the energy integral. Experimental results agree very well with theoretical results if the saturation of screens is low.

PERTURBATION FINITE ELEMENT METHOD

OF ELECTROMAGNETIC FIELD CALCULATION FOR EDDY-CURRENT COUPLING

Li Biao

The Air Force College of Engineering

Xi'an, Shaanxi, P.R.C.

ABSTRACT

This paper describes how the non-linear equation of the FEM for electromagnetic field calculation of eddy-current coupling can be turned directly into a set of linear perturbation equations by perturbation method. The set of perturbation equations can be solved directly without the need for iteration. For each and every perturbation equation, the coefficient matrices are identical and remain constant, with only the right hand column vectors being different from one another. The right hand column vectors of any perturbation equation can be calculated on the basis of the solutions of the preceding set of equations. Therefore, the method is simpler and more effective than the general FEM.

INTRODUCTION

The FEM is an effective method for electromagnetic field calculation. However, the equations are non-linear when non-linear problems are solved by the FEM. Usually, non-linear equations must be solved by iteration method. But the iteration methods now available (e.q. N-R method, Relax method, etc.) may prove to be ineffective: the iteration may converge very slowly, or it may not converge, or even diverge, thus consuming a lot of computer time even with no fruitful result. Therefore, it is considered urgent to find a more effective method.

The perturbation finite element method (PFEM) is an effective method for non-linear electromagnetic field calculation. In this paper, the non-linear discretization equation of FEM in electromagnetic field calculation for eddy-cuttent coupling is to be turned directly into a set of linear perturbation equations by the perturbation method, and, the calculation of the right hand column vectors of the set of perturbation equations is to be discussed in detail to see the features of the PFEM.

DISCRETIZATION EQUATION OF EDDY-CURRENT COUPLING

The boundary value problem of Poisson's equation for 2D magnetostatic field of eddy-current coupling is [1] :

$$\left.\begin{array}{l} \Omega : \dfrac{\partial}{\partial x}\left(\nu \dfrac{\partial A}{\partial x}\right) + \dfrac{\partial}{\partial y}\left(\nu \dfrac{\partial A}{\partial y}\right) = -J_f - s\Omega_0\sigma\left(x \dfrac{dA}{dy} - y \dfrac{dA}{dx}\right) \\ S_1 : A = A_0 \\ S_2 : \nu \dfrac{dA}{dn} = q \end{array}\right\} (1)$$

In which, A is the vector magnetic potential, ν is the magnetic reluctivity, J_f is the excitation current density, s is the slip, Ω_0 is the reference rotational anglular velocity, σ is the conductivity.

Equation (1) can be discretized by the weighted residuals method. As is the case with the general FEM, for every triangle element, we have

$$(\nu [k_1]_e + [k_2]_e)\{A\}_e = \{J_f\}_e + \{q\}_e \tag{2}$$

in which,

$$[k_1]_e = \frac{1}{4D}\begin{bmatrix} c_ic_i + b_ib_i & c_jc_i + b_jb_j & c_mc_i + b_mb_i \\ c_ic_j + b_ib_j & c_jc_j + b_jb_i & c_mc_j + b_mb_j \\ c_ic_m + b_ib_m & c_jc_m + b_jb_m & c_mc_m + b_mb_m \end{bmatrix} \tag{3}$$

$$[k_2]_e = \frac{s\Omega_0\sigma}{24}\begin{bmatrix} c_id_{xi} - b_id_{yi} & c_jd_{xi} - b_jd_{yi} & c_md_{xi} - b_md_{yi} \\ c_id_{xj} - b_id_{yj} & c_jd_{xj} - b_jd_{yj} & c_md_{xj} - b_md_{yj} \\ c_id_{xm} - b_id_{ym} & c_jd_{xm} - b_jd_{ym} & c_md_{xm} - b_md_{ym} \end{bmatrix} \tag{4}$$

in which,

$$d_{xi} = 2x_i + x_j + x_m,\quad d_{xj} = x_i + 2x_j + x_m,\quad d_{xm} = x_i + x_j + 2x_m,$$

$$d_{yi} = 2y_i + y_j + y_m,\quad d_{yj} = y_i + 2y_j + y_m,\quad d_{ym} = y_i + y_j + 2y_m,$$

$$\{A\}_e = \begin{Bmatrix} A_i \\ A_j \\ A_m \end{Bmatrix},\quad \{J_f\}_e = \begin{Bmatrix} J_f \cdot D/3 \\ J_f \cdot D/3 \\ J_f \cdot D/3 \end{Bmatrix},\quad \{q\}_e = \begin{Bmatrix} 0 \\ q \cdot 1/2 \\ q \cdot 1/2 \end{Bmatrix}.$$

where, D is the area of the triangle element.

Since B = f(H) is non-linear, ν is therefore non-linear parameter, and (2) is a non-linear equation.

SET OF PERTURBATION EQUATIONS

If we choose the perturbation parameter B_0, then ν can be represented by the power series of B_o :

$$\nu = \nu^{(0)} + \nu^{(1)} B_o + \nu^{(2)} B_o^2 + \ldots\ldots + \nu^{(n-1)} B_o^{n-1} \tag{5}$$

in which $\nu^{(0)} = \nu_o$ is the initial reluctivity, which is determined by the linear part of $B = f(H)$ curve.

Correspondingly, A, J_f, q are represented by the power series of B_o:

$$A = A^{(1)} B_o + A^{(2)} B_o^2 + \ldots\ldots + A^{(n)} B_o^n \tag{6}$$

$$J_f = J_f^{(1)} B_o + J_f^{(2)} B_o^2 + \ldots\ldots + J_f^{(n)} B_o^n \tag{7}$$

$$q = q^{(1)} B_o + q^{(2)} B_o^2 + \ldots\ldots + q^{(n)} B_o^n \tag{8}$$

Substitute equations (5) - (8) into equation (2), take out the terms which have identical power of B_o, then, the set of perturbation equations are obtained as follows :

$$\left.\begin{array}{l}
(\nu_o[k_1]_e - [k_2]_e)\{A^{(1)}\}_e = \{J_f^{(1)}\}_e + \{q^{(1)}\}_e \\
(\nu_o[k_1]_e - [k_2]_e)\{A^{(2)}\}_e = \{J_f^{(2)}\}_e + \{q^{(2)}\}_e + \{m^{(2)}\}_e \\
\vdots \\
(\nu_o[k_1]_e - [k_2]_e)\{A^{(k)}\}_e = \{J_f^{(k)}\}_e + \{q^{(k)}\}_e + \{m^{(k)}\}_e \\
\vdots \\
(\nu_o[k_1]_e - [k_2]_e)\{A^{(n)}\}_e = \{J_f^{(n)}\}_e + \{q^{(n)}\}_e + \{m^{(n)}\}_e
\end{array}\right\} \tag{9}$$

where

$$\{m^{(k)}\}_e = \sum_{i=1}^{k-1} \nu^{(k-i)} [k_1]_e \{A^{(i)}\}_e \tag{10}$$

here, the k is called perturbation order. When k=1, $\{m^{(k)}\}_e = 0$.

From the set of perturbation equation (9), we can observe that the coefficient matrices are identical and constant for any perturbation order. Therefore, it is not necessary to have them recalculated. But the right hand column vectors of all equations are different from one another, and we have only to recalculate them for any perturbation order.

Analogically, we obtain n sets of linear perturbation equations for the whole study field domain, and each set of equations contains m equations, m is equal to the total number of nodes for the whole study field domain. For example, the kth set of perturbation equations is :

$$\left[K^{(k)} \right] \left\{ A^{(k)} \right\} = \left\{ P^{(k)} \right\} \qquad (11)$$

where,

$$\left\{ P^{(k)} \right\} = \left\{ J^{(k)} \right\} + \left\{ Q^{(k)} \right\} + \left\{ M^{(k)} \right\}$$

RIGHT HAND COLUM VECTORS CALCULATION

The calculation of $J_f^{(k)}$ and $q^{(k)}$

When B_o, J_f, q are given, $J_f^{(k)}$ and $q^{(k)}$ can be calculated respectively in advance according to (7) and (8).

The calculation of $\left\{ m^{(k)} \right\}_e$

Since

$$\frac{dA}{dx} = \frac{d}{dx} \left(A^{(1)} B_o + A^{(2)} B_o^2 + \ldots\ldots + A_o^{(n)} B_o^n \right)$$

$$\frac{dA}{dy} = \frac{d}{dy} \left(A^{(1)} B_o + A^{(2)} B_o^2 + \ldots\ldots + A_o^{(n)} B_o^n \right)$$

and

$$B = \left[\left(\frac{dA}{dx} \right)^2 + \left(\frac{dA}{dy} \right)^2 \right]^{\frac{1}{2}}$$

therefore,

$$B = \left\{ \left[\frac{d}{dx} \left(A^{(1)} B_o + A^{(2)} B_o^2 + \ldots\ldots + A^{(n)} B_o^n \right) \right]^2 \right.$$

$$\left. + \left[\frac{d}{dy} \left(A^{(1)} B_o + A^{(2)} B_o^2 + \ldots\ldots + A^{(n)} B_o^{(n)} \right) \right]^2 \right\}^{\frac{1}{2}}$$

$$= \left[\left(B_x^{(1)} B_o + B_x^{(2)} B_o^2 + \ldots\ldots + B_x^{(n)} B_o^n \right)^2 \right.$$

$$\left. + \left(B_y^{(1)} B_o + B_y^{(2)} B_o^2 + \ldots\ldots + B_y^{(n)} B_o^n \right)^2 \right]^{\frac{1}{2}} \qquad (12)$$

For any given $B = f(H)$ curve, ν can be respresented by the power series of B :

$$\nu = \nu_o + a_1 B + a_2 B^2 + \ldots\ldots + a_{n-1} B^{n-1} \qquad (13)$$

Where a_1, a_2, ... a_{n-1} can be determined by B_1, B_2, ... B_{n-1} and its corresponding ν_1, ν_2, ... ν_{n-1}, which are obtained from the practical $B = f(H)$ curve.

On the other hand, according to (5), we have

$$\nu = \nu^{(0)} + \nu^{(1)} B_o + \nu^{(2)} B_o^2 + \ldots\ldots + \nu^{(n-1)} B_o^{n-1} \qquad (5)$$

Substitute (12) into (13) and compare the coefficient of $B_o^{(k-1)}$ in (13) with that in (5), obtain :

$$\left.\begin{array}{l} \nu^{(k-1)} = a_{k-1} \sum_{i=1}^{k-1} B_x^{(k-i)} B_x^{(i)} + B_y^{(k-i)} B_y^{(i)} \\ \qquad\qquad (k = 2,3, \ldots\ldots n) \end{array}\right\} \qquad (14)$$

where for any triangle element,

$$\left.\begin{array}{l} B_y^{(j)} = \frac{dA}{dx}^{(j)} = \frac{1}{2D} (b_i A_i^{(j)} + b_j A_j^{(j)} + b_m A_m^{(j)}) \\ B_x^{(j)} = \frac{dA}{dx}^{(j)} = \frac{1}{2D} (c_i A_i^{(j)} + c_j A_j^{(j)} + c_m A_m^{(j)}) \end{array}\right\} \qquad (15)$$

For the first perturbation, k=1, the first equation of (9) does not contain the non-linear part of ν, therefore, $A^{(1)}$ can be solved directly. For the second perturbation, k=2, $\nu^{(1)}$ can be calculated by $A^{(1)}$ according to (14) and (15). Analogically, for the kth perturbation, only $\nu^{(k-1)}$ needs to be calculated by $A^{(1)}$, $A^{(2)}$ ...... $A^{(k-1)}$, which have already been obtained before the kth perturbation, while $\nu^{(1)}$, $\nu^{(2)}$, ... $\nu^{(k-2)}$ have already been calculated respectively and keep constant during the 2th, 3th ,.. .... (k-1)th perturbation.

With $\nu^{(1)}$, $\nu^{(2)}$, ... $\nu^{(k-1)}$ obtained, the $\{m^{(k)}\}_e$ in kth perturbation can then be calculated according to (10). Therefore, the set of perturbation equations can be solved directly.

By solving n sets of perturbation equations one by one, we can obtain a more precise A.

The number n of the set of perturbation equations can be determined in advance. Obviously, the greater the number n is, the more precise the A will be. But from the (13) we can observe that the ν is precise enough for practical engineering when n is about five [3]. Therefore, the perturbation order n is small in practical use.

CONCLUSIONS

1. The non-linear discretization equation of FEM can be turned directly into set of perturbation equations by perturbation method.

2. The non-linear problem of electromagnetic field numerical calculation can be solved directly by the PFEM without the need for iteration.

3. The number of the sets of perturbation equations are finite and very small, and can be determined in advance.

4. Compared with the current iteration method, the PFEM is simpler, more effective and more time-saving.

So far, we have discussed the PFEM of 2D non-linear magnetostatic calculation for eddy-current coupling. As a numerical method, its use can be extended to 3D electromagnetic field calculation. It can be used not only in electromagnetic field calculation but also in the calculation of other fields. Therefore, it has a wide range of application and a broad perspective for development.

REFERENCES

1. LI BIAO : The Model of Electromagnetic Field Computation for Solid Rotor Eddy-Cuttent Coupling, Air Force College of Engineering Journal, No. 2, 1985.
2. Xi Qi Hua : Perturbation Finite Element Method of Electromagnet field, Air Force College of Engineering Journal, No.2, 1982.
3. Lin You-Yang : Application of Newton-Raphson Nethod to Numerical Solutions of Non-linear Magnetic Field Problems, in : " The Electromechanical Proceeding of China." Shanghai Society of Electrical Engineering, Shanghai, 1983.

MACRO - ELEMENT IN CARTESIAN COORDINATES

Dazhong Shen and Jean-Claude Sabonnadière

Laboratoire d'Electrotechnique de Grenoble
ENSIEG - INPG B.P. 46
38402 Saint-Martin-d'Hêres, France

Abstract. This paper presents the utilisation of macro-element in a Cartesian coordination system. Such macro-element has been introduced and applied to the straight overlap joint analysis. Their use increases the accuracy and avoids the new subdivision at different parameter values such as the overlap length. The harmonic number is analysed and a minimum number is given.

INTRODUCTION

The macro-element has been used with success in dynamic analysis of electric machines [1]. The relative change of rotor position with respect to the stator is taken into account. Using the macro-element increases the accuracy of the results and avoids the new subdivision in the air gap at each step of the rotor position. As is well known, the movement consideration is a great difficulty in the dynamic analysis. Some techniques can reduce either the new discretization only in the air gap or one new grid for many steps. But the elements of bad quality could not be avoided, so the accuracy is not ensured. In the macro-element, the accuracy depends only on the discretization of the element sides, not on the relative position between the discretization points. We can say that the macro-element has an analytical accuracy.

Such a distinctive element can only be used in the region which has the periodic boundary conditions as in the air gap of rotating machines. This condition limites their use in Cartesian coordinates because there are few periodic boundaries.

In the analysis of transformer overlapping joints, we have met the similar difficulties as in the rotating machines, namely the great thinness of the isolation layer. The reasonable accuracy cannot be easily obtained by using the classic element. On the other hand, in such analysis, the different overlapping lengths are necessary to investigate. The change of overlapping length makes a grave distortion of elements, so the new discretization is often necessary for each overlapping length.

Thanks to the periodic boundary conditions on the two extremes of the joint, the macro-element in Cartesian coordinate system is introduced in this paper.

OVERLAPPING JOINT ANALYSIS

The mitered overlapping joint is commonly used in the large power transformer cores. The overlapping length is an important parameter in such a joint design. Many experimental studies and numerical analysis have been made on the magnetic characteristics of the joint. In this paper, only the straight overlapping joint has been analysed. It is shown in Fig. 1 in a two-dimensional analysis. One pair of steel plates consists of one layer, which overlaps with after another in a deviation named the overlapping length. Since the flux distribution in each plate is symmetrical to the central line, the region surrounded by two neighbouring central lines is chosen as the finite element analysis region, the region ABDC in Fig. 1. If the boundaries AC and BD are chosen rather far from the centre of the joint, it is assumed that the periodic conditions exist on the boundaries AC and BD, that is to say, the corresponding points on these sides have the same vector potential values if we use the vector potential formulation.

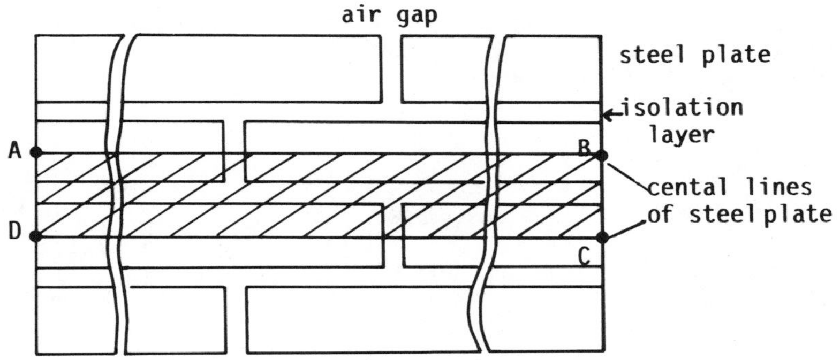

Fig.1 The straight overlap joint and the finite element region.

The steel plate has an ordinary thickness of 0.3mm or 0.35mm and an isolation layer distance of 0.01mm. The overlapping length is over the range from 0 to 5mm. The region length is chosen as at least 25mm.

For example, for getting the equilateral triangle elements in the isolation layer, there are about 5000 triangle elements! If we change the overlapping length. In Fig. 2, in the first case, the overlapping length is [1], the points situated on the lines have an uniform distribution; in the second case, the overlapping length is 1', on two sides, the points have a dense distribution, on the contrary, there are not sufficient points in the centre. In the case the overlapping extreme length is zero, the two central lines must be cancelled, on new discretization is compulsory. Such inconveniences and difficulties could be overcome by introducing the macro-element in this layer.

MACRO-ELEMENT

The complete analysis of macro-element has been made in 1 . In this paper we only give the results in Cartesian coordinates. One macro-element is shown in Fig.3.. It has m nodes on two sides in which n nodes are situated on the upper side and m-n on the lower side. The magnetic vector potential in it is :

$$A(x,y) = \sum \alpha_i(x,y)\, A_i$$

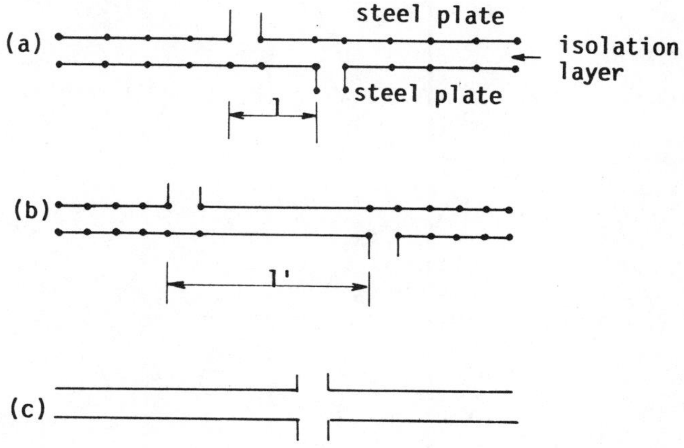

Fig.2 The distortion of point's distribution
(a) - the overlap length is l
(b) - the overlap length is l'
(c) - the overlap length is zero

where Σ is the sum over i from 1 to m. j A_i and α_i are the vector potential and the shape function at node i respectively. As the ordinary element, the contribution of the macro-element in the rigid matrix is :

$$S_{ij} = \frac{L}{4\mu_o} \frac{(H_2 - c') - (H_1 - c')}{(c - c')\,(e - e')} A_{io} A_{jo} +$$

$$\frac{L}{2} \sum \lambda_k \frac{\operatorname{sh} \lambda_k (H_2 - c') \operatorname{ch} \lambda_k (H_2 - e')}{\operatorname{sh} \lambda_k (c - c') \operatorname{sh} \lambda_k (e - e')} -$$

$$\frac{\operatorname{sh} \lambda_k (H_1 - c') \operatorname{ch} \lambda_k (H_1 - e')}{\operatorname{sh} \lambda_k (c - c') \operatorname{sh} \lambda_k (e - e')} (A_{ik} A_{jk} + B_{ik} B_{jk})$$

Where is the sum over all values of k and

$\lambda_k = \pm 2\pi k / L$. If i or j = 1, 2, ..., n, then

$c = H_1$ and $c' = H_2$; $e = H_1$ and $c' = H_2$.

If i or j = 1+n, 2+n, ..., m, then

$c = H_2$ and $c' = H_1$; $e = H_2$ and $c' = H_1$.

The coefficients A_{io} , A_{io} are

$$A_{io} = \frac{x_{i+1} - x_{i-1}}{L}$$

The coefficients A_{ik} , B_{ik} and A_{ik} , B_{ik} are given by

$$A_{ik} = -\frac{4}{L\lambda_k^2} A'_{ik} ,$$

$$A'_{ik} = \frac{1}{(x_i - x_{i-1})} \sin\left(\frac{(x_i + x_{i-1})\lambda_k}{2}\right) \sin\left(\frac{(x_i - x_{i-1})\lambda_k}{2}\right) + \frac{1}{(x_i - x_{i+1})} \sin\left(\frac{(x_{i+1} + x_i)\lambda_k}{2}\right) \sin\left(\frac{(x_{i+1} - x_i)\lambda_k}{2}\right)$$

$$B_{ik} = \frac{4}{L\lambda_k^2} B'_{ik} ,$$

$$B'_{ik} = \frac{1}{(x_i - x_{i-1})} \sin\left(\frac{(x_i - x_{i-1})\lambda_k}{2}\right) \cos\left(\frac{(x_i + x_{i-1})\lambda_k}{2}\right) + \frac{1}{(x_i - x_{i+1})} \sin\left(\frac{(x_{i+1} - x_i)\lambda_k}{2}\right) \sin\left(\frac{(x_{i+1} + x_i)\lambda_k}{2}\right)$$

Such a macro-element has been used in the straight overlapping joint analysis. For each overlapping length, the qeometry and the corresponding discretization do not have necessarily to be changed. In the Fig.4, the vector potential distribution is drawn for different overlapping lengths.

CHOISE OF HARMONIC NUMBER

The shape function of macro-element is based on the approximation of Fourier series. So the accurary depends also on the number of harmonics. The more the number of harmonic is reached, the more the accurary is obtained. At least one minimum number must be taken, it could be found in the following analysis.

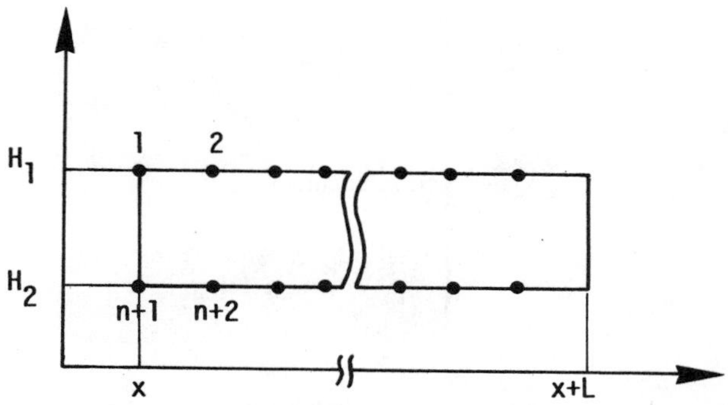

Fig.3 A macro-element in Cartesian coordinates.

Supposing the average interval between two neighbouring points is $\tau/2$, the region length or the macro-element length is L, and the ordinary element used in other parties outside macro-element is of prime importance. On the sides of macro-element, the shape function is a triangular function. Their Fourrier transformation function is

$$F(j\omega) = \frac{\tau}{2} \frac{\sin^2(\omega\tau)}{(\omega\tau)^2}$$

The shape function is a periodic function about L, then we have

$$\omega = \frac{2n\pi}{L}$$

If the harmonics belong to the first wave band of their frequency spectrum are taken into account, the sufficient accuracy is obtained. Therefore the minimum number of harmonic is at least:

$$n = \frac{L}{\tau}$$

CONCLUSION

The macro-element has been introduced in the Cartesian coordinate system and applied in the transformer straight overlapping joint. Their using increases the accuracy and avoids the new discretization at different parameter values such as the overlapping length. One minimum harmonic number is analysed and given. This paper complements the utilisation of macro-element.

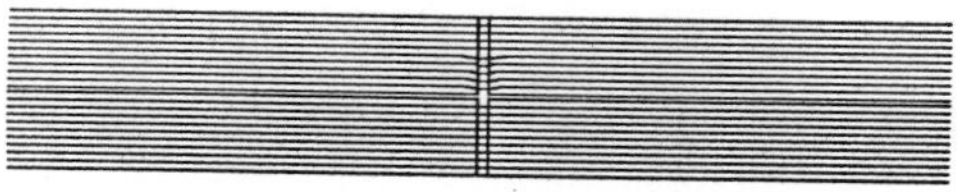

(a) the overlap length = 0

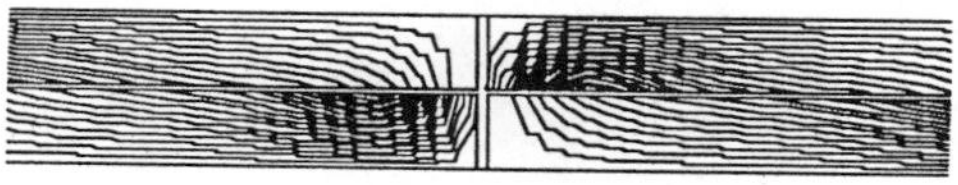

(b) the overlap length = 1 mm

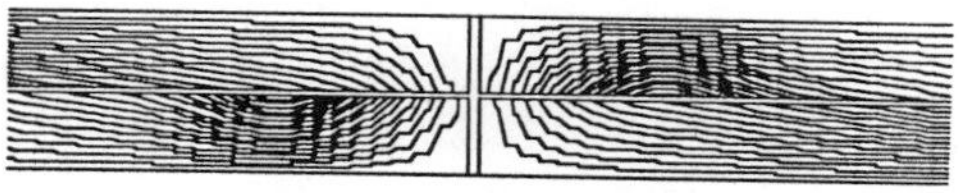

(c) the overlap length = 2 mm

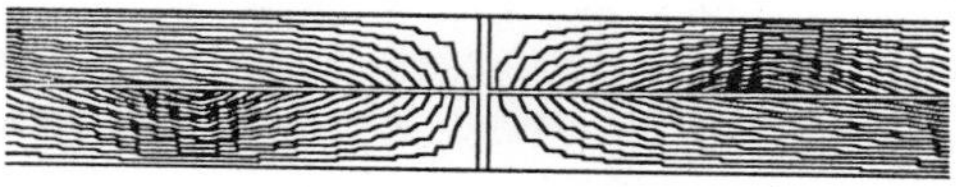

(d) the overlap length = 3 mm

Fig. 4 Vector potential distribution in the straight overlap joint

ACKNOWLEDGEMENT

The authors would like to thank the FRANCE TRANSFO and Mr. Sacotte for support in this work.

REFERENCES

1. A. Abdel-Razek, J.L. Coulomb, M. Féliachi, J.C. Sabonnadière "Conception of an air-gap element for the dynamic analysis of the electromagnetic field in electric machines". IEEE Trans. Magnetics, Vol. MAG-18, N°2, March 1982.

2. M.Féliachi : "Contribution au calcul du champ électromagnétique par la méthode des éléments finis en vue d'une modélisation dynamique de machines électriques". Thèse de Docteur-Ingénieur, Paris, 1981.

ANALYSIS OF TRANSIENTS IN ELECTRICAL CIRCUITS CONTAINING INITIALLY MAGNETIZED FERROMAGNETICS FOR IMPULSE EXCITATION

Sławomir Wiak

Institute of Electrical Machines and Transformers
Technical University of Łódź, Poland

INTRODUCTION

Many electrical devices have magnetic cores made of ferromagnetic materials with pronounced nonlinearity. Finding the proper values of equivalent circuit parameters for transient analysis, however, may be difficult or even virtually impossible, due to the nonlinear nature of the current skin effect in ferromagnetic cores. Beland and Gamach[1], Davidson[2],and MacBain[3] have studied experimentally and modelled numerically the transient processes in electrical machines or their parts, taking into account the nonlinearity of the magnetic media, but these investigations have been made with assumptions limiting the shape of the electric currents flowing in the circuit. Fridman[4] has proved the range of using an equivalent circuit of transformer type for fixed frequency of current flowing through the exciting winding and core thickness. Wiak[5,6] and Zakrzewski[5] have proved that it is possible to calculate transients in electrical circuits containing ferromagnetics, in many cases, without using an equivalent circuit idea. Instead, the contribution of the eddy-current field to the total impedance of the system is calculated throughout simultaneous solution of a nonlinear field equation in a ferromagnetic core, and the shape of the circuit is being corrected after each step of numerical iteration. In this paper, the author has successfully applied the "circuit-field method"[5] to the circuits containing ferromagnetics with DC initial magnetization of a core.

THE METHOD OF CALCULATION AND COMPARISON OF MEASUREMENTS

The elaborated method will be illustrated by an example circuit shown in Figure 1, which contains the non-linear element in the form of a solenoid with a tabular solid ferromagnetic core. The internal region (ferromagnetic core) is symbolised in the circuit diagram with the use of internal voltage u_i. Inductance L_e corresponds to the external flux. The exciting winding shows the additional resistance R_c. In the circuit shown in Figure 1 linear elements exist: resistance R_1 and inductance L. This approach, which takes full account of the non-linear eddy-current skin effect, is particularly beneficial when strong magnetic field impulses may occur in the system.

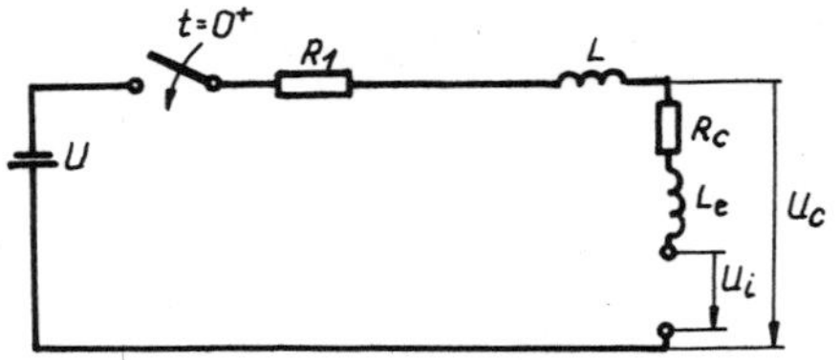

Fig.1. The circuit assumed for calculation

To serve as an example, the elaborated "circuit-field method" will be presented in the case of switching on the direct voltage to the circuit (Figure 1). In this case, the voltage equation is as follows:

$$u = (R_1 + R_c) \cdot i(t) + (L + L_e) \cdot \frac{di(t)}{dt} + u_i(t) \tag{1}$$

with the initial conditions

$$i(0) = 0 \quad \text{or} \quad i(0) = I_i \tag{2}$$

and in the ferromagnetic core

$$H(x,0) = 0 \quad \text{or} \quad H(x,0) = H_i \tag{3}$$

where: I_i - initial current in circuit, A,
H_i - initial value of magnetic field strength, A/m.

After the steady-state condition has been reached, the value of the current amounts to

$$I_{st} = \frac{u(t_{st})}{R_1 + R_c} \tag{4}$$

and magnetic field strength will equal

$$H(x,t_{st}) = H_{st}. \tag{5}$$

The transient magnetic field in the ferromagnetic core will be described by the diffusion equation. Thus, if the one-dimensional analytical model of the electromagnetic field is assumed, then

$$\frac{\partial^2 H(x,t)}{\partial x^2} = \sigma \cdot \mu_d \cdot \frac{\partial H(x,t)}{\partial t} \tag{6}$$

in solid iron. Nonlinearity of this equation arises from the fact that differential permeability is assumed to be a function of H through a non-linear B/H curve (Figure 2). The calculation of currents and voltages in the circuit requires the simultaneous solution of Equations 1 and 6 on every time step. At any time instant, the voltage on the terminals of the solenoid can be expressed as

$$u_c(t) = u_i(t) + R_c \cdot i(t) + L_e \cdot \frac{di(t)}{dt} = w \int_0^l E(0,t)dl + R_c \cdot i(t) + L_e \frac{di(t)}{dt} \tag{7}$$

and must be equal, with acceptable limits of accuracy, to

$$u_c(t) = u - R_1 \cdot i(t) - L \cdot \frac{di(t)}{dt} \;. \tag{8}$$

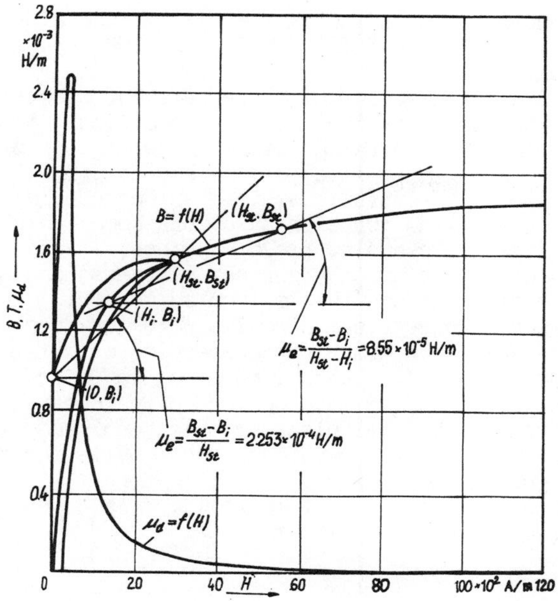

Fig.2. The primary and equivalent B/H curves of the core of solenoid and permeabilities for these curves

The electric field strength E(0,t) on the surface (Equation 7) may be found through the solution of diffusion equation (6) for which the finite difference method is used. For each time step in numerical solution a "prediction iterative correction" method is used. The values of the circuit current for time $(t+\Delta t)$ is initially extrapolated using values from previous instants. In order to check the accuracy of results of measurements with results of calculations, experimental investigations have been made (see Figure 3).

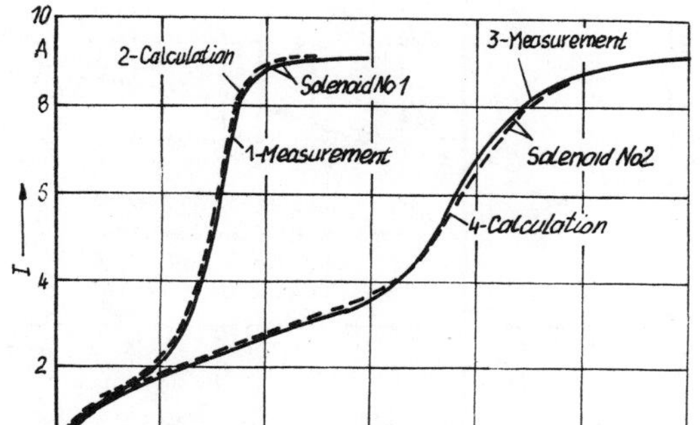

Fig.3. Comparison of current i(t) obtained from measurement and calculations

APPROXIMATE METHOD OF CALCULATING TRANSIENTS IN ELECTRICAL CIRCUITS
CONTAINING SOLENOIDS WITH DC INITIAL MAGNETIZATION OF CORES

The method described above and illustrated by specific examples shows way of calculating transients in nonlinear electrical circuits without the need for determining the equivalent circuits parameters of nonlinear elements. Having successfully applied this method to circuits with simple electromagnetic field configurations, this author considers that it provides an effective way of solving transients in more complicated non-linear circuits containing, for example, field windings, solenoids, one-phase transformers, electromagnets, eddy-current starters, and so on. Practically, it will not be possible to use this method successfully to calculate transient processes in electrical circuits with very complicated field configuration, because much time is required for calculating the electromagnetic field distribution on each time step. In these cases it is expedient to search for approximate methods of calculating transients. For this reason comparative calculations have been made. The comparative calculations of current curves have been made, while the following B/H curves are used for calculation:

- real B/H curve of solenoid core (Figure 1),
- substitute curve expressed by the formula:

$$B = C_1 \cdot H^{1/n} \tag{9}$$

  (for which $C_1 = 0.576$, and $n = 8$),
- linear B/H curve connecting two points (B_i, H_i) and (B_{st}, H_{st}) for which permeability is given by formula

$$\mu_e = \frac{B_{st} - B_i}{H_{st} - H_i} \ . \tag{10}$$

In order to provide the measurement for different initial values of magnetic field strength of the core, the additional exciting winding has been uniformely wound for each solenoid. Figures 4, 5 and 6 present current curves obtained from measurements and numerical calculations, These curves have been computed for different values of H_i and H_{st}, leading to different values of B_i and B_{st}.

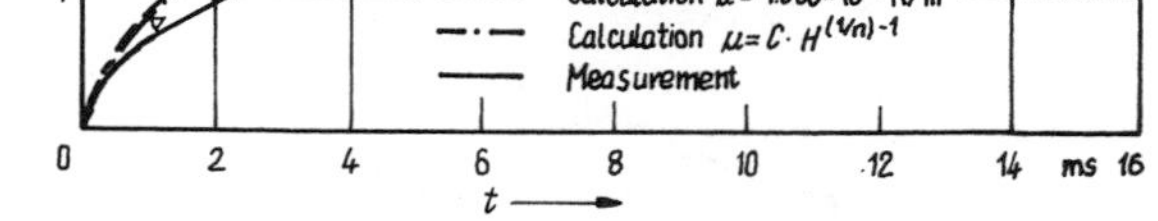

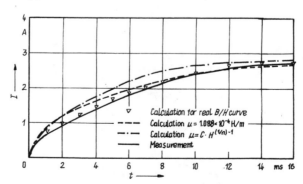

Fig.4. The current curves measured and calculated for different B/H curves in the case of switching on DC voltage to the circuit with solenoid "2". The process runs from H_i = 1483 A/m, until H_{st} = 3623 A/m

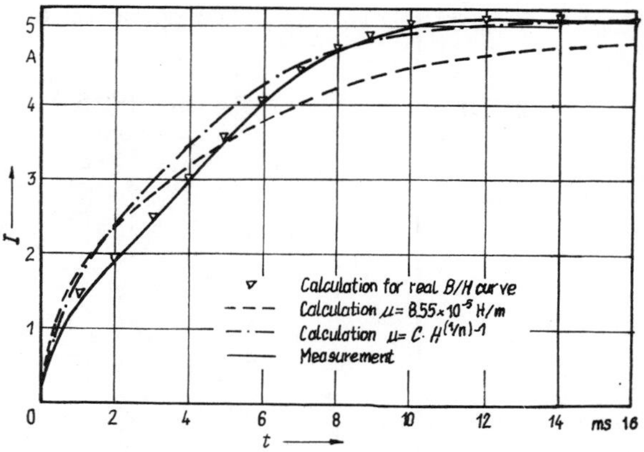

Fig.5. The current curves measured and calculated for different B/H curves in the case of switching on DC voltage to the circuit with solenoid "2". The process runs from H_i = 1566 A/m until H_{st} = 5482 A/m

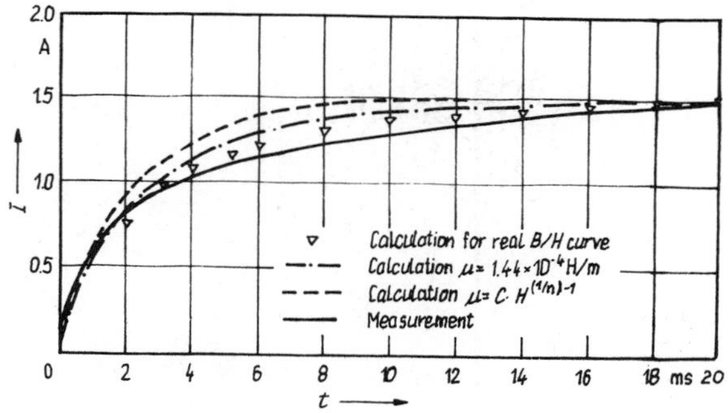

Fig.6. The current curves measured and calculated for different B/H curves in the case of switching on DC voltage to the circuit with solenoid "1". The process runs from H_i = 1475 A/m until H_{st} = 2582 A/m

CONCLUSION

The "circuit-field method" presented above can be fully adopted for computing transients in circuits containing elements, in which the electromagnetic field configuration in the elements is not complicated. Difficulties have appeared in the case of computing transients in circuits with elements of complicated field configuration. This is due to the necessity of solving, on each calculation step, algebraic equations system (resulting from the dividing cross-section of element into elementary subregions) and simultaneously circuit and field equations. The following method of solving the posed problem is proposed:

- calculating H_i and B_i distribution,
- finding a substitute B/H curve for each elementary subregion by connecting two points (B_i, H_i) and (B_{st}, H_{st}),

- finding equivalent permeability for each subregion or eventually a set of subregions or for the whole region,
- solving simultaneously circuit and field equations, on each time step for previously calculated magnetic permeability distribution not changing in the time period.

The proposed approximate method can provide the possibility of computing transients in electrical circuits, as in the case above, with a reasonable time spent in calculation. Such solutions will be presented in future papers.

APPENDIX

Numerical comparative calculations and measurements have been made for the following two solenoids:

Solenoid "1" - outer diameter D_{od} = 145.4 mm, inner diameter D_{id} = 140.6 mm, thickness d = 2.4 mm, height h = 100 mm, number of turns of exciting winding w = 350, number of turns of additional exciting winding for initial magnetization w_1 = 350.

Solenoid "2" - outer diameter D_{od} = 150.4 mm, inner diameter D_{id} = 139.8 mm, thickness d = 5.3 mm, height h = 100 mm, number of turns of exciting winding w = 350, number of turns of additional exciting winding for initial magnetization w_1 = 350.

The windings have been wound in such way that leakage flux associated with each winding is negligibly small. The core conductivity is $\sigma = 5.44 \cdot 10^6$ S/m.

REFERENCES

1. B. Beland and D. Gamach, The impedance of flat plate steel conductors, ICEM, Lausanne (1984).
2. J.M. Davidson and M.J. Balchin, Experimental verification of network method for calculating flux and eddy-current distributions in three dimensions, IEE Proc. Pt. A 7:242 (1981).
3. J.A. MacBain, A numerical analysis of time dependent two dimensional magnetic fields, Int. J. Num. Meth. Eng. 19:1033 (1983).
4. E.B. Fridman, The frequency characteristics of the resistance of the solenoid with conducting core and their applying for the calculation of the transient processes, Elektriczestwo 6:69 (1975).
5. S. Wiak and K. Zakrzewski, Numerical calculation of transients in electrical circuits containing elements with nonlinear eddy-current skin effect, IEE Proc. Pt. A 9:741 (1987).
6. S. Wiak, Analysis of transients in electrical circuits containing solenoids with initially magnetized cores for impulse excitation, Archiv für Elektr. 71:1 (1988).

3-D REACTANCE CALCULATION OF AIR-CORE COILS WITH MAGNETIC SCREENING

K. Zakrzewski* and M. Łukaniszyn**

* Technical University of Łódź, Poland

** High School of Engineering in Opole, Poland

INTRODUCTION

In the previous paper[5] a method for computing 3-D magnetic field of air-core coils has been proposed. These coils may be simulated by means of the four spatial solids (rectangular prisms) situated in the unbounded space as shown in Fig.1. The coils operating as air-core reactors may be suitably screened (partially or completely) with sheet iron. For other reasons they may also be located nearby the steel solids which may be considered as the magnetic screens as well. A magnetic screen is the more effective the less saturated than the sheet iron is ($\mu_s \gg \mu_0$). In the present paper 3-D field distribution as well as the reactance of an air-core coil are calculated having regard to the fact that nonsaturated magnetic screens can be located in a different way. In this connection this paper may be considered as a further development of the method described in the paper[5].

NUMERICAL MODEL

An assumption is made that each coil can be simulated by using the a.c. carrying bus-bars (Fig.1). Displacement currents are neglected.

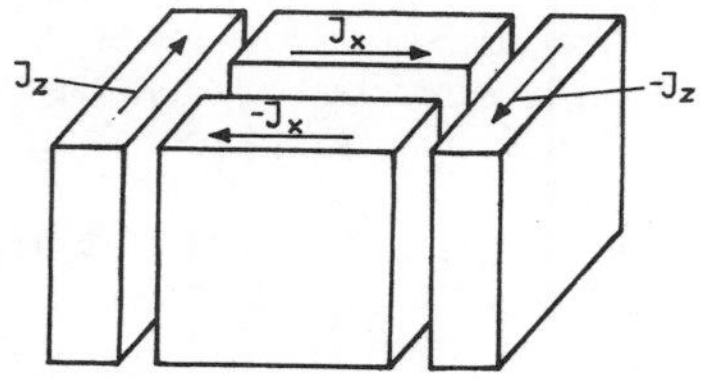

Fig.1. Calculating model of air-core coil

Solving the following partial differential equations yields the field distribution.

$$\text{rot}\,\vec{H} = \vec{J} \qquad (1)$$

$$\text{div}\,\vec{B} = 0 \qquad (2)$$

The magnetic field is described by the use of the vector potential $\vec{A}$ complete with the supplementary assumption

$$\operatorname{div} \vec{A} = 0 \tag{3}$$

Aforenamed vector potential $\vec{A}$ satisfies the following scalar Poisson equations

$$\nabla^2 A_x = -\mu J_x\,, \qquad \nabla^2 A_y = -\mu J_y\,, \qquad \nabla^2 A_z = -\mu J_z \tag{4}$$

with J_x, J_y, J_z being components of the current density vector $\vec{J}$.
Outside the windings (i.e. outside the bus-bars carrying an alternating current) the Laplace equation is satisfied. In an examined model one has $J_y = 0$ which implies that the vector potential has only two components A_x and A_z.
Components of the magnetic flux density are given as follows

$$B_x = \frac{\partial A_z}{\partial y}\,, \qquad B_y = \frac{\partial A_x}{\partial z} - \frac{\partial A_z}{\partial x}\,, \qquad B_z = -\frac{\partial A_x}{\partial y} \tag{5}$$

The air-core coil simulating the actual one has been for numerical purposes placed inside a rectangular prism with significantly larger volume as compared to that of the coil. On the face of the aforementioned polyhedron the homogeneous Dirichlet boundary condition holds ($\vec{A} = 0$) which implies in turn the field on these faces must vanish. Introducing a magnetic screen necessitates a boundary condition on its surface to be found. The screens are assumed to be made of an electric sheet being thin enough to neglect the influence of eddy currents due to static boundary conditions on the iron-dielectric interface. Assuming, hence, $\mu_r \to \infty$ a homogeneous Neumann boundary condition on the screen face $\partial\vec{A}/\partial n = 0$ is obtained.
The above boundary conditions remain valid also in case of thick magnetic screens being made up of electrical sheet packs in which the eddy currents interaction can be neglected (even if considerable saturation is taking place).

NUMERICAL SOLUTION

The numerical solution was obtained using the finite difference method (FDM).
For solving the three-dimensional Poisson differential equations the direct method consisting of the Fast Fourier Transform use was employed. For this purpose the software package FISHPACK[1] was used.
This package has been adapted and debugged on an Odra 1305 computer. Using an uniform finite difference grid two sets of algebraic equations were being solved. Computations were accomplished for seven variants of a magnetic screen location. The number of nodes being dependent on the specified variant ranged from $(17 \cdot 20 \cdot 21)$ to $(29 \cdot 27 \cdot 33)$ corresponding, therefore, to the comparatively coarse subdivision.
Numerical results for the inductance confirm the mesh was fine enough.
The diagrams of maximum inductance for the coil middle intersection are plotted in Cartesian coordinates in Figs.2,3,4,5, respectively. These diagrams concern distinct cases of the coil screening. The use of full screening, as shown in Fig.5 results in the extreme field amplification since the inner core having a form of a magnetic screen considerably reduces the permeance against the magnetic flux produced by the coil. Values of a coil inductance computed for $z = 50$ (z - number of turns) and $J = 5A$ (J - current flowing in the coil are given in Table 1). The distance between steel solids and the coil edge was assumed to be fixed and equal to 2 cm.

Tab.1. Measuring and calculating results

case	inductance mH meas.	inductance mH calc.	screen displacement
1	6.1	6.0	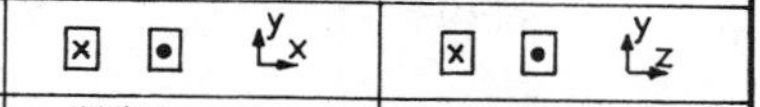
2	6.9	6.4	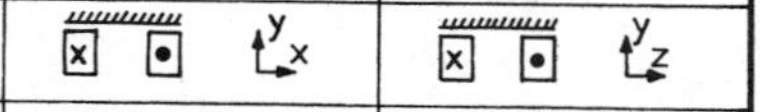
3	7.1	6.7	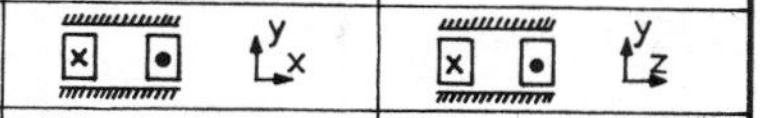
4	7.7	7.3	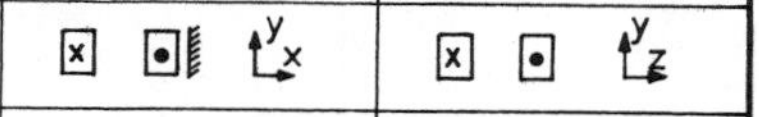
5	8.8	8.5	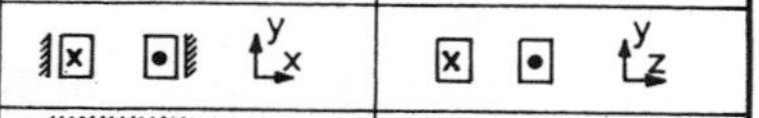
6	9.5	9.9	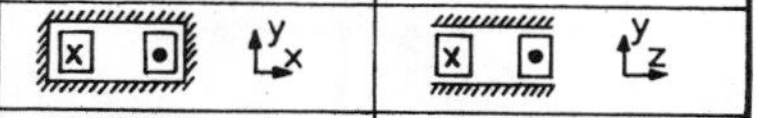
7	11.7	18.0	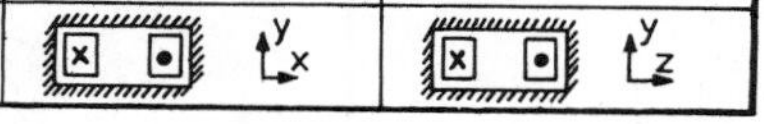

Integrating the energy of magnetic field stored inside the coil volume yields the inductance values.

$$W_m = \frac{1}{2\mu} \int_V \vec{B}\vec{B}\, dv \tag{6}$$

Hence it appears that

$$L = \frac{1}{\mu I^2} \int_V |B|^2\, dv \tag{7}$$

where $|B|^2 = |B_x^2 + B_y^2 + B_z^2|$ (8)

and $X = \omega L$ with $\omega = 2\pi f$ (9)

Obtained results were compared with experimental data. Results of comparisons are summarized in Table 1 for cases 1 ÷ 6 experimental and numerical results coincide very well. Significant discrepancy for the case 7 is evident. The full screening caused the magnetic flux (computed under the assumption $\mu_s \to \infty$) to grow nearly three times. On the ground of the saturation growth within a magnetic screen its permeance actually diminishes.

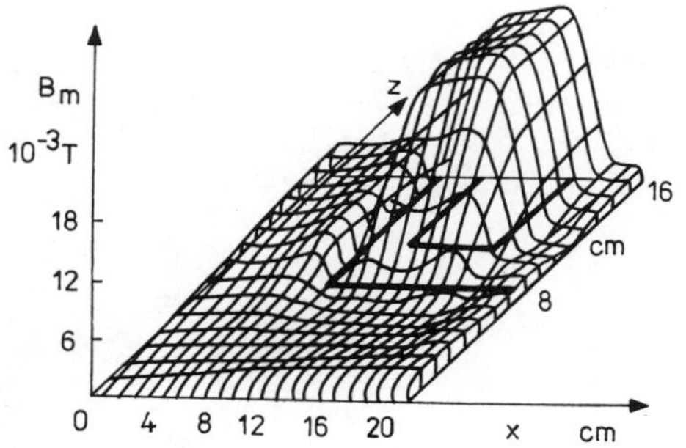

Fig.2. 3-D distribution of induction B in middle plane of air-core coil - case 4 in Tab.1.

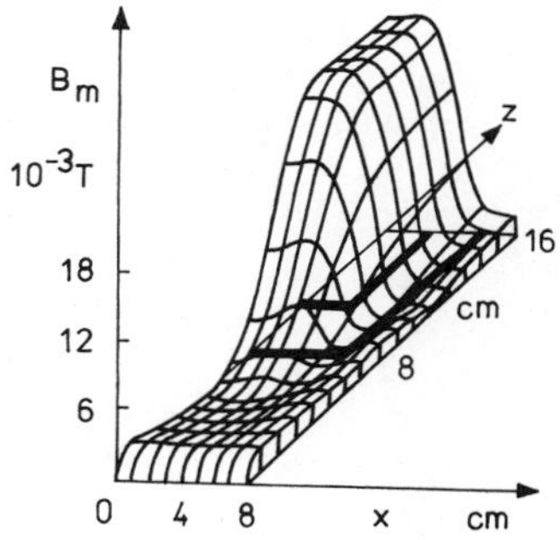

Fig.3. 3-D distribution of induction B in middle plane of air-core coil - case 5 in Tab.1.

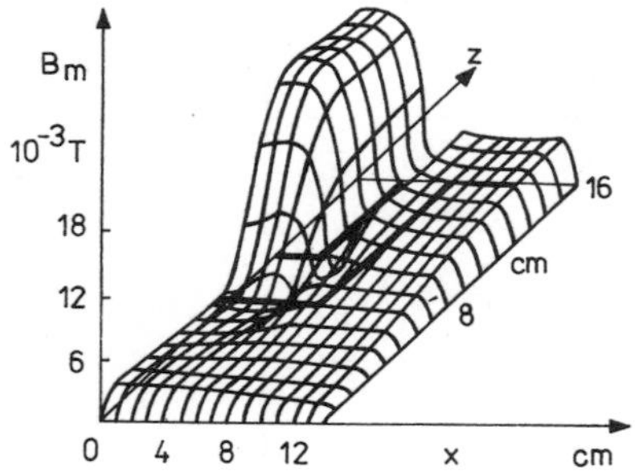

Fig.4. 3-D distribution of induction B in middle plane of air-core coil - case 3 in Tab.1.

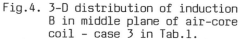

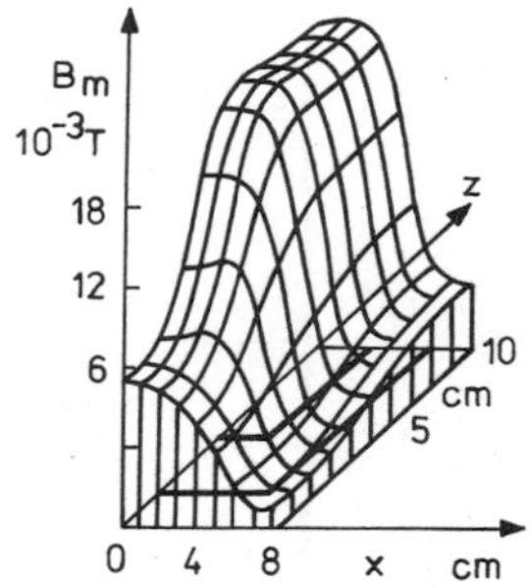

Fig.5. 3-D distribution of induction B in middle plane of air-core coil - case 7 in Tab.1.

Experimental results show that both magnetic flux and the inductance rised nearly twice in the range from non-screened state (case 1) to full - screening (case 7).

CONCLUSIONS

Presented results of experimental investigations confirm the method elaborated by the present authors is applicable for calculating both the field and the inductance (reactance) of screened coils without magnetic saturation and with eddy currents interaction being negligible. To make allowance for the magnetic saturation phenomenon the present method should be modified by applying an iterative choice of the magnetic permeability $\mu_s \neq \infty$. Simultaneously the Neumann boundary condition on the screen face should be corrected to the nonhomogeneous one $\partial \vec{A} / \partial \vec{n} \neq 0$ as well.

REFERENCES

1. J. Adams, P. Schwartztrauber, R. Sweet, "FISHPACK a Package of Fortran Subprograms for the Solution of Separable Elliptic Partial Differential Equation". Research Report, NCAR, Colombo, (1979).
2. B. Buzbee, G. Golub, C. Nielson, "On direct Method for Solving Poisson a Equations". SIAM J., Numer. Anal., vol.7, Nr 4 (1970).
3. P. Schwartztrauber, "A direct Method for discrete Solution of Separable elliptic Equations". SIAM J. Numer. Anal., vol.11 (1974).
4. J. Turowski, "Technical Electrodynamics". WNT, Warsaw (1968), (in polish).
5. K. Zakrzewski, M. Łukaniszyn, "Calculation of the 3-Dimensional Field of the Air-core Coils of Rectangular Section". Int. Symp. on Electromagnetic Fields in Electrical Engineering ISEF´85, Warsaw (1985).

3. TRANSFORMERS

Introductory remarks

T. Nakata

Department of Electrical Engineering
Okayama University
Tsushima, Okayama 700, Japan

Eight papers from six countries are included in this chapter. Most of them are concerned with two-dimensional numerical analysis of flux and eddy current distributions. Forces, stray losses and temperature distributions are calculated from those results. Experiments have been done to verify the results obtained by numerical analysis. These researches are useful for the optimum design of electric machines.

An outline of each paper is introduced here.

D. Kerenyi proposes two simple approximation methods for calculating the stray losses produced in the yoke-beams of transformers. The influences of the geometry and material on the stray losses are examined. It is found that the stray losses in the yoke-beam and those in the tank wall have mutual effect each other. For example, when the losses in the yoke are increased, the losses in the tank are decreased. The results calculated are verified by measurements.

J.A. Tegopoulos and his group present an analytical method for determining the distribution of current and force in two parallel rectangular non-magnetic conductors with equal dimensions carrying equal and oppositely directed sinusoidal currents.

A. Savini and J. Turowski examine the distance between the tank wall and the transformer windings. The leakage flux distribution is analyzed by using the reluctance network method and the finite element method. They evaluated what is called "the critical tank wall distance". Experimental verification has been also done.

J. Takehara and his group analyze inrush currents in transformers by using the finite element method taking into account three-dimensional magnetic fields and residual fluxes. Analyzed results for a single phase transformer show good agreement with experimental ones.

Z. Valkovic tries to decrease core losses by using highly grain oriented silicon steel. The influence of construction of T-joints, core dimensions, core forms, number of laminations per stagger layer, corner overlap length, etc. on the loss reduction are investigated experimentally.

M. Jablonski's group investigates the transient magnetic field in converter transformers during faults using the finite element method.

T. Janowski and R. Goleman estimate additional losses in the windings of magnetic frequency triplers. The influence of input current distortion on the winding losses is examined. The results obtained can be used for the optimum design of magnetic triplers.

Finally A. Di Napoli and his group propose an evaluation method of stray losses in large power transformer tanks. The linear diffusion equation is solved using the finite element method. Tank losses are calculated from the eddy current distribution under several hypotheses. The results obtained by numerical calculation are compared with experimental results.

A METHOD FOR LOSSES EVALUATION
IN LARGE POWER TRANSFORMER TANKS

A. Babare[1], A. Di Napoli[2], E. Santini[2], and G. Scendrate[1]

[1]Nuova Industrie Elettriche di Legnano
Legnano (MI)

[2]Dipartimento di Energia Elettrica
Universita' di Roma

ABSTRACT

The aim of this paper is to show how it is possible to calculate, with a good approximation, the eddy current losses in metal transformer parts. The calculation method is firstly described: it consists in the calculation of vector potential, starting from the diffusion equation discretized by means of the Finite Elements Method, with the ferromagnetic material of the tank taken as linear; the eddy current losses are then calculated. Finally, the method is applied to the analysis of two machines, respectively of 350 kVA and 370 MVA; it is also shown that in order to achieve good results, the above method requires several basic hypotheses which vary from case to case.

INTRODUCTION

Bearing in mind the effect of eddy currents in ferromagnetic structures, the study of the magnetic field in a transformer may be performed beginning from the diffusion equation written in terms of vector potential $\vec{A}$:

$$\frac{\delta}{\delta x}\frac{1}{\mu}\frac{\delta \vec{A}}{\delta x} + \frac{\delta}{\delta y}\frac{1}{\mu}\frac{\delta \vec{A}}{\delta y} + \frac{\delta}{\delta z}\frac{1}{\mu}\frac{\delta \vec{A}}{\delta z} = - J - \frac{1}{\tau}\frac{\delta \vec{A}}{\delta t} \qquad (1)$$

with μ magnetic permeability and τ electric resistivity.

The performance of a three-dimensional study, in addition to increasing difficulties in calculation algorithms, entails a considerable increase in occupation of the computer memory and an increase in calculation time, with a resulting increase in truncation errors, with no adequate increase in precision of the results; it should be borne in mind that the measurements of primary interest in the output are often integral quantities, and that also if the quantity is local it is sought at points where the effect due to neglecting the three-dimensional nature is small. On the other hand, this may be checked easily by studying the machine, both as regards plane and cylindrical geometry.

In this way, since vector potential A is always in the same direction ($\vec{A} = A_z$ or $\vec{A} = A_\theta$), at each point of the area in question, it may be considered as a scalar quantity.

Since the field which involves the tank is the leakage field, it is sufficient to consider the space outside the window, excluding the core which may be considered as having infinite permeability; along the surface of air column separation, the condition $\delta A/\delta n = 0$ may be fixed.

In addition, the field outside the tank is practically zero both because it decreases with the square of the distance from the source; because of the shielding effect of the structure of the tank, it is possible to fix $A = 0$ at a small distance from the tank.

Lastly, it may be noted that the behaviour of the machine is, both owing to the arrangement of the windings and to their electric behaviour, frequently symmetrical with respect to the half-way line, and it is therefore possible to fix the condition $\delta A/\delta n = 0$ on this line and study only half the machine.

FORMULATION ACCORDING TO THE F.E.M.

It has been shown that the correct solution of (1) is always the one which minimizes the energy functional in the region being examined. It is therefore possible to write the energy functional, discretizating the region into triangular elements of the first order and thus assuming the density of flux to be constant inside each triangle, differentiating the energy functional with respect to the vector potential, it is possible to obtain for each element a matrix equation of the type [1]:

$$[S]\,[A] + [Z]\,\frac{\delta A}{\delta t} = -\,[W]\,[J] \qquad (2)$$

Supposing that the machine works in such a way that in the windings is flowing alternating current at a frequency of 50 Hz, it is possible to carry out the derivative of the vector potential in the time domain according to the Kennelly and Steinmetz well-known rules, obtaining the following matrix equation:

$$[S][A] + [Z]\,j\,\omega[A] = -[W][J] \qquad (3)$$

Determining matrices [S], [W] and [Z] for all the triangular elements, it is possible to derive an equation of type (3) written for the entire region being studied:

$$[P][A] + [Q]\,j\omega\,[A] = -[T][J] \qquad (4)$$

with [P], [Q] and [T] as square matrices of order n, and [A] and [J] vector of the potentials and of the current densities; they are also of order n, where n represents the number of the nodes in the discretized region.

One important problem is found in performing the discretization. Many studies have been carried out to determine the best mesh, and it was seen that for this the highest real part of the eigenvalues of the matrix [P] of equation (3) must be greater than a function of the values of the parameters μ and τ. Not only is it difficult to respect this condition, but at times, as in the case where a ferromagnetic material with non-linear features is present, it is virtually impossible. It has, however, been noted that the solution of (4), for a fixed discretization, diverges when in the materials where the eddy current phenomenon is present the thickness of the element in the field direction is greater than d/2, in which the penetration depth d is obtained from the expression

$$d^2 = \frac{\tau}{\omega \mu} \tag{5}$$

CALCULATION OF THE LOSSES

Once the distribution of the vector potential is known, it is possible to calculate eddy current losses in the tank:

$$P_v = E^2/\tau \tag{6}$$

Using $E = - \delta A/\delta t$, it is possible to calculate these losses per unit of volume. To obtain total losses, therefore, (6) must be integrated into the whole volume. If a plane study is conducted, (6) may be integrated using the following expression

$$P = \int_S (E^2/\tau)\, dS \tag{7}$$

dS being the infinitesimal area element; the losses calculated in this way are losses per unit of depth, i.e. in W/m.

Bearing in mind that the study is performed in the frequency domain, E in magnitude thus equals:

$$| E | = \omega | A | \tag{8}$$

in which $| A |$ is the vector potential module, obtained starting from the values of A' and A", which are. respectively, the value of the vector potential A in its two components in phase and in quadrature. Thus (7) may be discretized as follows:

$$P = \Sigma_N ((E_i)^2/\tau)\Omega_i = \Sigma_N ((\omega A_i)^2/\tau)\Omega_i \tag{9}$$

E_i being the average R.M.S. value of the electric field in the i-th triangle, Ω_i the area of the i-th triangle, N the total number of triangles.

All that has been said so far is valid for any section of the transformer made on the plane (x,y). To obviate this, a parametric evaluation was performed, with respect to angle θ, of the losses for different values of the same angle (Tab.1), for the transformer (A) in the following.

Table 1 - Losses in relation to angle

θ (degrees)	Losses (W)
0	4461
10	4253
20	3939
30	3503
45	2520

The possibility of finding a law of variation of the losses in relation to θ, allows one to pass from the losses calculated in the section where the tank is nearer to the windings, to the average losses along the perimeter and thus to derive the total losses by multiplying the average losses on the transformer perimeter. One may find from the values of Table 1 that the law of variation (6) is close to the following expression

$$P(\theta) = P_M \cos^2\theta \qquad (10)$$

where P are the maximum losses for the unitary depth.

If (10) is accepted, the average value of the losses is found to be:

$$P_m = P_M \frac{4}{\pi} \int \cos^2\theta \, d\theta = 0.82 \, P_M \qquad (11)$$

Thus the total value of the losses will be given by:

$$P_t = P_m \, L \qquad (12)$$

where L is equal to the perimeter of the transformer.

APPLICATIONS OF THE CALCULATION METHOD

In order to verify the congruence of the results obtained by the application of the proposed method with the measured values, two transformers were analysed presenting different features, as to power, voltage and layout of the windings. In the following, the two transformers will be identified as (A) and (B).

The main features of the transformer (A) are: nominal power 700/350/350 kVA, voltages 11000/400/400 V, frequency 50 Hz, connections of the windings ⅄ - ⅄ - ⅄, three-phase core with three wounded columns. The layout of the windings is shown in Fig. 1a.

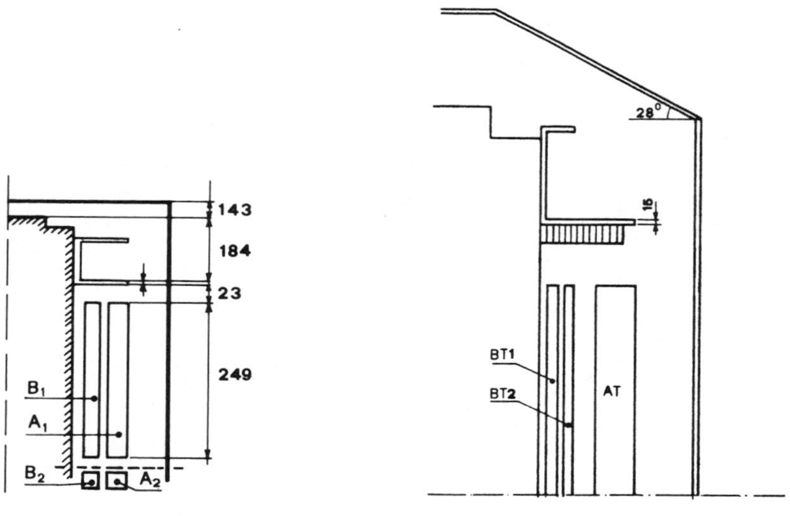

a) b)

Fig.1 - a) Geometric scheme of transformer (A); b) idem for transformer (B).

The main features of the transformer (B) are: nominal power 370 MVA, voltages 400/20 kV, frequency 50 Hz, connections of the windings ⅄ - Δ , three-phase core with five columns, three wounded columns. The layout of the windings is shown in Fig. 1b.

The calculations were performed with the following numerical values of the electromagnetic parameters:

- iron resistivity τ: $0.23\ 10^{-6}\ \Omega/m$;
- magnetic permeability of the iron constituting the magnetic shields: $800 \cdot \mu_o$.

For the transformer (A) the operating conditions were considered as shown in the upper row of Tab. 2.

With those connections, the leakage flux presents a high tranversal component; thus, the integral value of the losses due to the relevant eddy currents is considerable (as a matter of fact, the losses due to the normal axial flux are generally negligible).

For the transformer (B), the operating condition of H.V. winding feeded and L.V. winding short-circuited was examined.

For the determination of the numerical value of the losses in the non-active ferromagnetic materials, the measured total losses were subtracted from the losses in the windings.

COMPARISON WITH EXPERIMENTAL RESULTS

For an overall evaluation of the eddy current losses one ought to take into account not only the tank but also the presence of all the other ferromagnetic parts (end frames, shields, etc.); the respective losses were calculated separately, considering an unitary depth in the direction orthogonal to the plane under investigation. Taking P_i (i=1,2,7) as eddy current losses in the tank and P_j (j=3,4,5,6) as eddy current losses in the end frames (unitary depth), the total losses were obtained as follows (see Fig. 2):

$$P_t = P_i\ 0.82\ L + P_j\ L' \qquad (13)$$

where L is the perimeter of the transformer, L' the length of the end frames.

Table 2 - Partial losses in the transformer (A)

feed open wind. short cir.	A1 B1,B2 A2	A2 B1,A2 B2	B1 A1,A2 B2
P_1 [W/m]	6.11	6.2	6.2
P_2	2785	2420	2420
P_3	739	749	749
P_4	2761	3081	3081
P_5	2538	2933	2933
P_6	392	392	392
P_7	.05	.05	.05

For transformer (A), the disposition of the tank and of the end frames is shown in Fig. 2a. In Tab. 2 the partial losses for each block of ferromagnetic material referring to

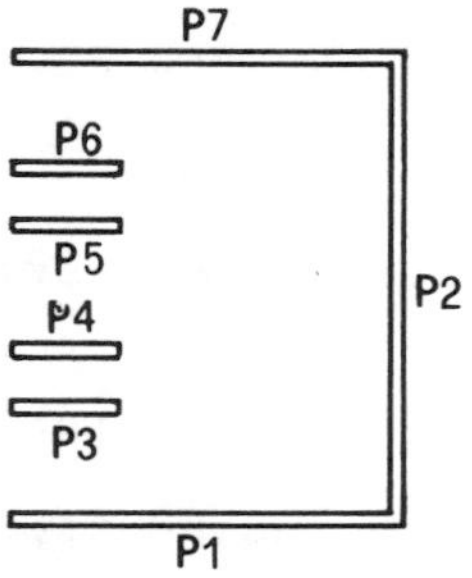

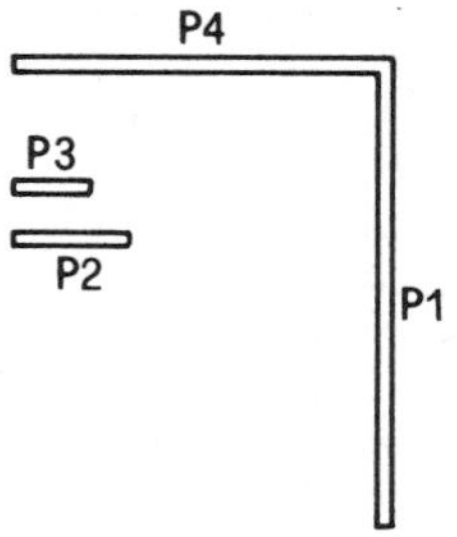

Fig.2 - a) Disposition of the parts involved in the eddy current of transformer (A); b) idem for transformer (B).

different working conditions are shown. From the study of this table, it may be noted that the losses in the lid and base of the tank are far lower than those in the end frames and the walls, as may also be justified by the distance between these and the windings.

For transformer (B), the disposition of the tank and of the end frames is shown in Fig. 2b, where the half wall and lid of the tank are schematically represented with P_1 and P_4; P_2 and P_3 represent the end frames.

In the shields no account was taken of the eddy current losses, in that they were constructed with sheets isolated from each other; in addition, to make the model plane adopted to conform to reality, it was necessary to consider the shield contiguous to the core. In such a way, a great part of the flux from the shield crosses the core, thus unloading the end frames. Tab. 3 shows, for the two transformers, a comparison between the overall results obtained with the method proposed, the experimental results, and the results obtained with a traditional analytical method [1].

Table 3 - Comparison between values calculated, by numerical method and analytical method, and the values measured for the two transformers.

		losses (W)		
transf.	config.	num.	anal.	meas.
A 350kVA	A1/A2	20160	18500	21000
A	B1/B2	20500	18500	20000
A	A1/B2	20500	18500	20000
B 370MVA		70000	80000	61000

REFERENCES

1. V.K. Chari: Finite Element Solution of the Eddy Current Problem in Magnetic Structures, IEEE P.A.S., p.62, April 1973.
2. A. Di Napoli, R. Paggi: Sizing of Magnetic Shields in Saturated Conditions for Large Transformers Calculated by menas of the Finite Element Method, IEEE P.A.S., A 77 663-8, July 1977.

THE FEM ANALYSIS OF MAGNETIC FIELD IN CONVERTER TRANSFORMERS DURING FAULTS

Michał Jabłoński* and Ewa Napieralska-Juszczak**

* Institute of Electrical Machines and Transformers
Technical University of Lodz, Poland

** Institute of Informatics Techn.University of Lodz, Poland

INTRODUCTION

The rectifier transformers during internal faults in valve sets undergo strong submagnetization with DC, saturating its core. Simultaneousely the nonsymetrical short circuit conditions produce large amount of currents in both windings. The core in instants of high saturation loses its "magnetic mirror" properties and the main flux finds its way as well in the nonmagnetic space out of the core. The configuration of leakage flux varies in function of time and contains the DC component fast increasing with time. This conditions greatly differ with classic understanding of leakage flux in usual transformers operating in networks during faults. The above phenomena can influence the electromagnetic field in transformer space, as well as the field of dynamic forces and the mechanical behaviour of windings. Evidently they deserve a more detailed analysis. It must consider the magnetic properties of steel and should explain the difference of field distribution in convertor transformers in comparison to usual network units.

ELECTROMAGNETIC FIELD GOVERNING EQUATIONS

The electromagnetic field in transformer space is assumed in first approximation to be flat-parallel having two components of flux density $\overline{B}$ in directions x and y, as well as one component of the current density J in direction z.
These quantities are described in Maxwell´s equations:

$$\mathrm{rot}\, H = \gamma \cdot E \qquad \mathrm{rot}\, E = -\frac{\partial B}{\partial t} \qquad H = \nu \cdot B \tag{1}$$

where ν (B) - reluctivity $(\frac{1}{\mu})$ of considered space varying in function of magnetic flux density
H - magnetic field density
B - magnetic flux density
E - electric fields strenght
Introducing the vector potential to Maxwell´s equation and assuming $A = A_z$; $A_x = 0$; $A_y = 0$ it can be stated that:

$$\frac{\partial}{\partial x}\left(\nu_x(B)\frac{\partial A}{\partial x}\right)+\frac{\partial}{\partial y}\left(\nu_y(B)\frac{\partial A}{\partial y}\right) = \gamma\frac{\partial A}{\partial t} - \tilde{q}^{B}(J_o) \qquad (2)$$

The bounds-the conditions of Dirichlet or Neuman can be assumed. Dirichlet, if the bounds are far, and Neuman it they are near to the transformer core. The criterion of critical distance depends on saturation and is not yet formulated as Savini and Turowski stated[1];

γ - electric conductivity of the material

J_o - current density of the forcing current in particular windings computed on the basis of equivalent circuit equations.

Expression (2) in its $\gamma\frac{\partial A}{\partial t}$ takes into account the eddy currents.

In the case of a dry transformer with laminated core the share of eddy currents is negligable. The authors wish to form a general model allowing the computation of eddy current reaction in constructive parts (screens, tank walls etc.) the $\gamma\frac{dA}{dt}$ is generaly considered. In this paper it can be and consequently is neglected.

Considering next a variational formulation, the functional Π is defined such that when involving the stationarity of Π, the governing differential equation (2) is obtained

$$\Pi = \int \frac{1}{2}\left(\nu_x(B)\left(\frac{\partial A}{\partial x}\right)^2 + \nu_y(B)\left(\frac{\partial A}{\partial y}\right)^2\right) dv - \int_V A\, q^B\, dv \qquad (3)$$

where $q_B = \tilde{q}_B - \gamma\frac{\partial A}{\partial \tau}$

$\tilde{q}_B$ - the current density in particular windings of the transformer

Introducing the interpolation function h_i we obtain the general matrix equation in following form.

$$\underline{K}_c(A)\cdot\left\{A(t)\right\} + \underline{C}\cdot\left\{\dot{A}(t)\right\} = \left\{C_o\right\} \qquad (4)$$

where $\underline{K}_c$ - the reluctivity matrix depending on vector potential; $\underline{C}$ - the matrix of eddy currents; $\left\{C_o\right\}$ - the vector of knownforcing currents, computed with circuit equations.

The considered space D is divided into elements with 3 to 8 nodes.

3 nodes elements are applied in the model of the iron core with interlacing angle $\varepsilon \neq \frac{\pi}{2}$, eg $\varepsilon = \frac{\pi}{4}$. Such system of nodes allows the consideration of anisotropic properties of steel. In this paper the anisotropy and interlacing are not considered. A new paper (in preparation) takes both into account.

For transient analysis a numerical time integration scheme must be employed. In this paper a family of one step methods is used in which we assume.

$$^{t+\alpha\Delta t}\dot{A} = \left(^{t+\Delta t}A - {}^{t}A\right) / \Delta t$$
$$^{t+\alpha\Delta t}A = (1-\alpha)\,^{t}\underline{A} + \alpha\,^{t+\Delta t}\underline{A} \qquad (5)$$

where $\alpha = 0$ - Euler forward method

$\alpha = 1/2$ - trapezoidal rule

$\alpha = 1$ - Euler backward method

To solve the set of above equations Napieralska[2] applied a modified method of Gauss for symmetrical band matrices. The full matrix was divided into submatrices, and due to this subdivision it was possible to calculate the field distribution for large objects, discretised with several thousands of elements on small Polish made computer Odra 1305. Larger computers can allow one to solve three dimentional problems.

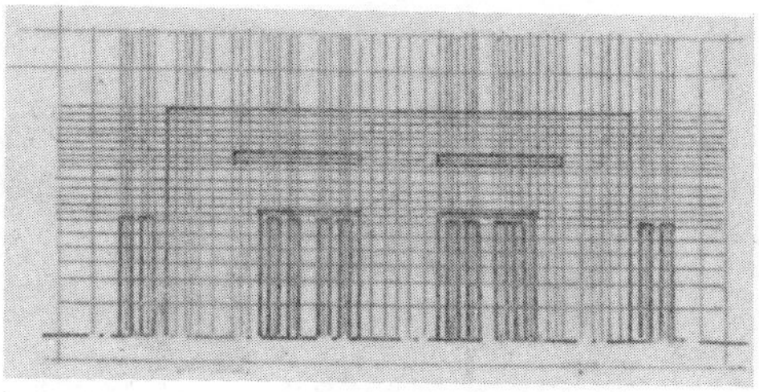

Fig.1. The 3 phase transformer space divided into elements

CIRCUIT EQUATIONS

The forcing quantity for the field analysis in transformer space is the set of instantaneous values of primary and secondary currents in all phases. The computation of such currents during the operation of a convertor set, specially for transients and faults needs an adequate equivalent circuit with a variable structure. The circuit has to allow the consideration of nonlinear magnetic permeability of the main core as well as a smoothing reactor in the receiver branch and the natural nonsymmetry of the core and the properties of diodes, thyristors and fuses. It should be taken into account that the faults in the valve set can cause a strong undirectional submagnetizing of the core with the following consequences:

a) the instantaneous oversaturation ($B > 1.95$ T) ousts the flux from the main core to a nonmagnetic space of the transformer

b) the core surface loses in this instants the usually considered property of "magnetic mirror"

c) the saturation of the core yields a large increase of magnetizing ampere turns in the network winding, increasing greatly the effective value of primary current

d) a variable flux distribution appears in the leakage space of the transformer in consecutive instants providing an intrinsic influence on short circuit forces and their harmonics.

In this paper, as an example, a three phase diode bridge is considered supplied with a Yy transformer. An important fault, a break through of one diode is taken into account, as shown on fig.2a.

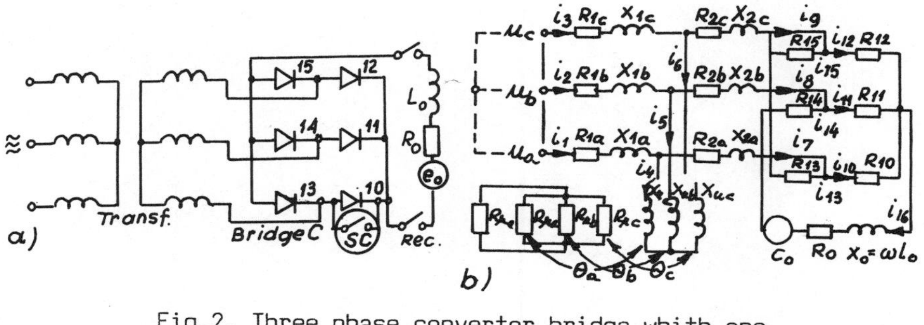

Fig.2. Three phase convertor bridge whith one diode short-circuited
2a. conection sketch
2b. the equivalent circuit

The prime forcing quantity in this set is a three phase rigid symmetrical and sinusoidal voltage with f = 50Hz supplying the primary winding. The equivalent circuit (fig.2b) allows to obtain 19 equations with nonlinear

magnetizing inductances. Every valve is represented with resistance R. During the period of conduction or when losing the rectifier property it obtains the value of dynamic forward resistance R_F; during the reverse or blocking period it is equal to the backward resistance R_B. The left bottom side of the sketch 2b represents the magnetic circuit coupled with electric one through the magnetomotive force (ampereturns $\theta = iz$, where z - number of turns).
The detailed information how to prepare the particular elements of the equivalent circuit and the set of equations is given by Jabłoński[3]. The solution of this nonlinear implicit equations with numerical methods yields the primary and secondary currents needed for the field analysis with FEM.
Several other quantities can be found, like magnetizing currents containing an increasing unidirectional component (DC), main fluxes (DC + AC) in particular phases etc. The numerical example was done for a specially prepared model transformer supplying 6D (or 6T) bridge with parameters $S_n = 23.65$ kVA; $U_2 = 167.2$ V; $J_2 = 81.65$ A; $\vartheta = 1$; $U_x = 6.5\%$; $U_R = 4\%$; $U_z = 7.63\%$; on the DC side $U_{do} = 225.8$ V; $J_{dn} = 100$ A; the diodes, have $R_F = 0.01\,\Omega$; $R_B = 2\,k\Omega$ where S_n - the rated kVA; U_2 = the rated secondary voltage; J_2 - the rated secondary current; U_x = short circuit reactance; ϑ - transformation ratio; U_R - short circuit resistance in %; U_2 - short circuit impedance in %; U_{dn} - no load DC voltage; J_{dn} - rated DC current.
Nowicki[4] prepared the detailed description of the transformer and initial experiments performed on it. For equivalent circuit purposes the short circuit reactance was divided into internal and external windings in ratio 1:2. The magnetizing curves and corresponding nonlinear reactances X_μ were computed separatly for external phases and for the internal one, to consider the natural nonsymmetry of the 3-phase core. According to the need, both $\mu_{dyn} = \frac{dB}{dH}$ (for voltage equations) and $\mu = \frac{B}{H}$ (for flux equations) were applied. Beginning B = 1.98 T the $\mu_{dyn.r.}$ falls to 1.
The smoothing reactor is practically linear up to 120 A (L = 0.01 H). Over this value it begins to saturate and by 500 A its dynamic inductance achieves only $L_{dyn} = 0.0005$ H (the cases of short circuit on DC side).
The circuit computation with given parameters for the above mentioned case of break through of one diode gives results shown on fig.3. Five periods that means 0.1 s are enough to obtain full saturation of the core with DC

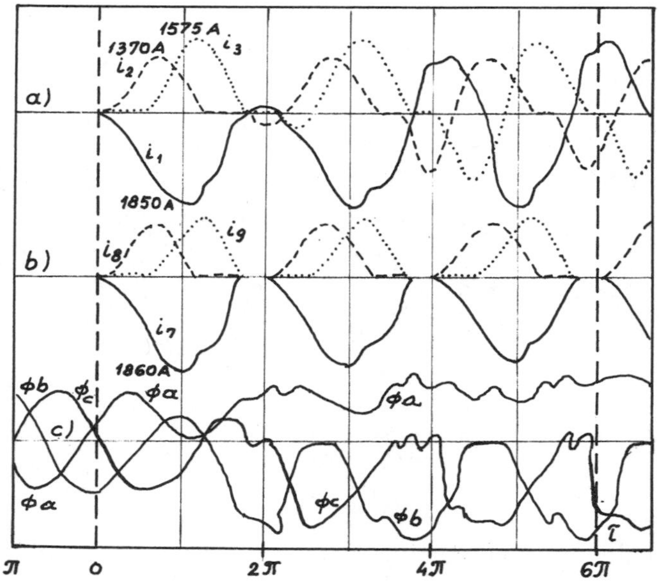

Fig.3. The transient after s-c of diode no 10, 3a. primaty currents; 3b. secondary and magnetising currents; 3c. main fluxes.

MAGNETIC FIELD IMAGES

The limited volume of this paper does not allow us to show a series of plottings representing subsequent distributions of flux in the core and leakage space. Fig.4 represents only one case from the series, in the instant, when the saturation is at its highest and secondary currents are zero in the 4 th period of fault The analysis done in this report did not take the saturation of ferromagnetic tank into account.

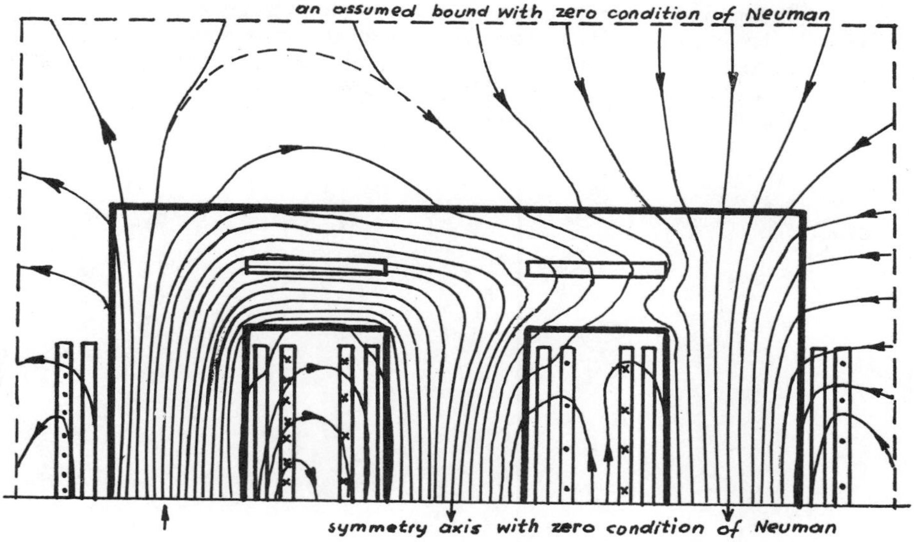

Fig.4. The magnetic field image in the instant τ on fig.3

The feed back between the field and circuit models after every step of computation could allow the fluent correction of the circuit model because of the non uniform saturation of the core. That all has not yet been done and remains to be the next step in this research.

CONCLUSIONS

The DC increasing submagnetizing of the transformer core during faults in the convertor set provides a peculiar effect for the field distribution in leakage space. The usual methods applied to the analysis of leakage S.C. flux in power transformers are not correct in the case of convertor transformers. The field distribution varies with in function time, the peak of the short circuit current in the damaged phase is larger than during the symmetrical short circuit, and the core becomes in some instants over saturated.
It all shows clearly the necessity of careful consideration of magnetic fields in convertor transformers during the faults. The method described in this report allows us to obtain proper results of field analysis with its consequences for dynamic forces consideration.
First experiments just performed by J. Nowicki on the test transformer equipped with large number of sensors and the research described by Ławnicki[5] confirm this statement.

REFERENCES

1. A. Savini, J. Turowski, "Computer analysis of critical distance of tank wall in power transformers". ISEF 87, Pavia (1987).
2. E. Napieralska-Juszczak, "Method of modeling the coupled electromagnetic and thermal fields in electrical machines and apparatus". Modeling, Simulation Control A, 12:4 (1987).
3. M. Jabłoński, E. Napieralska-Juszczak, "Model matematyczny do badania stanów dynamicznych w mostkowych układach tyrystorowych". Materiały konferencyjne. Konferencja Obwodów Nieliniowych, Poznań (1987).
4. J. Nowicki, "Model fizyczny do badania stanów dynamicznych pola elektromagnetycznego w transformatorze przekształtnikowym". Przegląd Elektrotechniczny (in preparation).
5. A. Ławnicki, "Analiza przebiegów prądów pobieranych z sieci przez mostkowy trójfazowy zespół przekształtnikowy przy zwarciu wewnętrznym w przekształtniku". Doctor thesis Technical University of Łódź (1974).

ADDITIONAL LOSSES IN FREQUENCY TRIPLER WINDINGS

Tadeusz Janowski and Ryszard Goleman

Technical University of Lublin, Poland

INTRODUCTION

Magnetic frequency triplers under construction have mega-Watt outputs[1] and accurate determination of their winding losses is essential. Methods established for rotating machines and transformers[2,4,5] are inadequate for frequency multipliers which have highly distorted primary current and higher frequency output current. The additional loss caused by the magnetic flux in the cross-sections of the windings is the sum of the losses from each flux harmonic. The direct analytical method is used to calculate the losses in each wire of the windings. This method consists in solving Helmholtz's equation with boundary conditions defined by the distribution of the magnetic field acting on the wire[3]. This solution makes possible calculation of the additional loss factors in the tripler windings.

POWER LOSS IN A WIRE OF THE WINDING

Let us now consider concentric primary and secondary windings placed on both legs of the core. Each half of the windings consists of m_1 layers with n_1 turns in the primary and m_2 layers with n_2 turns in the secondary, respectively (Fig. 1). Windings of magnetic frequency triplers are similar to those in transformers and therefore the present analysis is carried out with the assumption that the magnetic field intensity in the winding region contains only an axial component. On the whole, the eddy-current loss caused by the radial component of the magnetic field may be neglected . This will be low in the slender windings typical of magnetic frequency triplers and other magnetic converters. Distribution of the magnetic fields in the region of tripler windings is different from the well-known trapezoidal distribution for transformers because of the various frequencies of the primary and secondary magnetomotive force (Fig. 1). The magnetic field produced by the winding carrying the input current appears in the regions of both windings, however, in the case of the inner winding the third harmonic of magnetic field intensity generated by the secondary current appears only in this winding (Fig. 1). For the case shown, Fig. 1. the electric field of pulsation ω in the wire "pq" of a winding with conductivity γ and permeability μ_0 can be determined from the equation:

$$\frac{d^2\underline{E}_2\ (x)}{dx^2} = \Gamma_w^{\ 2}\ \underline{E}_z\ (x) \qquad (1)$$

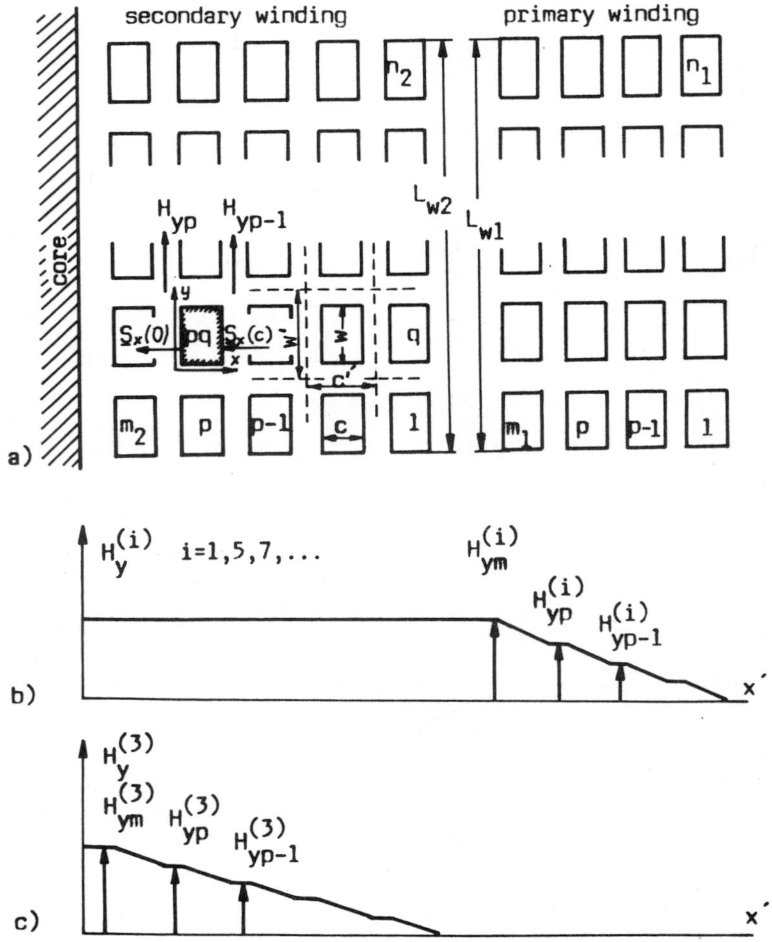

Fig. 1. Denotation used in the calculation of winding losses of a magnetic frequency tripler: a) windings location and their cross-sections; b) distribution of the magnetic field intensity harmonics produced by the current of the primary winding; c) distribution of the third harmonic of magnetic field intensity produced by the secondary winding.

where

$$\Gamma_w^2 = j\omega\mu_0\gamma\frac{w}{w'}$$

its solution may be written as:

$$\underline{E}_z(x) = \underline{A}_1 e^{\Gamma_w x} + \underline{A}_2 e^{-\Gamma_w x} \qquad (2)$$

According to Maxwell's second equation the magnetic field intensity is given by the relation:

$$\underline{H}_y = \frac{\Gamma_w}{j\omega\mu_o}(\underline{A}_1 e^{\Gamma_w x} - (\underline{A}_2 e^{-\Gamma_w x}) \tag{3}$$

The unknown constants $\underline{A}_1$ and $\underline{A}_2$ may be obtained from boundary conditions as components of the magnetic field intensity on the conductor surface (Fig. 1). After some algebra, we obtain:

$$\begin{bmatrix} \underline{A}_1 \\ \underline{A}_2 \end{bmatrix} = \frac{j\omega\mu_o}{2\Gamma_w \mathrm{sh}\,\Gamma_w c} \begin{bmatrix} H_{yp-1} - H_{yp} e^{-\Gamma_w c} \\ H_{yp-1} - H_{yp} e^{\Gamma_w c} \end{bmatrix} \tag{4}$$

The active power in the wire of length l may be found as the difference between the power streams penetrating the flanks of the wire:

$$\Delta P_{pq} = \frac{1}{2} \mathrm{Re}\left[\frac{w'}{w} (1 \int_o^w \underline{E}_z(c)\underline{H}_y^*(c)dy - 1 \int_o^w \underline{E}_z(o)\underline{H}_y^*(o)dy)\right] \tag{5}$$

The power loss expressed by (5) has only been derived for the harmonic component of pulsation ω. Linearity of the windings environment permits us to use dependences in the form of (5) in superimposing losses from harmonics of any pulsation. The total power losses, taking into account the magnetic fields of both windings, may be expressed [3] in the form:

$$\Delta P_{pq} = \frac{1}{cw\gamma} \sum_{i,r} \Big\{ I_i^2 + \frac{1}{2}(w')^2[(H_{1yp}^{(i)} - H_{1yp-1}^{(i)})^2 \varphi_1(\xi^{(i)}) + H_{1yp}^{(i)} H_{1yp-1}^{(i)} \Psi(\xi^{(i)})] + \frac{1}{2}(w')^2[(H_{2yp}^{(r)} - H_{2yp-1}^{(r)})^2 \varphi_1(\xi^{(r)}) + H_{2yp}^{(r)} H_{2yp-1}^{(r)} \Psi(\xi^{(r)})]\Big\} \tag{6}$$

where: I_i - effective value of the i-th current harmonic in the wire, $H_{1yp}^{(i)}$, $H_{1yp-1}^{(i)}$ - amplitudes of the i-th harmonics of magnetic field intensity produced by the considered winding, $H_{2yp}^{(r)}$, $H_{2yp-1}^{(r)}$ - amplitudes of the r-th harmonics of magnetic field intensity generated by the current in the adjacent winding,

$\xi^{(n)} = c\sqrt{\dfrac{n\omega\mu_o\gamma w}{2w'}}$ - reduced dimensionless thickness of the wire for the n-th harmonic of the magnetic field intensity, $n \in (i,r)$,

$$\varphi_1(\xi) = \xi \frac{\mathrm{sh}2\xi + \sin 2\xi}{\mathrm{ch}2\xi - \cos 2\xi} - 1 \quad \text{and} \quad \Psi(\xi) = 2\xi \frac{\mathrm{sh}\xi - \sin\xi}{\mathrm{ch}\xi + \cos\xi}$$

The first term $1(\gamma cw)^{-1}\sum_i I_i^2$ in eqn. (6) represents the I^2R loss in the considered wire in the absence of skin effect. The other terms represent the additional losses dependent on magnetic field harmonics. The relation (6) is the basis for calculating the winding losses of a frequency tripler. The total loss in the wire as well as in the whole winding may be written in the form:

$$\Delta P = (1 + \varkappa)\Delta P_{I^2R} = k\Delta P_{I^2R} \tag{7}$$

where k - additional loss factor

$\varkappa$ - additional loss index $\left(\varkappa = \dfrac{\Delta P_{add}}{\Delta P_{I^2R}}\right)$

ADDITIONAL LOSSES IN THE PRIMARY OUTER WINDING

In the case of on outer primary winding, the magnetic field intensity in its vicinity contains odd harmonics (1,5,7,...) except those of the zero sequence (3,9,...). Their amplitudes have linear distribution on the conductors surfaces (Fig. 1b). Thus, from (6) and (7) after some manipulation, we obtain the expression for the additional loss factor in the primary winding:

$$k = 1 + \varkappa \tag{8}$$

where

$$\varkappa = \frac{\sum\limits_{i=1,5,7,\dots} h_i^2 \alpha_I^2 \left(\frac{w_1'}{w_1}\right)^2 \left[\varphi_1(\xi^{(i)}) + \frac{1}{3}(m_1^2 - 1)\,\psi(\xi^{(i)})\right]}{1 + h_5^2 + h_7^2 + \dots}$$

and

$$\alpha_I = \frac{n_1 w_1}{L_{w1}}\,; \qquad h_i = \frac{I_i}{I_1}$$

In this case the factor (8) has been calculated taking into account harmonics up to the 7-th order. Harmonics of the input current higher than the 7-th have relatively small amplitudes under commonly occurring loads of magnetic triplers and can be neglected. Variation of higher harmonics in the input current causses a change in the index of additional losses ($\varkappa$) which may even reach 100% under leading and lagging load, as compared with a resistive load. By an analysis of equations (7), (8) it can be proved that winding losses are directly proportional to the expression defined by the ratio of the additional loss factor and reduced thickness of the wire for the fundamental harmonic:

$$\frac{k}{\xi^{(1)}} = \frac{1}{\xi^{(1)}} + \frac{1}{\xi^{(1)}(1+h_5^2+h_7^2+\dots)} \sum_{k=1,5,7,\dots} h_i^2\left[\varphi_1(\sqrt{i}\,\xi^{(1)}) + \right.$$
$$\left. + \frac{1}{3}(m_1^2 - 1)\,\psi(\sqrt{i}\,\xi^{(1)})\right] \tag{9}$$

By calculating $\xi^{(1)}$ which corresponds to the minimum of the function (9), we can express the critical thickness of the wire. The results of numerical solution are shown in Tabl. 1. In the case of layer windings, if $m_1 \geqslant 3$, the critical thickness of the wires determined by the analytical solution can be used in practice, namely:

$$c_{cr} = a\left(\frac{L_{w1}}{n_1 w_1}\right)^{\frac{1}{2}} \left[F(h_i)(m_1^2 - 0{,}2)\right]^{-\frac{1}{4}} \tag{10}$$

Table 1. Critical dimensionless thickness of the wire $\xi^{(1)}_{cr}$

Number of layers m_1	Higher harmonics content			
	$h_5 = 0,4$	$h_5 = 0,5$	$h_5 = 0,6$	$h_5 = 0$
	$h_7 = 0,3$			$h_7 = 0$
1	1,580	1,590	1,590	1,570
2	0,630	0,590	0,568	0,960

where

$$F(h_i) = \frac{1 + 5^2 h_7^2 + 7^2 h_7^2}{1 + h_5^2 + h_7^2}$$

the parameter "a" for aluminium and copper conductors is 1,75 10^{-2} m and 1,37 10^{-2} m, respectively.
The additional loss factor of a single layer winding corresponding to the critical size of the wire is about 2 in the case of usual input current distortion. If the number of layers in the winding grows this factor decreases and its values for two, three and four layers are 1,38-1,46; 1,33-1,39 and 1,32-1,38 respectively.

ADDITIONAL LOSSES IN THE SECONDARY INNER WINDING

Magnetic fields produced by both the input and output current appear in the region of the secondary winding (Fig. 1c). From relation (6), it can be shown that the additional loss factor for the whole winding contains terms which depend on the stray fields of both windings:

$$k = 1 + \varkappa_1 + \varkappa_2 \qquad (11)$$

where

$$\varkappa_1 = m_2 \alpha_{II}^2 \left(\frac{w_2'}{w_2}\right)^2 \sum_{i=1,5,7,\dots} s_k^2 \,\Psi\left(\sqrt{\tfrac{i}{3}}\ \xi^{(3)}\right)$$

$$\varkappa_2 = \alpha_{II}^2 \left(\frac{w_2'}{w_2}\right)^2 \left[\varphi_1(\xi^{(3)}) + \tfrac{1}{3}(m_2^2 - 1)\,\Psi(\xi^{(3)})\right]$$

and

$$\alpha_{II} = \frac{n_2 w_2}{L_{w2}}; \qquad s_i = \frac{I_i}{I_2}$$

I_i - effective value of the i-th harmonic of the primary current referred to the secondary, I_2 - output current, $\xi^{(3)}$ - reduced thickness of the wire for the third harmonic of the magnetic field intensity.
By analysing the function "k/ $\xi(3)$", we can determine the critical thickness of the wire for the secondary winding. After some simplification of relation (11) we obtain the analytical solution in the form:

$$c_{cr} = a\left(\frac{L_{w2}}{n_2 w_2}\right)^{\frac{1}{2}} \left[F(s_i)m_2^2 + 3(m_2^2 - 0,2)\right]^{-\frac{1}{4}} \qquad (12)$$

where

$$F(s_i) = s_1^2 + 5^2 s_5^2 + 7^2 s_7^2$$

the parameter "a" for aluminium and copper conductors is 1,38 10^{-2} m and 1,33 10^{-2}m respectively.
Equation (12) can be used for both single layer and multilayer windings. The additional loss factor corresponding to the critical thickness of the wire for usual ratios of stray fields produced by the primary and secondary windings amounts to 1,32-1,37. The above analyses apply for the field distribution with secondary winding nearest to the core. For the converse case a modified analysis is necessary.[3]

CONCLUSIONS

The additional primary winding losses due to distortion of the input current are several times greater than those due to the first harmonic alone. Due to the large difference in the m.m.fs. of the two windings, 75-80% of the total winding losses occur in the primary winding. The magnetic field caused by current of the primary (outer) winding determines in principle the additional loss in the secondary internal winding. The higher harmonics in the input current waveform and the higher frequency of the output current determine the critical thicknesses of the wires. For the primary this is 40-100% and for the secondary 20-58% of the values which would be calculated for normal transformer action.

REFERENCES

1. P.P. Biringer, J.D. Lavers, Recent advances in the design of large frequency changers, IEEE Trans. Magn., MAG-12, No. 6 (1976).
2. W. Dietrich, Berechnung der Wirkverluste von Transformatorenwicklungen unter Berucksichtigung des tatsachlichen Steufeldverlaufes, Arch. Elektrotech.H. 4 (1961).
3. R. Goleman, Losses in a transformer frequency tripler, (in Polish), Doctorate Thesis, Lublin Technical University (1983).
4. J. Jezierski, Transformers, (in Polish), WNT, Warszawa (1983).
5. J. Turowski, Electromagnetic calculation of elements of electrical machines and devices, (in Polish), WNT, Warszawa (1982).

STRAY-LOAD LOSSES IN YOKE-BEAMS OF TRANSFORMERS

D. Kerényi

Ganz Electric Works
Budapest
Hungary

INTRODUCTION

The yoke-beams, serving for clamping the yoke of the transformer core and for supporting the windings, are structural parts of complicated shape. The beams are in the stray magnetic field of the windings and the stray flux can produce considerable eddy current losses in them. Due to the large surface and efficient cooling of the beam hot spots generally do not form, however, the loss can reach or even exceed the tank losses. Thus it is advisable to assess the stray losses in the yoke-beams as well when designing the transformer. However, it is impracticable to accurately simulate such a complicated construction in the calculation and to pay attention to the non-linearity as well as the reaction of induced eddy currents at the same time. Calculation methods based on simplified models and on simplifying conditions are required. In the following two simple models are presented for the approximate calculation of the losses in yoke-beams. The losses calculated are compared to those measured on a transformer.

NUMERICAL CALCULATION

The model for calculation is shown in Fig.1. The cross-section given in the figure is a plane going through the axis of the windings and perpendicular to the tank wall. The stray field can be sub-divided into three parts as shown in the figure. One part of the stray flux is linked with the core, the other part with the yoke-beam and the third part with tank wall. Eddy-current losses of considerable magnitude are generated only by the two latter, therefore only losses in the yoke-beams and in the tank wall will be dealt with in the following. These losses have been calculated using a two-dimensional computer program based on the finite-difference method [1]. The calculation considers the stray field of the windings in the plane of Fig.1 and currents normal to this plane are calculated. Using this model the calculated losses and their proportion will differ somewhat from the real ones. The reason is that in the model the minimum distance between the outer winding and the tank wall is considered though the tank wall is at almost all places farther away

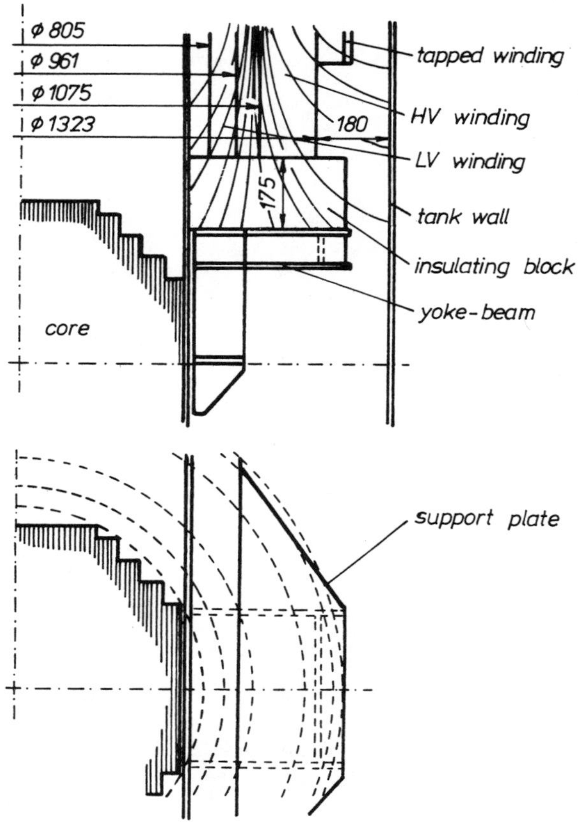

Fig. 1. Yoke-beam-design for a 63 MVA, 120 kV transformer.

from the winding consequently, the tank loss calculated is higher than the real one. However, the distance between the bottom of the winding and the support plate of the yoke-beam is constant and equal to the distance considered in the model. In addition, flux generated by the other phase windings penetrates into the beam sections not under the windings. Thus calculated losses in the yoke-beams are lower than the real ones. Despite, the method enables both the comparison of the various designs and assessing the effect of the factors influencing the phenomena as well.

Fig. 1 shows the cross-section of a three-phase 63 MVA, 120 kV transformer. The beam-design in the figure is one of the most frequently used ones. For this transformer with the dimensions given in the figure and assuming 170 kA Ampere-turns in the HV and LV windings /with no current in the tapped winding/ the calculation gave a tank loss of P_t = 19,7 kW and a yoke-beam loss of P_y = 10,2 kW. In the following we will consider in what manner the position, the shape and the material of the yoke-beam influence the losses. Losses in the layout of Fig. 1 will be regarded as unity.

Fig. 2 gives the losses as a function of the distance between the bottom of the windings and the yoke-beam's support plate. At distances larger than that in the basic layout /d_y = 175 mm/ the losses in the yoke-beams decrease considerably while the tank-losses increase slightly. Reducing the distance

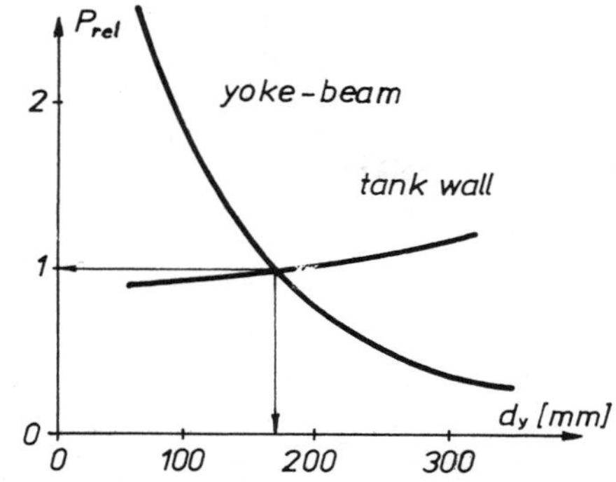

Fig. 2. Losses as functions of distance d_y.

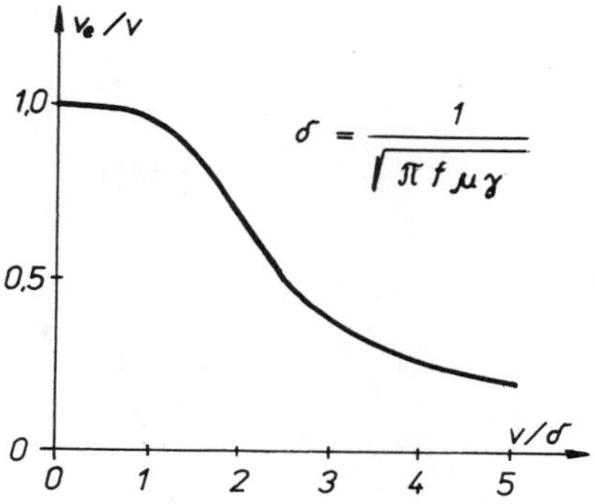

Fig. 3. Equivalent thickness as a function of δ.

between the bottom of the windings and the yoke-beam the losses in the yoke-beam increase rapidly while the tank losses practically do not change.

Table 1 gives the calculated losses for 7 yoke-beams of different shapes. Losses in the yoke-beams can be reduced to less than half of their original value by a more suitable design, as shown in the Table. The magnitude and the rate of change of the tank losses are lower than the values given in the table for the reason mentioned above.

Table 2 shows the effect of the material an the losses. Considering the data given in the Table aluminum yoke-beams may be advantageous if the tank losses are reduced by magnetic shunts /flux diverters/. The aluminium makes the structure lighter, but one must think of the fact that it is more expensive than the magnetic steel and its welding is also more costly. Nonmagnetic steel is not recommended: it increases the losses, it is far more expensive than the magnetic steel and its machining is rather difficult.

One important result of the calculation is the observation that, even though to a less extent than calculated, the losses produced in the yoke-beam and in the tank wall have a mutual effect on each other.

A SIMPLE APPROXIMATE FORMULA

The biggest part of the losses is produced in the upper horizontal support-plate of the yoke-beam in Fig. 1. Therefore, for an approximate calculation of the losses in this layout the yoke-beam can be substituted by the upper plate. Thus a plate perpendicular to the axis of the windings will be used as our second model. The losses can be calculated using a simple formula obtained as a result of an analytical calculation [2]. The losses of a plate with dimensions r[m] x s[m], having a thickness v[m] and an electric conductivity γ[A/Vm] and a magnetic permeability μ[Vs/Am], are given by the relationship

$$P = \gamma f^2 \left(\mu k_H H_m\right)^2 r^3 s \frac{1}{1+\left(\frac{r}{s}\right)^2} v_e \; [W], \qquad /1/$$

Table 1. Losses in the yoke-beam $/P_y/$ and in the thank wall $/P_t/$ related to those for the basic layout

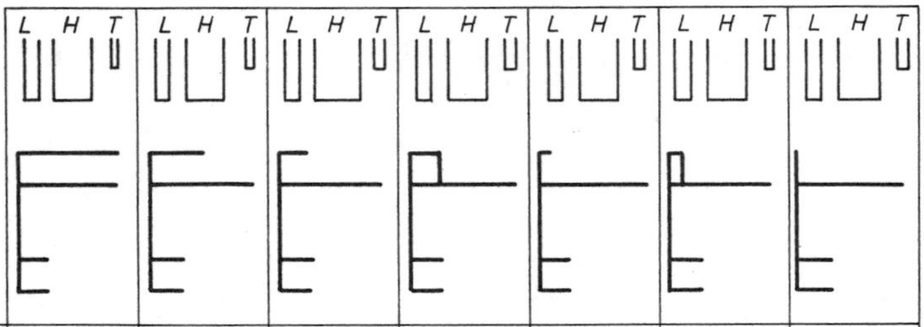

P_y	1	0,73	0,66	0,61	0,53	0,49	0,43
P_t	1	1,22	1,38	1,31	1,55	1,31	1,51

in which $f\,[1/s]$ is the frequency, $H_m\,[A/m]$ is the peak value of the field strength normal to the plate on the surface, k_H is a factor depending on the distribution of the field strength and v_e is the equivalent thickness depending on the depth of penetration of the plates. v_e can be obtained from the curve of Fig. 3. Assuming a field strength varying along r as a sine half wave and being constant along s, $k_H = 1$. Applying relationship /1/ to the calculation of the losses in the yoke-beam the following are assumed

a/ The dimensions r and s are equal to those given in Fig. 4. /Losses are produced in the hatched part of the plate./.

b/ The flux distribution along the width r is described by a sine half wave instead of the usual trapezium form $/k_H = 1/$

c/ The stray-flux density $B_s = \mu_o \frac{\Theta}{l_w}$ $/\frac{\Theta}{l_w}$ = Ampere turns per length in the windings/ must be multiplied by a factor k_y depending of the distance between the bottom of the windings and the plate and obtained experimentally /see Fig.5/.

d/ The equivalent thickness is $v_e = 0{,}003$ m $/\gamma = 7.10^6$ A/Vm and $\mu = 100\,\mu_o = 1{,}256.10^{-4}$ Vs/Am/. With these the equation /1/ can be reduced to the following form

$$P_y = 6{,}3.10^7\, n\, /k_y\, B_s/^2\, r^3\, s\, \frac{1}{1+\left(\frac{r}{s}\right)^2}\ [W] \qquad /2/$$

where n is the number of the yoke-beam parts hit by the stray field. /For three-limb transformers n = 6./ Yoke-beam losses

Table 2. Relative values of losses in yoke-beams of different material

	Steel	Aluminium	Nonmagnetic steel
P_y	1	0,28	1,17
P_t	1	2,04	1,65

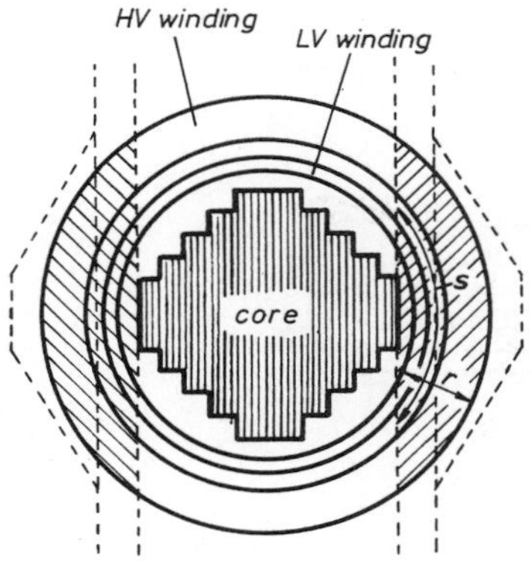

Fig. 4. Cross section illustrating r and s.

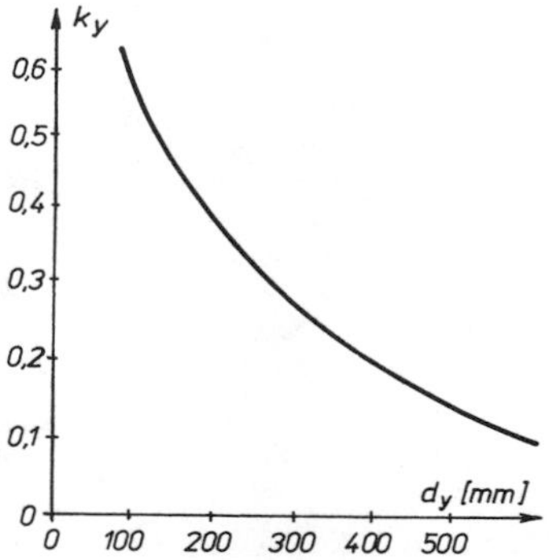

Fig. 5. Factor k_y as a function of d_y.

obtained from this approximate formula are higher than the real ones, because the flux-diverting effect of the tank wall has been neglected.

Yoke-beam losses in the aforementioned 63 MVA 120 kV transformer obtained from equation /2/ are P_y = 18 kW /n = 6, k_y = 0,4, B_o = 0,148 Vs/m^2, r = 0,259 m and s = 0,85 m/.

MEASUREMENTS

The stray losses generated in the yoke-beam and the tank wall of the transformer, mentioned in sections 2 and 3, were determined with short circuited secondary winding. A large number of thermocouples and search coils for measuring the flux density were fitted onto the surface of the yoke-beam and the tank wall. The surface loss density was calculated from the initial rate of rise of the temperature measured at the given spot after switching on by means of the relationship

$$p = c\varrho \left.\frac{\Delta\vartheta}{\Delta t}\right|_{\Delta t \to 0}$$

where p [W/cm] is the surface loss density, c [Ws/g°C] is the specific heat of the component investigated, ϱ [g/cm^3] is the density, $\Delta\vartheta$ [°C] is the measured temperature rise of the spot in question in time Δt [s], thus $\frac{\Delta\vartheta}{\Delta t}$ is the initial rate of rise

Table 3. Computed and measured values of the losses

	Numerical analysis	Approximate formula	Measurement
P_y [kW]	10,2	18	13
P_t [kW]	19,7	-	15

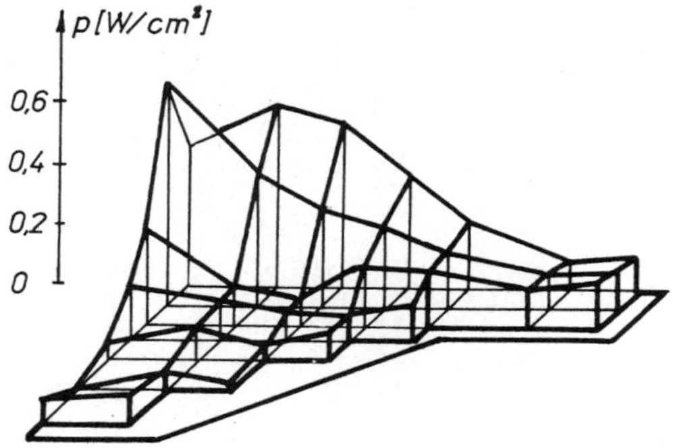

Fig. 6. Surface loss density distribution on the support plate.

of the temperature ϑ /t/. Fig.6 shows the distribution of the loss density measured on the half part of one of the support plates. On the basis of this the total loss of the yoke-beams amounted to P_y = 13 kW whereas the value of the tank loss was P_t = 15 kW.

The calculated and the measured losses are shown in Table 3. Yoke-beam losses calculated numerically are lower, those obtained from the approximate formula are higher than the measured ones as can be expected.

CONCLUSION

Simple approximate calculation can give information on the stray load losses produced in the yoke-beams of transformers. The losses in the yoke-beams and those in the tank wall mutually influence each other. Using the results of the calculation the designer of the transformer can decide if it is necessary to reduce the stray-load losses, by using magnetic shunts for example.

REFERENCES

1. I.Kézér, S.Márkon and L.Szabó: „Computation of two dimensional eddy current problems with application to the calculation of stray losses in power transformers". International Symposium on Electrodynamics. Lodz, 1979. pp 21-32

2. D.Kerényi: „Analytisches Näherungsverfahren zur Berechnung von Wirbelströmen in Metallplatten". Acta Technica Academiae Sc. Hungaricae, 93/1-2/ pp 77-99 /1981/

INFLUENCE OF STRUCTURE GEOMETRY, SCREENS AND EDDY CURRENTS ON THE CRITICAL DISTANCE OF TANK WALL IN POWER TRANSFORMERS

A. Savini — University of Pavia, Italy

J. Turowski — Technical University of Lodz, Poland

ABSTRACT

It is generally known that at small distance of a tank wall from transformer windings the state and position of tank wall strongly effects the leakage field configuration. At larger distance this influence is practically negligible and e.g. measurements of stray losses in a tank are possible with the short-circuit test "with and without tank". Such a "critical" tank wall distance has been investigated with both the reluctance network and the finite element method for normal large transformers with various structures and screens. The existence of such a characteristic distance and its influence on leakage field phenomena has been investigated and evaluated.

INTRODUCTION

In spite of the amount of progress made in computer modelling and simulation of the magnetic leakage field in transformers, simple constructional indications and syntetic quality criteria are still necessary to simplify practical conclusions and laboratory test. One of such practical

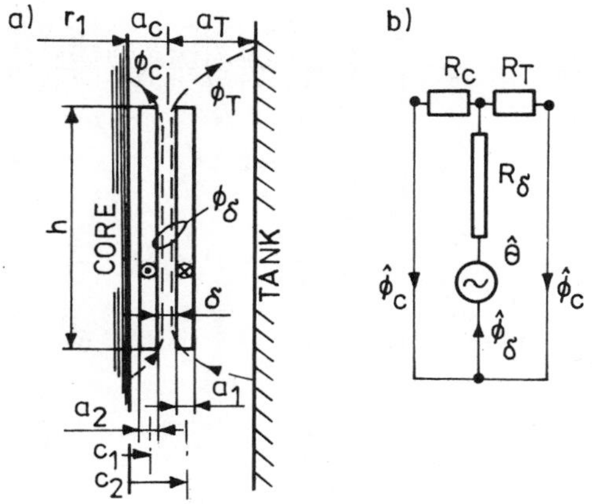

Fig.1. Simplified model of leakage flux distribution between iron core (ϕ_C) and tank wall (ϕ_T): a) physical model, b) equivalent magnetic circuit.

criteria is the so called "critical distance" of the tank wall from windings in transformers[1,2]. Many important phenomena in transformers, like additional losses in windings and other structural parts, electrodynamic forces, short circuit voltages and so on are strongly connected with leakage field distribution. This distribution depends on the conditions on the tank surface, on core saturation, and on the ratio of the distances of tank wall and core from the interwinding gap.

In the extremally simplified model (Figure 1) of leakage flux distribution between the iron core (ϕ_C) and the steel wall (ϕ_T) the principal dependence can be expressed[1] as

$$\phi_T/\phi_C \simeq a_C/a_T \quad ; \quad \phi_T+\phi_C \simeq \phi_\delta \qquad /1/$$

Assuming that, due to the big gap reluctance, ϕ_δ is constant one can conclude that ϕ_C/ϕ_δ decreases when a_T/a_C decreases. However, when a_T/a_C grows, ϕ_C/ϕ_δ increases only up to a certain limited distance, called "critical distance" $a_{CR}=(a_T/a_C)_{CR}$ beyond which ϕ_C/ϕ_δ becomes constant, even when a_T/a_C continue to grow. Following this idea the critical distance a_{CR} of a tank wall has been defined[1] as "the distance from the gap axis at which the part ϕ_T of stray flux closed by the tank is equal to the $flux_{air}$ which would close through open, unlimited air space when the tank is completely removed."

In transformers with the distance relation $(a_T/a_C) \geqslant a_{CR}$ the part ϕ_C of leakage flux closed through core and structural elements of removable parts of transformer is practically independent on presence or absence of the tank (not having Cu or Al screens).

In this case e.g. additional losses in the tank can be measured by the elimination method of short-circuit tests "with and without tank".
Cu or Al screens have, however, much stronger influence, which have been additionally proved in this work.

TANK WALL WITH MAGNETIC SCREENS

In the works 1 and 2 the simplified analytical formula, for the simple model in the Fig.1a, has been developed:

$$\left(\frac{a_T}{a_C}\right)_{CR} = \frac{\left| \ln \dfrac{1+(\frac{2c_2}{h})^2}{1+(\frac{2c_1}{h})^2} + 2\left(\dfrac{2c_2}{h}\,\mathrm{artg}\,\dfrac{h}{2c_2} - \dfrac{2c_1}{h}\,\mathrm{artg}\,\dfrac{h}{2c_1}\right) \right|}{2\pi\dfrac{\delta}{h_R}\cdot\left| \ln \dfrac{1+(\frac{2c_2}{h})^2}{1+(\frac{2c_1}{h})^2} + 2\left(\dfrac{2c_2}{h}\,\mathrm{artg}\,\dfrac{h}{2c_2} - \dfrac{2c_1}{h}\,\mathrm{artg}\,\dfrac{h}{2c_1}\right) \right|} \qquad /2/$$

or for large transformers, where $h/a_C > 10$.

$$(a_T/a_C)_{CR} \simeq \pi / (8a_C/h) - 1 \qquad /2a/$$

The simplified formula /2a/ can be used for large power transformers with slim proportions. For instance for h/a_C=10 to 12, a_{CR}=3 to 3.5.

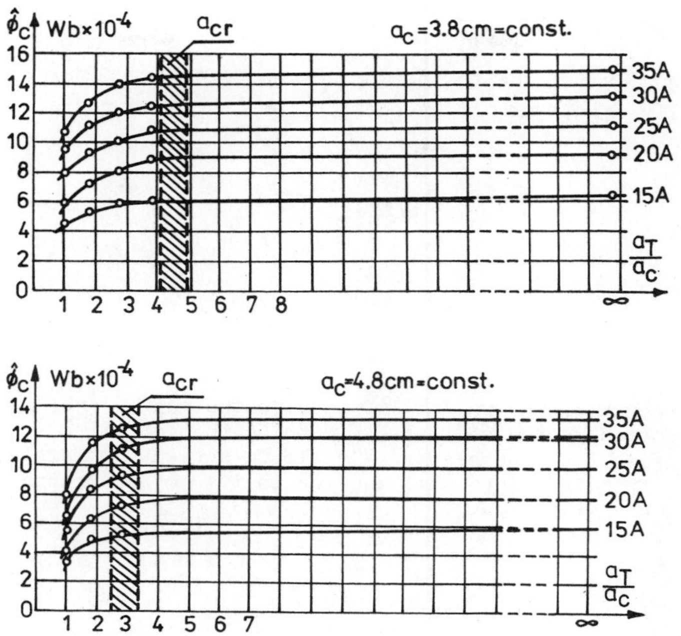

Fig.2. Experimental verification of formula /2/[2] a_{CR} - critical relative distance of steel wall from windings calculated with formula /2/.

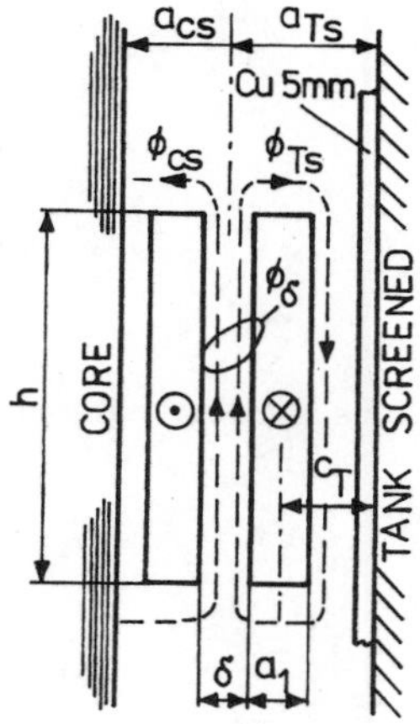

Fig.3. Simplified model of leakage flux distribution; tank wall with electromagnetic screen

Formula /2/has been checked experimentally.[2] Fig.2 confirms the conclusion that at $(a_T/a_C) > a_{CR}$, the part ϕ_C of total leakage flux ϕ_δ closing through the core is practically constant and independent on the distance of tank wall from windings.

TANK WALL WITH ELECTROMAGNETIC SCREENS

The total gap flux ϕ_δ,due to the dominating role of gap reluctance R_δ/Fig.1/, is practically constant, independently on the changes of a_C and a_T. Therefore

$$\phi_{TS}/\phi_{CS} \approx R_C/R_{TS} \tag{/3/}$$

where

$$R_C \approx R_{CS} \approx 2a_{CS}/(\mu_o \beta h) \tag{/4/}$$

$2a_C$ is the length of the way of ϕ_{CS} outside the gap and β= 0,6 to 0,8 is the cross-section factor of flux ϕ_{CS} outside the gap .

If we assume that the electromagnetic Cu screen completely repulses the leakage flux from the tank wall and forces it to go through the gap with width c_T, we can write that

$$R_{TS} \approx (h+a_{TS})/(\mu_o c_T) \tag{/5/}$$

and from the equivalent circuit /Fig.1b/ we have

$$\phi_\delta = \phi_{TS}+\phi_{CS} = \phi_{TS}(1+R_{TS}/R_C) = \phi_{TS}(R_C+R_{TS})/R_C \tag{/6/}$$

from where

$$\phi_{TS}/\phi_\delta = R_C/(R_C+R_{TS}) \tag{/7/}$$

Considering /4/ and /5/ we have analogically

$$\phi_{CS}/\phi_\delta = R_{TS}/(R_C+R_{TS}) = \frac{(h+a_{TS})/c_T}{2a_{CS}/(\beta h)+(h+a_{TS})/c_T} =$$

$$= \frac{h/a_{CS} + a_{TS}/a_{CS}}{(a_{TS}/a_{CS})2a_{CS}/(\beta h)-(a_1+\delta)/(\beta h)+h/a_{CS}+a_{TS}/a_{CS}} \tag{/8/}$$

In order to find the "critical" distance we are looking for the particular ratio $(a_{TS}/a_{CS})_{CR} = x$ at $a_{CS} = a_C$, for which flux ϕ_{CS} going towards the core, after the screening, will be the same as the flux ϕ_C existing in the non-screened transformer at the critical distance /2a/. Considering /1/ and /2a/ we have

$$\phi_{CS}/\phi_\delta \approx \phi_C/\phi_\delta = 1/(1+x) = 1/(1+\pi/(8a_C/h)-1) = (8/\pi)a_C/h \qquad /9/$$

After equalization of /8/ and /9/ we have

$$\frac{h/a_C + x}{(h/a_C-(a_1+\delta)/\beta h)+(1+2a_C/\beta h)x} = (8/\pi)a_C/h \qquad /10/$$

i.e. an equation of the type $\frac{a+x}{b+cx} = d$, from where $x = (a-bd)(cd-1)$, hence

$$(a_{TS}/a_{CS})_{CR} = \frac{h/a_C-(h/a_C-(a_1+\delta)/\beta h)(8/\pi)(a_C/h)}{(1+2a_C/\beta h)(8/\pi)(a_C/h)-1} \qquad /11/$$

In large transformers $h/a_C >> (a_1+\delta)/\beta h$ and $1 >> 2a_C/\beta h$. Therefore the "critical" distance of screened tank wall /11/ can be expressed with the simple formula

$$(a_{TS}/a_C)_{CR} \approx h/a_C \qquad /12/$$

It means that only at the distance ratio $a_{TS}/a_C > h/a_C$ the presence of tank walls has no influence on the flux ϕ_C turning towards the core.

In large transformers $h/a_C = 10$ to 12, which gives $a_{TS}/a_C = 10$ to 12 as well. Because in practice a_{TS}/a_C is of the order of 1 to 5, we can conclude that the screening of tank walls has always an influence on the change of leakage field distribution. As a result, it gives the increase of ϕ_C and decrease of ϕ_T. It causes also the increase of axial forces and additional losses from the radial component in internal winding.

These considerations concern the two-dimensional model. For real transformers three-dimensional and three-phase fields should be considered as well as the influence of yokes, yoke beams etc.

The influence of yoke beams and yokes can be evaluated on the basis of mirror images /Fig.4/.

A qualitative consideration /Fig.4/ shows that yoke and yoke-beams have not important influence on leakage field distribution according to /1/ and /3/.

The experimental verification[3] in principle has confirmed the conclusions presented above /11 /as well as the small influence of yoke-beams /Fig.5 and 6/.

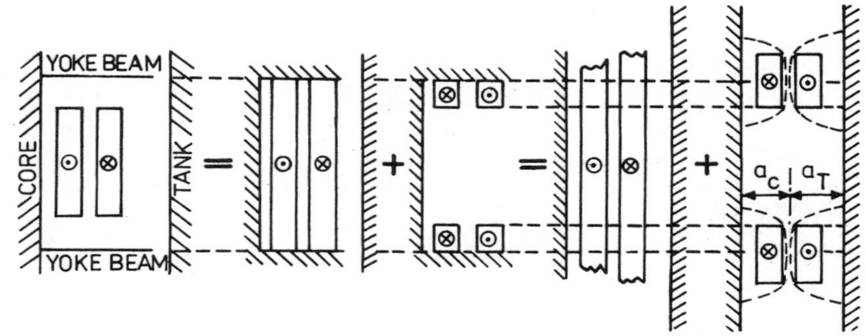

Fig.4. Checking of influence of yoke beams on the leakage field distribution.

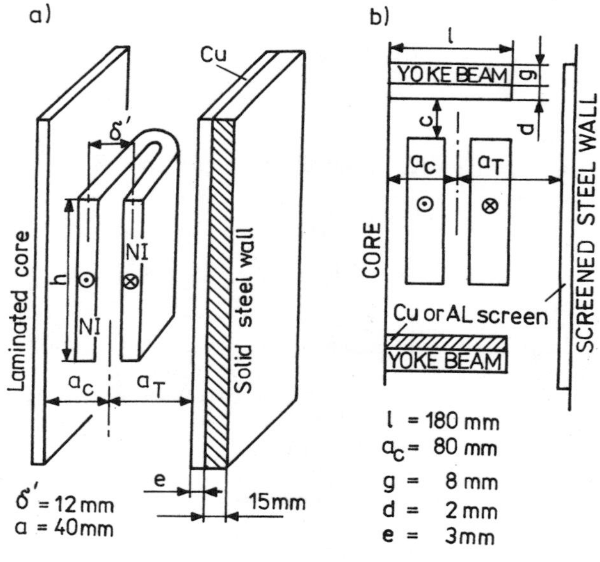

Fig.5. Model for checking of influence of Cu screens and yoke-beams.

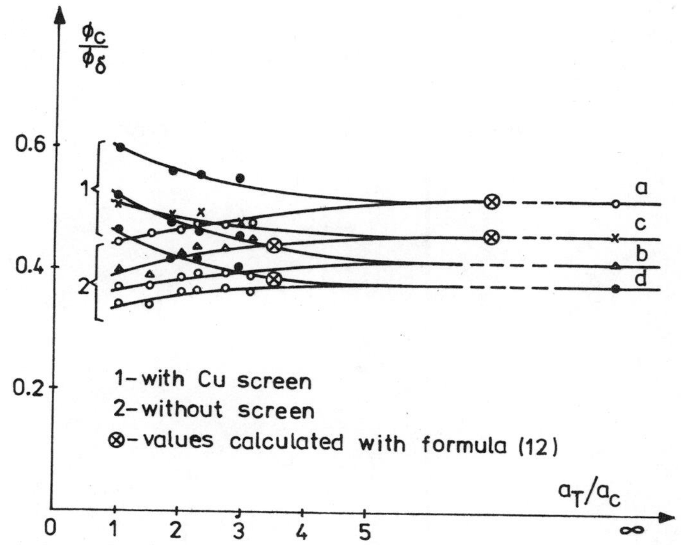

Fig.6. Experimental verification of the simplified formula /12/.

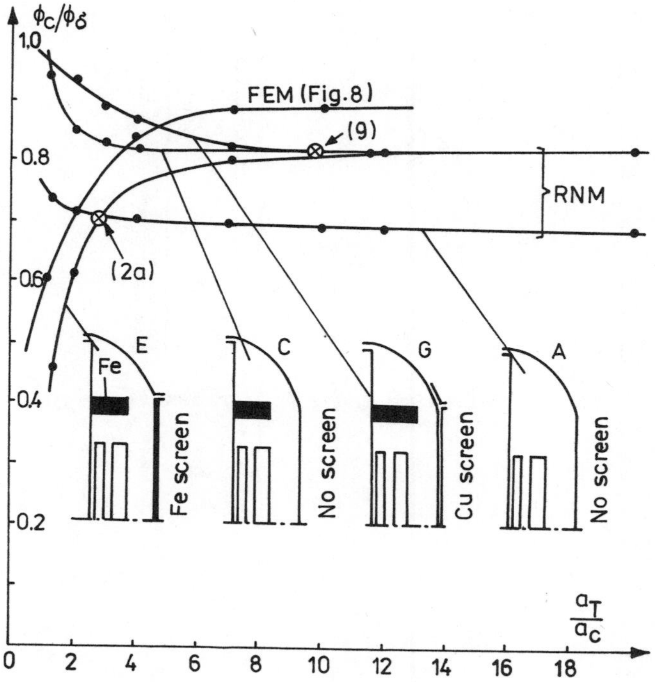

Fig.7. RNM analysis of screens and yoke-beams influence in 240 MVA transformers: E,C,G, - with yoke beams ($\mu=\infty, \gamma=0$); C,A - solid steel tank wall; FEM analysis - like A but with magnetic screens ($\mu=\infty, \gamma=0$)

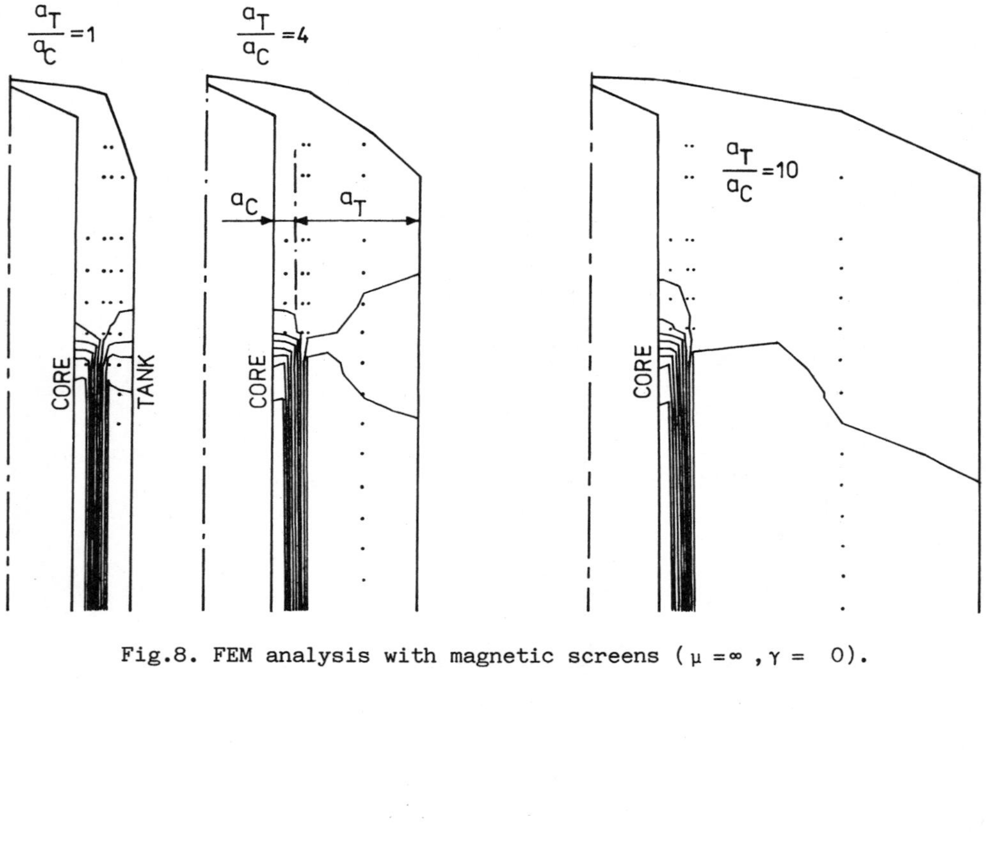

Fig.8. FEM analysis with magnetic screens ($\mu = \infty$, $\gamma = 0$).

EXPERIMENTAL VERIFICATION

The experimental verification of a structure with Cu screens has been made on the model described in[1], -see Fig.5.

Measurements[3], made for various ratios a_T/a_C and distances c, without and with Cu screens (Fig.5 and 6), have proved that practically $\phi_C/\phi_\delta = f(c) \simeq$ const. for c = 0,6, 12 and ∞ cm.

MORE COMPLICATED STRUCTURES

Complicated structures have been examined by means of the quasi-three-dimensional cylindrical reluctance network RMS[4] and the 2 D finite element method FEM[5] /Fig.7,8/. The results confirm that the "critical" tank wall distance also exists in more complicated structures except for walls made of solid steel, without yoke-beams /Fig.7,A/, where eddy currents compensate the effect.

CONCLUSION

The analysis presented confirms in principle the existence and practical usefulness of the tank wall "critical" distance, except for walls made of solid steel without any screens. The influence of saturation should be next examined[6].

REFERENCES

1. Turowski J., "Elektrodynamika Techniczna", WNT, Warszawa (1968).
2. Janowski T., Turowski J., Kriterium wyboru metody pomiaru strat mocy w kadziach transformatorow, Rozpr Elektr. 16: 205 (1970).
3. Turowski J., Sykulski J., Sykulska E., Określenie krytycznej odległości kadzi ekranowanej od uzwojeń. Inst.T.M.i. AE Polit Lodzk. Int. rep. No 52, (1974).
4. Turowski J., Turowski M., A network approach to the solution of stray field in large transformers, Rozpr Elektr, 31: 405 (1985).
5. Bassi E., Gobetti A., Savini A., Finite element magnetic field calculation taking into account saturation and hysteresis, Compumag, Grenoble. (1978).
6. Jablonski M., Napieralska-Juszczak E., The FEM analysis of magnetic field in convertor transformers during faults, ISEF'87, Pavia (1987).

NUMERICAL ANALYSIS OF INRUSH CURRENTS IN TRANSFORMERS

J. Takehara and M. Kitagawa

Chugoku Electric Power Co.

Hiroshima, Japan

Nakata and N. Takahashi

Dept. of Electrical Engineering
Okayama University
Okayama, Japan

ABSTRACT

The finite element method for analysing inrush currents in transformers connected to constant voltage sources taking into account three-dimensional magnetic fields and residual magnetism is developed.

The principle and the finite element formulation of the method are described, and an example of application is shown.

INTRODUCTION

Inrush currents are due to the over-saturation of flux density in a transformer core. Therefore, the flux distribution should be examined to understand the phenomenon exactly. In conventional analysis[1-3], this fundamental behaviour has been ignored, and an equivalent circuit with electrical circuit constants L, M and R has been solved. However, in this method, it is very difficult to take into account the construction of the winding and the core, the quality of core material, the three-dimensional leakage fluxes, etc.

If the finite element method is used for the analysis of inrush currents, it is not necessary to estimate the circuit constants, because the flux distribution can be obtained from material constants and the geometrical shape of the transformer. As the inrush currents are unknown, the conventional finite element method[4] cannot be applicable. Therefore, a new method for calculating the exciting currents directly from the given terminal voltage[5] should be introduced in the analysis of inrush currents. The hysteresis of the core material should be taken into account in order to investigate the effects of residual magnetism on the behaviour of inrush currents. As the leakage flux from the core is distributed three-dimensionally, 3-D analysis is necessary.

In this paper, a new finite element method for analysing the inrush currents has been developed by combining the finite element method for analysing electrical machinery connected to a constant voltage source[5], the technilque for treating hysteresis characteristics[6] and the approximate method for calculating 3-D magnetic fields[7]. As an example of the application, the inrush currents have been analysed taking into account 3-D construction and residual magnetism.

METHOD OF ANALYSIS

Outline of the Method

The relationship between the flux density B and the magnetic field intensity H in the presence of residual magnetism Br is shown in Fig.1. In order to take into account such a hysteresis phenomenon, the hysteresis characteristic should be represented in terms of the magnetization $\mathbb{M}$ as follows[6]:

$$\mathbf{B} = \mu_0 \mathbb{H} + \mathbb{M} \tag{1}$$

where μ_0 is the permeability of vacuum.

It is assumed that the transformer core is composed of limbs with axisymmetric sections and yokes with rectangular sections as shown in Fig.2(a). Then, the approximate method for calculating 3-D magnetic fields in magnetic circuits composed of axisymmetric and rectangular regions may be introduced. Using this method, 3-D magnetic fields can be calculated with only a small increase of computer storage and computing time compared with the 2-D analysis.

Magnetic fields in the axisymmetric regions a-b-d-c-a and c-d-f-e-c in Fig.2(b) can be written as

$$\frac{\partial}{\partial r}\left(\frac{\nu_0}{r}\frac{\partial A}{\partial r}\right) + \frac{\partial}{\partial z}\left(\frac{\nu_0}{r}\frac{\partial A}{\partial z}\right) = -\left(\frac{nI}{Sc} + Jm_\theta\right) \tag{2}$$

where A is the vector potential, ν_0 is the reluctivity of vacuum and r is the radius. n and Sc are the number of turns and the cross-sectional area of the winding respectively. Jm_θ is the θ-component of the equivalent magnetizing current density $\mathbb{J}m$, and is given by

$$Jm_\theta = \nu_0\left(\frac{\partial Mr}{\partial z} - \frac{\partial Mz}{\partial r}\right) \tag{3}$$

where Mr, Mz are the r- and z-components of $\mathbb{M}$.

Magnetic fields in the rectangular region b-g-h-f-d-b in Fig.2(b) can be written as

$$\frac{\partial}{\partial x}\left(\frac{\nu_0 \pi}{t^*}\frac{\partial A}{\partial x}\right) + \frac{\partial}{\partial y}\left(\frac{\nu_0 \pi}{t^*}\frac{\partial A}{\partial y}\right) = -\left(\frac{nI}{Sc} + Jmz\right) \tag{4}$$

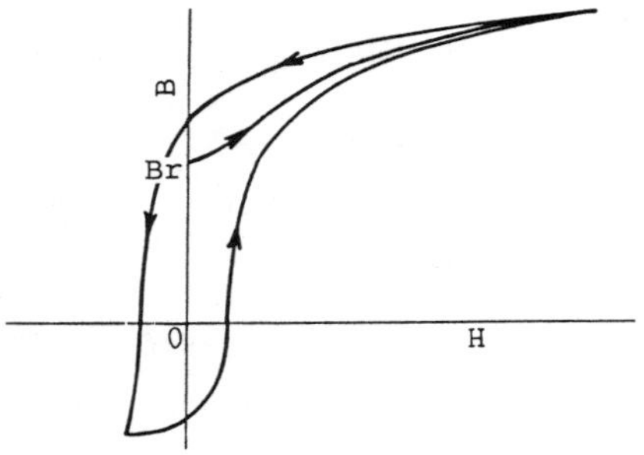

Fig.1 Hysteresis curve.

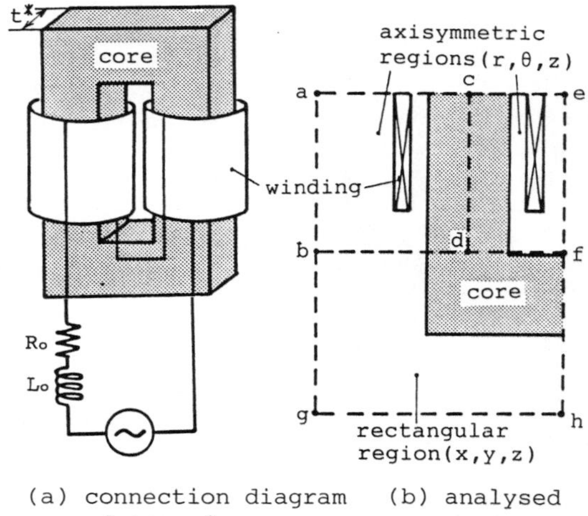

(a) connection diagram of transformer (b) analysed region

Fig.2 Analysed model.

where t^* is the thickness of the yoke. Jmz is the z-component of $\mathbb{J}m$, and is given by

$$J_{mz} = \nu_0 \left(\frac{\partial M_y}{\partial x} - \frac{\partial M_x}{\partial y} \right) \tag{5}$$

where Mx an My are the x- and y- components of $\mathbb{M}$.

In order to calculate inrush currents, Poisson's equations (2) and (4) and the following equation obtained from Kirchhoff's law for the electric circuit in Fig.3 should be solved simultaneously by treating the vector potentials and the currents as unknown variables.

$$V = \frac{\partial}{\partial t} \int_c A\, ds + L_o \frac{\partial I}{\partial t} + (R_o + R_c)\, I \tag{6}$$

where Lo and Ro are the inductance and the resistance of the power source, and Rc is the resistance of the winding. c is the contour along the winding in the finite element region. ds is a unit vector tangent of the contour c. Though there is only one relationship of Eq.(6) for the case of Fig.2, three kinds of relationships similar to Eq.(6) exist in the case of a three-phase transformer.

Finite Element Formulation

The following equation can be obtained at an instant t by using the Galerkin method from Eqs.(2) and (4).

$$G_i^{rz} = \sum_{e=1}^{NE^{rz}} \pi \alpha^{(e)} \left\{ rc^{(e)} \nu_0^{(e)} \sum_{k=1}^{3} S_{ike}^{rz} A_{ke}^{t} - \frac{\Delta^{(e)} n I}{3 S_c} - \frac{\nu_0^{(e)}}{2} \left(Mr^{(e)} d_i^{rz(e)} - Mz^{(e)} c_i^{rz(e)} \right) \right\} \tag{7}$$

$$G_i^{xy} = \sum_{e=1}^{NE^{xy}} \pi \left\{ \frac{\pi}{t^*} \nu_0^{(e)} \sum_{k=1}^{3} S_{ike}^{xy} A_{ke}^{t} - \frac{\nu_0^{(e)}}{2} \left(Mx^{(e)} d_i^{xy(e)} - My^{(e)} c_i^{xy(e)} \right) \right\} \tag{8}$$

$$G_i = G_i^{rz} + G_i^{xy} \tag{9}$$

where NE^{rz} and NE^{xy} are the numbers of elements in the axisymmetric and rectangular regions respectively. $rc^{(e)}$ is the r-coordinate of the center of gravity of an element e. A_{ke}^{t} is the vector potential at a node ke. $\alpha^{(e)}$ is unity when the element e is on the left-hand side of the center line of the axisymmetric region, and is equal to -1 when e is on the right-hand side. S_{ike}^{rz} and S_{ike}^{xy} are defined by

$$\left.\begin{aligned} S_{ike}^{rz} &= \frac{c_i^{rz(e)} c_{ke}^{rz} + d_i^{rz(e)} d_{ke}^{rz}}{4\Delta^{(e)}} \\ c_{ie}^{rz} &= -(z_{je} - z_{ke}) / rc^{(e)} \\ d_{ie}^{rz} &= -(r_{ke} - r_{je}) / rc^{(e)} \end{aligned}\right\} \tag{10}$$

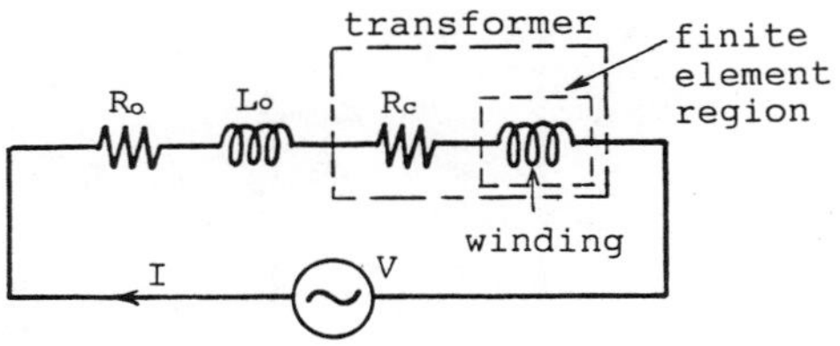

Fig.3 Equivalent circuit.

$$\left.\begin{array}{l} S_{ike}^{xy}=\dfrac{c_i^{xy(e)}c_{ke}^{xy}+d_i^{xy(e)}d_{ke}^{xy}}{4\Delta^{(e)}} \\ c_{ie}^{xy}=y_{je}-y_{ke} \\ d_{ie}^{xy}=x_{ke}-x_{je} \end{array}\right\} \qquad (11)$$

where $\Delta^{(e)}$ is the area of the element e.

Let us define F as follows:

$$F=\frac{\partial}{\partial t}\int_c A\,ds+Lo\frac{\partial I}{\partial t}+(Ro+Rc)\,I-V \qquad (12)$$

By approximating the differentiation ∂/∂t in Eq.(12) by the finite difference, the following equation can be obtained:

$$F=\pi\sum_{e=1}^{Nc}\frac{n\Delta^{(e)}}{3\Delta t\,Sc}\sum_{j=1}^{3}(A_{je}^{t}-A_{je}^{t-\Delta t})+\frac{Lo}{\Delta t}(I^{t}-I^{t-\Delta t})+(Ro+Rc)I^{t}-V^{t} \qquad (13)$$

where Δt is the time interval. Nc is the number of elements in the winding.

If Eqs. (7)-(9) and (13) are solved simultaneously, the vector potentials and the inrush currents can be directly obtained.

In the nonlinear analysis using the Newton-Raphson iteration technique, the increments $\{\delta A_i\}$ and δI are obtained from the following equation.

$$\begin{bmatrix} \left[\dfrac{\partial G_i}{\partial A_j}\right] & \left\{\dfrac{\partial G_i}{\partial I}\right\} \\ \left\{\dfrac{\partial F}{\partial A_j}\right\} & \dfrac{\partial F}{\partial I} \end{bmatrix} \begin{Bmatrix} \{\delta A_i\} \\ \delta I \end{Bmatrix} = \begin{Bmatrix} \{-G_i\} \\ -F \end{Bmatrix} \qquad (14)$$

AN EXAMPLE OF APPLICATION

The inrush current of a single-phase transformer shown in Fig.4 is

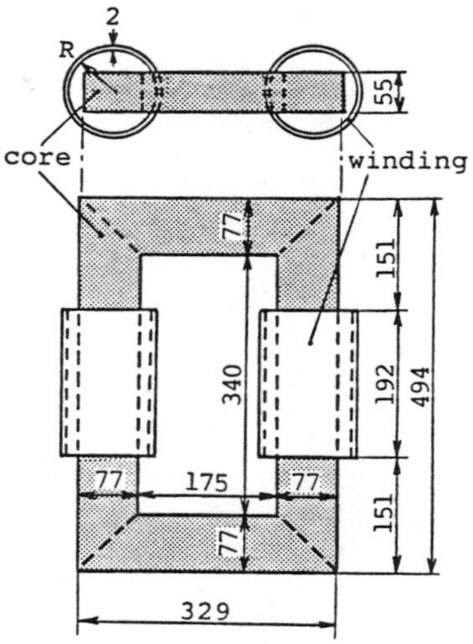

Fig.4 Dimensions of the transformer.

analysed using the new method. The specifications are shown in Table 1. The residual magnetism Br is assumed to be 0.8(T).

Figure 5 shows the calculated inrush currents in the case when the innner radius R of the winding is 58(mm). Results obtained using the equivalent circuit method and experiments are also shown in Fig.5. In the equivalent circuit method, it is assumed that the core is represented by the equivalent magnetic resistance having the mean length of the magnetic circuit of the core.

Figure 6 shows the inrush currents in the case when R is increased to 109(mm). Inrush currents are smaller than the case when R=58(mm), due to the leakage flux between the core and the winding. When the leakage flux is increased as in the case of Fig.6, our method is especially effective.

CONCLUSIONS

The method proposed in this paper enables us to analyse precisely the inrush currents of transformers.

The application of our method to a three-phase transformer will be shown in another paper[8].

Table 1 Specifications of the transformer.

Core material	M-5
Working flux density (T)	1.7
Number of primary turns/limb	30
Sum of primary winding resistance and source impedance (Ω)	2+j0
Frequency (Hz)	60

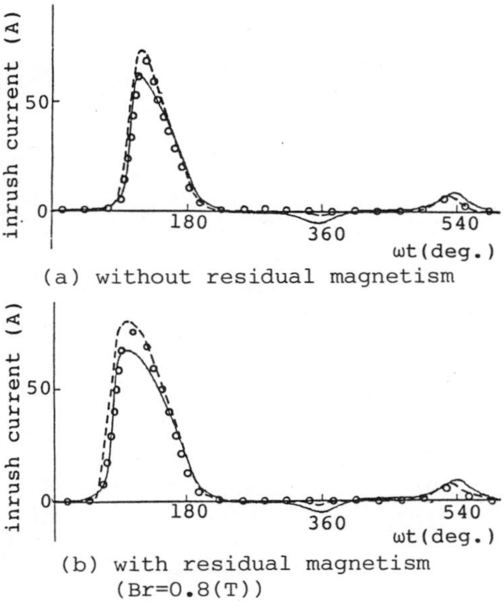

Fig.5 Waveforms of inrush currents (R=58(mm)).

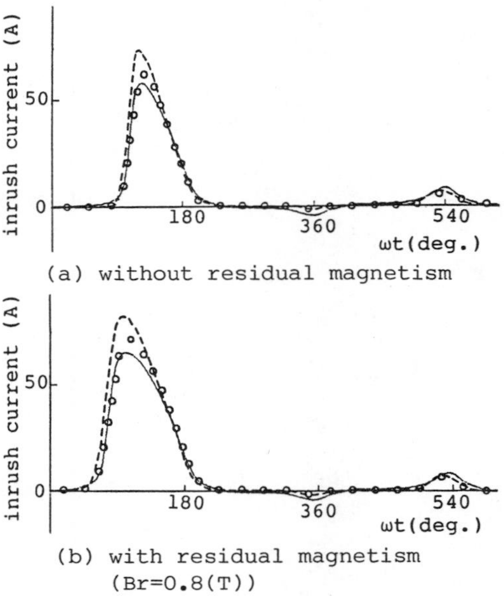

Fig.6 Waveforms of inrush currents (R=109(mm)).

o :new FEM
----:equivalent circuit method } calculated
——:measured

REFERENCES

1. K. Okuyama and K. Inagaki:"Exciting Inrush Current of Three-Phase Core-Type Transformer with Three Legs", Trans. of IEE Japan, 91, 2, 337 (1971).
2. A.A. Shaltout: "New Method for Calculating Transient Current of Transformers", Proceedings of ICEM, 1, 99, Lausanne (1984).
3. R. Yacamini and A. Abu-Nasser:"The Calculation of Inrush Current in Three-Phase Transformers", Proc. IEE, 133, B, 1, 31 (1986).
4. T. Nakata and N. Takahashi:"Finite Element Method in Elecrical Engineering", Morikita Shuppan, Tokyo (1982).
5. T. Nakata and N. Takahashi:"Direct Finite Element Analysis of Flux and Current Distributions under Specified Conditions", IEEE Trans. on Magnetics, MAG-18, 2, 325 (1982).
6. T. Nakata , N. Takahashi and Y. Kawase: "Finite Element Analysis of Magnetic Fields Taking into Account Hysteresis Characteristics", ibid., MAG-21, 5, 1856 (1985).
7. T. Nakata , N. Takahashi, Y. Kawase and H. Funakoshi: "Finite Element Analysis of Magnetic Circuits Composed of Axisymmetric and Rectangular Regions", ibid., MAG-21, 6, 2199 (1985).
8. J.Takehara, M. Kitagawa, T. Nakata and N. Takahashi: "Finite Element Analysis of Inrush Currents in Three- Phase Transformers", ibid., MAG-23, 5 (1987).

IMPROVEMENT OF TRANSFORMER CORE LOSS BY USE OF LOW-LOSS ELECTRICAL STEEL

Zvonimir Valković

Institut Rade Končar
Zagreb, Yugoslavia

ABSTRACT

The results of an experimental investigation of the influence of high permeability grain-oriented (HGO) steel on transformer core loss are presented. The power losses of cores assembled with HGO material of grade M2H and with conventional grain-oriented material of grade M5 are compared. The aim of the investigation was to determine the influence of a number of parameters (e.g. single-phase and three-phase core forms, different corner and T-joint designs, core proportions, core dimensions, a number of laminations per stagger layer, overlap length) on the amount of loss reduction in the cores of HGO material.

INTRODUCTION

A trend of reduction of transformer core losses in the last ten years is related to a considerable increase of energy price. One of the ways to reducing core losses is to use better grade material the so called high permeability grain-oriented (HGO) electrical steel having about 15% lower specific core loss compared with the conventional grain-oriented (CGO) steel. However, it has been noticed[1,2] that the use of this new material in transformers does not result in as great a reduction of core losses as could be expected from the improvement on the material. This discrepancy has been neither sufficiently explained nor investigated enough as yet.

The purpose of this work is to investigate the influence of a number of different core parameters on the efficiency of the HGO material.

PROCEDURE

Investigation was carried out on scale models of transformer cores, made parallelly of HGO (grade M2H) and CGO (grade M5) materials. The sheet thickness in both cases was 0.3 mm. After slitting, the laminations were stress-relief annealed. In order to increase the reliability of investigation, five identical cores were made for each variant. The results given in this paper are an average of measurements on all the five cores. All the models were, if not stated otherwise, assembled in the usual way with two laminations per stagger layer. Each core was assembled with 120 laminations per stack.

RESULTS ON MODELS

Influence of Core Design

For the first part of the research, nine different core variants were made (Table 1., Fig. 1-3). On these models we investigated how the core loss reduction was influenced by the core form (single-phase or three-phase core, Fig. 1.), corner design (45° mitred overlap joint or 90° butt and lap joint, Fig. 2.), form of T-joint in three-limb core (V-45° joint, 45-90° joint, and staggered joint[3], Fig. 3.), and core proportions (models with lamination width of 40 or 70 mm).

Table 1. Basic data of models

Model No.	Core form	Corner form	T-joint form	w(mm)	l(mm)	h(mm)	d(mm)	V_c
1	Fig. 1a	Fig. 2a		40	290	290	5	0.14
2				70	250	250	5	0.28
3		Fig. 2b		40	290	290	5	0.14
4				70	250	250	5	0.28
5	Fig. 1b	Fig. 2a	Fig. 3a	40	140	230	5	0.20
6				70	140	230	5	0.36
7			Fig. 3a	60	140	220	10	0.31
8			3b	60	140	220	10	0.31
9			3c	60	140	220	10	0.31

The loss reduction (ΔP) achieved due to the application of HGO material, compared with CGO material, in those nine variants is summarized in Table 2.

We can notice that the loss reduction greatly depends on the core induction and the core form. Better results are obtained at a higher induction and on simpler core forms (single-phase). On the single-phase cores of 45° mitred overlap corners nearly the same loss reduction was achieved as on Epstein samples. The reduction achieved on the three-phase models was more than two times smaller than on the Epstein samples.

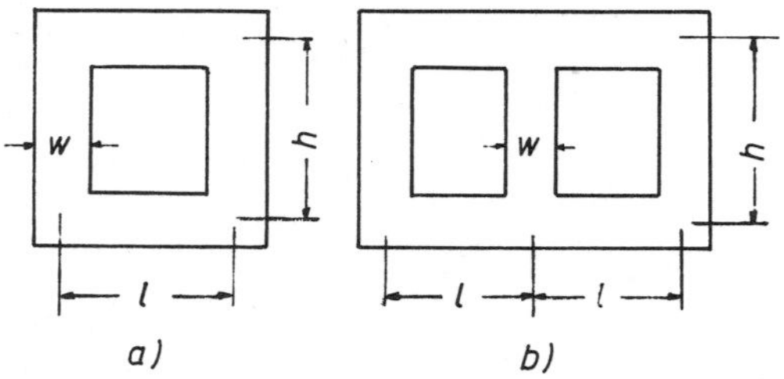

Fig.1. Core forms. (a) single-phase core, (b) three phase core.

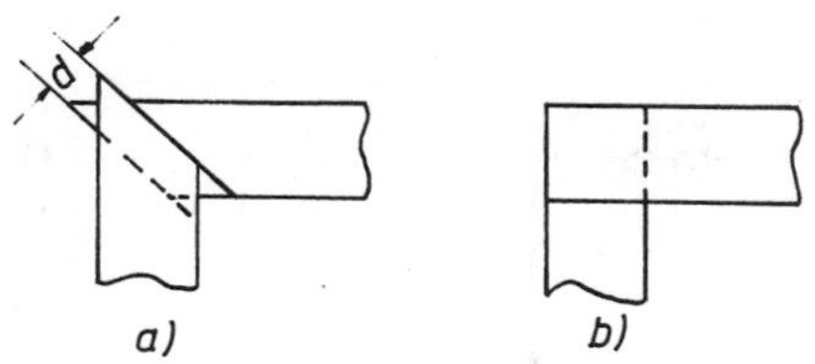

Fig. 2. Corner designs. (a) 45° mitred joint, (b) 90° butt and lap joint.

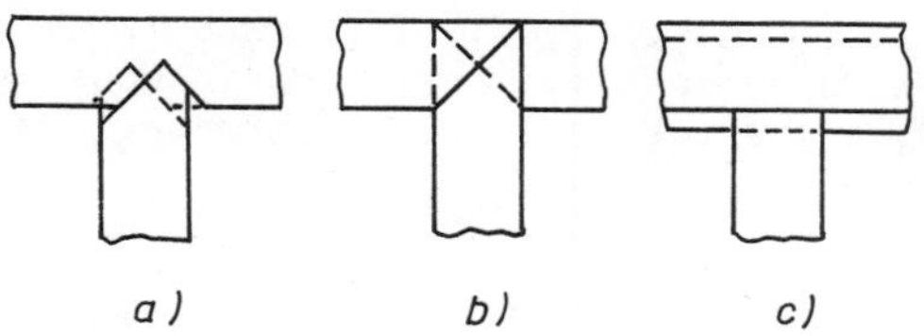

Fig. 3. T-joint designs. (a) V-45° joint, (b) 45-90° joint, (c) staggered joint.

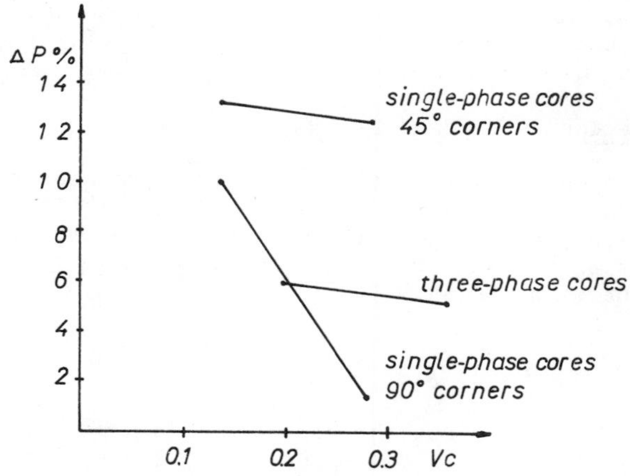

Fig. 4. Power loss reduction in models with HGO material versus core proportions (V_c), at 1.7 tesla.

Table 2. Loss reduction (in %) achieved with HGO material

Model No.	Induction (T) 1.5	1.7
1	7.9	13.2
2	6.9	12.5
3	6.0	10.0
4	1.6	1.4
5	2.0	6.0
6	0.1	5.2
7	2.0	5.7
8	2.7	8.8
9	1.4	6.2
Epstein samples	8.6	14.9

The core proportions have a small influence if the T-joints are 45°, while on models with 90° joints the influence is considerable. This can be seen from Fig. 4. where the influence of proportions is related with parameter V_c. Parameter V_c is the ratio of the volume of all corners and T-joints and the total core volume.

All three investigated T-joints have more or less equal influence on the loss reduction.

Influence of Number of Laminations

In all previous experiments the cores were assembled with two laminations per stagger layer. In order to check how the number of laminations influences the efficiency of HGO steel, some models were reassembled with one, and then with three laminations per layer. These experiments were carried out on models No. 1, 2, 5 and 6 (Table 1.). The loss reduction at 1.7 tesla in relation to the number of laminations is shown in Fig.5. It is evident that the results are better in the case of a smaller number of laminations, but this influence is relatively small.

Influence of the Corner Overlap Length

Some single-phase models (models 1 and 2) were reassembled with a larger overlap, of 10 and then of 15 mm. The results obtained for 1.7 tesla are given in Fig.6. We can see that the smaller the overlap, the more efficient HGO steel is, but even for a three times larger overlap the difference is not significant.

RESULTS ON LARGE CORES

In order to determine how the core dimensions influence the efficiency of aplication of HGO steel, a few experiments were carried out with cores of normal dimensions. These were three-limb three-phase cores of power transformers of 20 to 150 MVA. The corners and T-joints were according to Fig. 2a and Fig. 3a respectively.

The percentage reduction of losses on large cores is approximately two times bigger than on the scale models, and similar to that obtained on single-phase models. These results point to the conclusion that core dimensions have a considerable influence on the efficiency of application of HGO material, although this investigation has not given the possibility of determining precisely the relation between dimensions and core loss reduc-

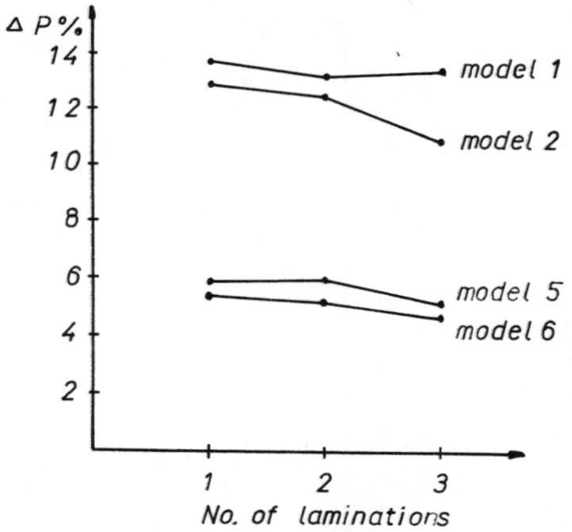

Fig. 5. Power loss reduction in models with HGO material versus number of laminations per stagger layer, at 1.7 tesla.

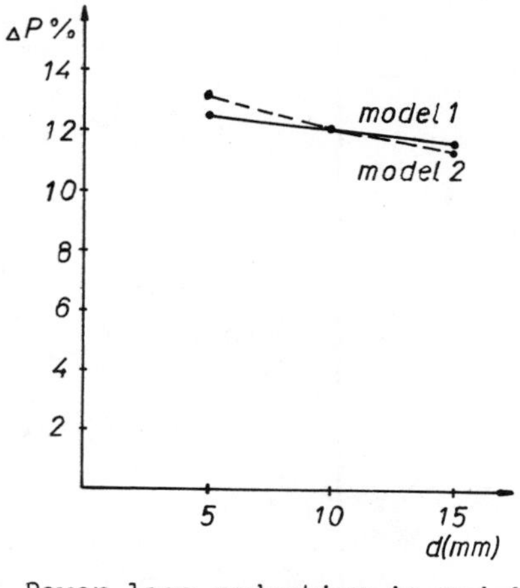

Fig. 6. Power loss reduction in models with HGO material versus overlap length, at 1.7 tesla.

Table 3. Loss reduction (in %) achieved with HGO steel

Transformer capacity (MVA)	Induction (T) 1.5	1.7
20	5	9
40	7	14
40	4	10
150	10	14
Average	7	12

tion. It is not quite clear why there is such a strong influence of dimensions, but it can be assumed that in a core of larger dimensions, the space distribution of magnetic flux in the core with HGO steel is somewhat more favourable compared to small models. This assumption necessitates, naturally, some further verification.

CONCLUSIONS

The influence of core form, proportion and dimensions, corner and T joint design, number of laminations per stagger layer and corner overlap length on the reduction of power loss in the cores with HGO electrical steel has been investigated on a number of scale models and on a few large cores.

It has been found that the efficiency of HGO material depends on the core form. The best results have been obtained on single-phase models with 45° mitred overlap corners. The reduction of losses on small three-phase three-limb models is about two times smaller than on Epstein samples. On large three-phase cores the results are better and similar to that on single-phase small models.

The influence of other parameters investigated in this experiment is not so significant.

REFERENCES

1. Z. Valković, Improvement of transformer magnetic properties by use of high permeability grain-oriented silicon steel, Digests of INTERMAG Conference, Florence, 1978.
2. A.J. Moses, Problems in the use of high permeability silicon iron transformer cores, J. Magn. Mat., vol. 19, No. 1-3, 1982.
3. Z. Valković, Influence of transformer core design on power losses, IEEE Trans. Magn., vol. MAG-18, No.2, 1982.

CURRENTS AND FORCES IN CONDUCTORS OF RECTANGULAR CROSS SECTION

P.P. Yannopoulos, J.A. Tegopoulos, and M.P. Papadopoulos

National Technical University of Athens
42, October 28th str.,
106 82 Athens, Greece

ABSTRACT

In this paper an analytical method is given for the determination of current density and forces in two parallel rectangular conductors of equal dimensions carrying equal and oppositely directed sinusoidal currents. The solutions are given in integral form and a method has been developed for the numerical treatment of the solutions on a digital computer. An application is also given.

INTRODUCTION

The determination of the current distribution in two parallel rectangular conductors of equal dimensions is of particular interest in studies of foil wound transformers, substation busbars, etc. The calculation of the forces that develop in such constructions is particularly important especially during short-circuit. From the determination of eddy current distribution the loss in these conductors may also be obtained.

The two conductors considered are of equal width c and thickness d and they are placed symmetrically parallel to each other as shown in Fig. 1. $(y_0\text{-}d)$ is the air gap between them. The conductivity σ and the magnetic permeability μ of the conductors are also constant. Both conductors carry equal and oppositely directed currents of magnitude I and angular frequency ω. Displacement currents are neglected. The length of the conductors is considered to be infinite so that the problem is two dimensional. The current density is directed along their length (z-axis) and varies both with the x and y directions. In this problem both skin and proximity effects are encountered.

ANALYSIS

If O(0,0) is the origin of the coordinates and M(x,y) an arbitrary point in conductor 1 we consider the closed ciruit OMM'O'O made up of the line OM, parallel to the z-axis MM', another parallel OO', both of unit length and closed by the line O'M'. Then, the flux density, dB_{11}, at point M due to a current density $J_1(x',y')$ in an infinite rectangular wire of infinitesimal cross section dx'dy' at point P, is given by

$$dB_{11}(x,y) = \frac{\mu}{2\pi} \cdot \frac{J_1(x',y')}{r} dx'dy'$$

whereas the magnetic flux, $d\Phi_{11}$, through the closed circuit OMM'O'O is

$$d\Phi_{11} = \int_{r=r_o}^{r} dB_{11}(x,y)dr = \frac{\mu}{2\pi} J_1(x',y')dx'dy' \ln \frac{r}{r_o}$$

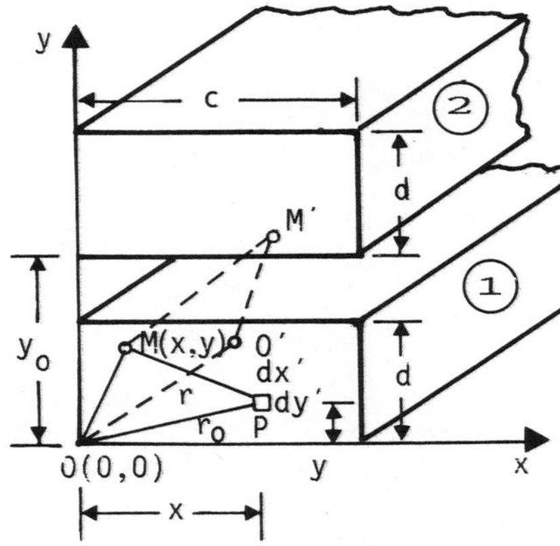

Fig. 1. Two infinitely long conductors of rectangular cross-section.

If the total magnetic flux through the area OMM'O'O due to all elemental currents $J_1(x',y')dx'dy'$ in conductor 1 is Φ_{11} and due to all elemental currents $J_2(x',y')dx'dy'$ in conductor 2 is Φ_{12}, then the total flux, Φ_1, through area OMM'O'O in conductor 1 due to all elemental currents of both conductors is

$$\Phi_1 = \Phi_{11}+\Phi_{12} = \int_{x'=0}^{c} \int_{y'=0}^{d} \frac{\mu}{2\pi} J_1(x',y')f(x,y,x',y')dx'dy' - \int_{x'=0}^{c} \int_{y'=y_o}^{y_o+d} \frac{\mu}{2\pi} J_2(x',y')f(x,y,x',y')dx'dy' \qquad (1)$$

where $f(x,y,x',y') = \ln \sqrt{\frac{(x-x')^2+(y-y')^2}{x'^2+y'^2}}$

Furthermore, if we apply Faraday's law to the area OMM'O'O

$$J_1(x,y) - J_1(0,0) = -j\omega\Phi_1\sigma \qquad (2)$$

considering there is no current flowing along O'M' and OM and by substituting (1) into (2) we get

$$J_1(x,y)-J_1(0,0) = \lambda \int_{x'=0}^{c} \int_{y'=0}^{d} J_1(x',y')f_1(x,y,x',y')dx'dy' - \lambda \int_{x'=0}^{c'} \int_{y'=y_o}^{y_o+d'} J_2(x',y')f_1(x,y,x',y')dx'dy' \quad (3)$$

where $\lambda = -j\omega\sigma\mu/2\pi$.

To obtain the equation for $J_2(x,y)$, the current density in conductor 2, subscripts 1 and 2 in (3) are interchanged. Then,

$$J_2(x,y)-J_2(0,y_o+d) = \lambda \int_{x'=0}^{c} \int_{y'=y_o}^{y_o+d} J_2(x',y')f(x,y,x',y')dx'dy' - \lambda \int_{x'=0}^{c} \int_{y'=0}^{d} J_1(x',y')f(x,y,x',y')dx'dy' \quad (4)$$

Equations (3) and (4) represent a system of two simultaneous double integral equations for $J_1(x,y)$ and $J_2(x,y)$ which may be solved numerically. The reference current densities $J_1(0,0)$ and $J_2(0,y_o+d)$ which are also required for this solution may be determined.

NUMERICAL TREATMENT

Integral equations (3) and (4) can be solved by the application of Simpson's rule for double integrals. For this purpose we divide the width of each conductor into m equal intervals each of length $h = c/2m$ and the thickness of each conductor into n equal intervals each of length $k = d/2n$. Then, with

$$x_i = ic/m, \quad y_j = jd/n \qquad \text{cond.1} \qquad 0 \leq i \leq 2m$$

$$x_i' = ic/m, \quad y_j' = y_o+d-j\frac{d}{2n} \qquad \text{cond. 2} \qquad 0 \leq j \leq 2n$$

Equation (3) gives

$$J_1(x,y)-J_1(0,0) = \frac{\lambda hk}{9} \sum_{i=0}^{2m} \sum_{j=0}^{2n} c_{ij}(J_1(x_i,y_j)f(x,y,x_i,y_j)-J_2(x_i',y_j') f(x,y,x_i',y_j')) \quad (5)$$

where : $c_{ij} = w_i u_j$ with $w_i = \begin{cases} 1 \text{ if } i=0 \text{ or } 2m \\ 4 \text{ if } i=\text{odd} \\ 6 \text{ if } i=\text{even} \end{cases}$

Similarly for u_j. Equation (4) gives

$$J_2(x,y)-J_2(0,y_o+d) = \frac{\lambda hk}{9} \sum_{i=0}^{2m} \sum_{j=0}^{2n} c_{ij}[J_2(x_{i2},y_{j2})f(x,y,x_{i2},y_{j2}) - J_1(x_{i2}',y_{j2}') f(x,y,x_{i2}',y_{j2}')] \quad (6)$$

where $x_{i2} = ic/2m, \quad y_{j2} = -\frac{jd}{2n} + y_o+d$

$x'_{i2} = ic/2m, \quad y'_{j2} = jd/2n$

Finally, (5), (6) five for $\xi=i$, $\tau=j$ but $(x_\xi,y_\tau)\neq(0,0)$, $(x'_\xi,y'_\tau)\neq(0,y_o+d)$

$$J_1(x_\xi,y_\tau)-J_1(0,0) = \frac{\lambda hk}{9} \sum_{i=0}^{2m} \sum_{j=0}^{2n} c_{ij}[J_1(x_i,y_j)f(x_i,y_j,x_i,y_j) - J_2(x'_i,y'_j)f(x'_i,y'_j,x'_i,y'_j)] \tag{7}$$

and

$$J_2(x'_\xi,y'_\tau)-J_2(0,y_o+d) = \frac{\lambda hk}{9} \sum_{i=0}^{2m} \sum_{j=0}^{2n} c_{ij}[J_2(x'_i,y'_j)f(x'_i,y'_j,x'_i,y'_j) - J_1(x_i,y_j)f(x_i,y_j,x_i,y_j)] \tag{8}$$

Equations (7) and (8) constitute a system of $2[(2m+1)(2n+1)-1]$ simultaneous equations with $2[(2m+1)(2n+1)]$ unknowns $J_1(x_i,y_j)$ and $J_2(x'_i,y'_j)$ for $0\leq i\leq 2m$, $0\leq j\leq 2n$. The extra two equations required for the solution of the system are given by the conditions that the total current through each conductor is equal to I.

Thus,

$$\int_{x=0}^{c} \int_{y=0}^{d} J_1(x,y)dxdy = I \quad \text{(a)} \quad \text{and}$$

$$\int_{x=0}^{c} \int_{y=y_o}^{y_o+d} J_2(x,y)dxdy = I \quad \text{(b)} \tag{9}$$

Integration of (9)a and (9)b over x,y and substitution of (5) and (6) into the resulting integrals respectively give

$$I-J_1(0,0) = \frac{\lambda hk}{9} \sum_{i=0}^{2m} \sum_{j=0}^{2n} c_{ij}[J_1(x_i,y_j)F_1(c,d,x_i,y_j)-J_2(x'_i,y'_j) F_1(c,d,x'_i,y'_j)] \tag{10}$$

$$I-J_2(0,y_o) = \frac{\lambda hk}{9} \sum_{i=0}^{2m} \sum_{j=0}^{2n} c_{ij}[J_2(x'_i,y'_j)F_2(c,d,x'_i,y'_j)-J_1(x_i,y_j) F_2(c,d,x_i,y_j)] \tag{11}$$

where

$$F_1(c,d,x_i,y_j) = \frac{1}{2}(c-x_i)(d-y_j)\ln[(c-x_i)^2+(d-y_j)^2]+ \frac{1}{2}(c-x_i)y_j\ln [(c-x_i)^2+y_j^2]+ \frac{1}{2}\, x_i(d-y_j)\ln[x_i^2+(d-y_j)^2]+ \frac{1}{2}\, x_iy_j\ln [x_i^2+y_j^2]-cd\ln\sqrt{[x_i^2+y_j^2]}+ \frac{1}{2}(c-x_i)^2\tan^{-1} \left[\frac{d(c-x_i)}{(c-x_i)^2-y_j(d-y_j)}\right]+ \frac{1}{2}\, x_i^2\tan^{-1}\left[\frac{x_id}{x_i^2-y_j(d-y_j)}\right]+ \frac{1}{2}\, y_j^2 \tan^{-1}\left[\frac{y_jc}{y_j^2-(c-x_i)x_i}\right]+ \frac{1}{2}(d-y_j)^2\tan^{-1}\left[\frac{c(d-y_j)}{(d-y_j)^2-(c-x_i)x_i}\right]$$

and $F_2(c,d,x_i',y_j') = F_1(c,d,x_i',y_j'-y_o)$.

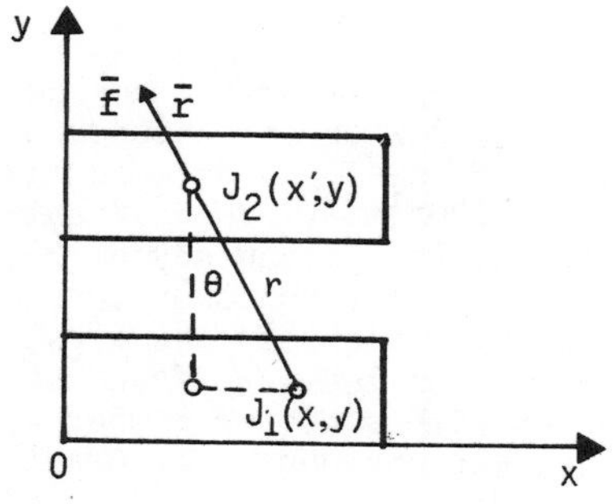

Fig. 2. Calculation of force.

CALCULATION OF FORCE

The force between the two elements at (x,y) and (x',y') is

$$d\bar{f} = \frac{\mu}{2\pi} \cdot \frac{[J_1(x,y)dxdy][J_2(x',y')dx'dy']}{r}\,\bar{r} \qquad (12)$$

This force lies on the straight line connecting the two elements (fig. 2) and has the components in x and y direction given by

$$\begin{pmatrix} df_x \\ df_y \end{pmatrix} = \frac{\mu}{2\pi} \cdot \frac{[J_1(x,y)dxdy][J_2(x',y')dx'dy']}{[(x-x')^2 + (y-y')^2]} \begin{pmatrix} (x-x') \\ (y'-y) \end{pmatrix} \qquad (13)$$

From (13) the force between the two conductors has the components :

$$\begin{pmatrix} f_x \\ f_y \end{pmatrix} = \frac{\mu}{2\pi} \int_{x=0}^{c} \int_{y=0}^{d} J_1(x,y) \{ \int_{x'=0}^{c} \int_{y'=y_o}^{y_o+d} J_2(x',y')$$

$$\frac{dx'\ dy'}{[(x-x')^2+(y-y')^2]} \begin{pmatrix} (x-x') \\ (y'-y) \end{pmatrix} \} dxdy \tag{14}$$

which may finally be written

$$\begin{pmatrix} f_x \\ f_y \end{pmatrix} = \frac{\mu}{2\pi} (\frac{hk}{9})^2 \sum_{i=0}^{2m} \sum_{j=0}^{2n} c_{ij} J_1(x_i,y_j) \{ \sum_{k=0}^{2m} \sum_{\ell=0}^{2n} c_{ij}\ J_2(x'_k,y'_\ell)$$

$$\frac{1}{[(x_i-x'_k)^2+(y_j-y'_\ell)^2]} \begin{pmatrix} (x_i-x'_k) \\ (y'_\ell-y_j) \end{pmatrix} \} \tag{15}$$

EXAMPLE AND CONCLUSIONS

As an example two sets of non-magnetic conductors were considered. The first with thin, the second with thick conductors. Dimensions of their cross sections are shown in Figs. 3 and 4 respectively. The conductivity of the material was $\sigma = 57.10^6 \Omega^{-1} m^{-1}$. The current flowing in each conductor for both cases was 300 A(rms) at a frequency of 50 Hz. In carrying out the numerical solution for the set of thin conductors a grid of 9 points for the width and 3 points for the thickness was first considered. Then a grid of 11x3 points was tried and the two were found to have a deviation in current density of ±5%. Grids with a higher number of points along the width of the thin conductor gave current densities practically coinciding with the 11-point grid. Current densities were insensitive to the variation of the distance between the two thin conductors. This was valid unless the distance was very short. Results are given in Fig. 3.

For the thick conductors the grid was made 9x9 points. Results are given along the width at three levels, 1,2 and 3, Fig. 4 and along the thickness at positions 1 and 2, Fig. 5.

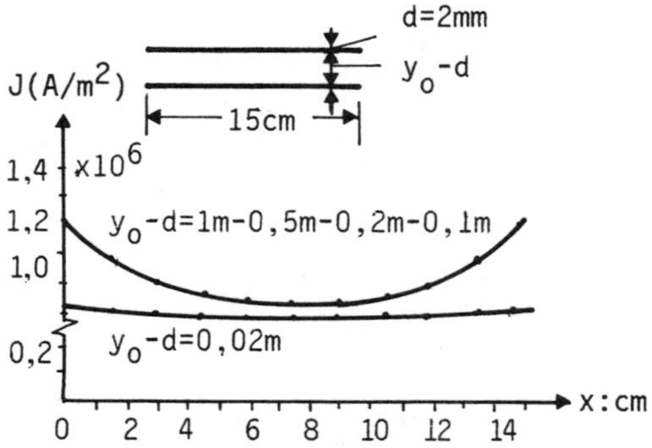

Fig. 3. Current densities along the width for thin plates.

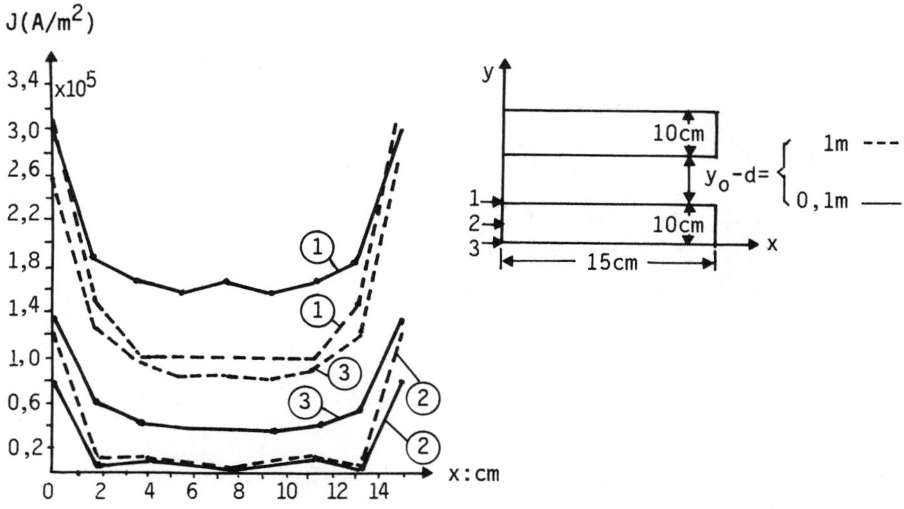

Fig. 4. Current densities along the width for thick plates.

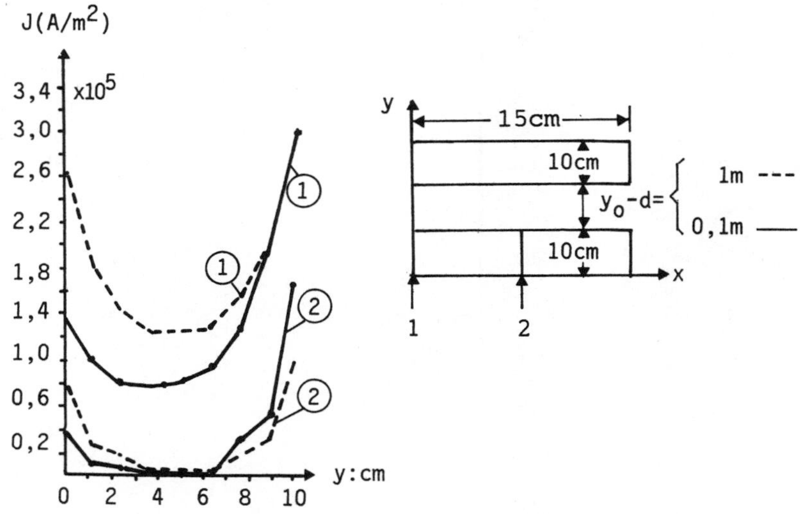

Fig. 5. Current densities along the thickness for thick plates.

REFERENCES

1. Manneback C. : "An integral equation for skin effect in parallel conductors". Journal Mathematics and Physics, pp. 123-146 (1922).
2. Gradshteyn I.S., Ryzhik J.W. : "Table of Integrals series and products". Academic Press, (1965).

4. ELECTRICAL MACHINES – Synchronous Machines, DC and Linear Motors

Introductory remarks

T. Sliwinski

Department of Fundamental Research
Instytut Elektrotechniki
Warsaw, Poland

The permanently growing possibilities of computers enable one to use more and more complicated and accurate methods of analyzing the electromagnetic field in electrical machines and of calculating the machine parameters. Two approaches can be followed. The first one consists in setting up field equations and boundary conditions for a machine part, in solving these equations in an analytical way and in using the computer only in the last stage of calculation. The second one starts with the discretization of the analyzed area into surface elements by means of special programs and in the numerical solution of a large set of equations. This last approach is represented by the finite difference method and by the finite element method, based on variational principles. Another numerical method called the boundary element method consists in the discretization of the area boundaries instead of the area itself and in discribing the field by boundary integral equations converted into a numerical form. It is also possible to represent the separate parts of the magnetic and electrical circuits of an electrical machine by lumped elements and to put them together in a network, which may be solved afterwards, in an usual way.

In the case of electrical machines the evaluation of magnetic and/or electric field distribution is only half-way, the final aim being the calculation of windings inductances and, if power losses exist, of windings impedances. In the course of these computations flux linkages with separate winding space elements must be taken into account. Further complications in analysing the magnetic field in electrical machines arise because of changings in the position of the rotor in relation to the stator, of variable permeability of ferromagnetic parts, of mutual reaction of different windings, etc. The magnetic field in stator and rotor cores may be treated as two-dimensional, but in the case of the end connections a three-dimensional approach is inevitable, which causes further difficulties. The solution of the enumerated problems for different kinds of electrical machines requires great efforts and invention and has been done until now in a limited degree.

In this chapter five papers are presented, which deal with several kinds of electrical machines: synchronous generators and motors, direct current motors and linear induction motors. They represent different cases of analysis problems and different kinds of approach. Static and dynamic problems are solved using both the finite difference and the finite element method, the equivalent network method and analytical methods. All these papers are aimed at better knowledge of the physical phenomena and at more accurate calculation of machine parameters. They may contribute to more common use of advanced calculation methods in the design of electrical machines and in the prediction of their static and transient performance.

MAGNETIC SHIELDING OF TURBO-GENERATOR OVERSPEED TEST TUNNELS

E. Ch. Andresen, and W. Müller

Institut für Elektrische Energiewandlung
Technische Hochschule Darmstadt
D - 6100 Darmstadt

ABSTRACT

Turbo-generator rotors induce high eddy current losses in the surrounding steel walls when turning in an overspeed test facility with the field current switched on. By means of cylindrical shielding the losses are reduced to a very small fraction. The interesting field quantities, the power losses and the currents per pole are numerically calculated depending on the rotation frequency, on the number of poles and on the shield dimensions.

INTRODUCTION

Alternating or moving magnetic fields occurring in power apparatus e.g. in electrical machines, linear motors, nuclear fusion magnets, overspeed test tunnels etc. must be shielded in many cases in order to protect the surroundings against heating or magnetizing and to avoid power losses. The field of electrical machines is normally shielded by the laminated stator core necessary anyway to guide the flux and to limit the need of a field excitation current. The stator core represents a passive magnetic shield. Synchronous generators with a superconducting field winding may be shielded by passive or active elements. The passive shield is normally preferred, because the flux density at the back of the stator winding is below the iron saturation. In case of higher flux densities an active shield from conducting material is needed. The magnetic field is kept away from the outer space by eddy currents induced by the field itself.

In this contribution the active magnetic shielding of turbogenerator rotors in overspeed test tunnels is treated. Overspeed test tunnels are often used for testing both the generator and the turbine rotor. They are constructed from solid steel to enable test runs under vacuum thus reducing the friction losses of the turbine rotor. Turning the excited generator rotor inside the steel tunnel induces high eddy current losses, because the big permeability of the steel brings about a very small penetration depth of the currents. So the solid steel must be protected against the rotating field by a nonmagnetic shield of high electric conductivity, e.g. by an aluminum cylinder or cage.

First of all the important electrical and magnetical data of a given overspeed test facility are determined by a two-dimensional finite difference method [1]. Interesting quantities represent the flux densities inside and outside the shield, the current density, the current per pole and the power losses. The influence of the number of pole pairs and of the frequency is also considered. For the benefit of an optimal shield design the paper shows how the power losses and the current density depend on the thickness of the shield wall and on the diameter of the shield cylinder. For a comparison the power losses in the unshielded steel container are determined by non-linear iterative calculation [1].

METHOD OF CALCULATION

Assumptions

1) The field quantities do not depend on the coordinate z perpendicular to the plane of calculation.
2) The magnetic vector potential and the exciting current density have z-components only.
3) All field quantities vary sinusoidally with time, only the first harmonics are taken into account.

The complex amplitude A of the vector potential A_z fulfills the well known diffusion equation in polar coordinates r and φ

$$\frac{1}{r}\frac{\partial}{\partial r}\frac{r}{\mu}\frac{\partial A}{\partial r} - \frac{1}{r}\frac{\partial}{\partial \varphi}\frac{1}{r\mu}\frac{\partial A}{\partial \varphi} + j\omega\varkappa A = J - \varkappa\frac{\partial \Phi}{\partial z} \qquad /1/$$

where ω, $\varkappa$ denote the circular frequency and the electric conductivity respectively. Φ is a scalar electric potential due to the external voltage source and J denotes the complex amplitude of the current density which may be given arbitrarily within all nonconducting domains. In case of ferromagnetic material the permeability μ is regarded as the proportionality factor in the relation $B = \mu H$ between the vectors of the complex amplitudes of the magnetic flux density and the magnetic field strength. It can be evaluated from a modified magnetization characteristic

$$\hat{B} = \hat{B}(\hat{H}) \qquad /2/$$

where $\hat{B}$, $\hat{H}$ denote the maximum values of $Re(\underline{\hat{B}}e^{j\omega t})$, $Re(\underline{\hat{H}}e^{j\omega t})$ during one time period. $\hat{B}$ is determined by harmonic analysis of the flux density curve excited by a sinusoidal field strength. The calculation of the nonlinear field is started with a rough estimate of the local distribution of μ. The resulting linear field problem /1/ is solved numerically by the finite difference method using a special block-SOR technique described in [1]. After some cycles the μ's are recalculated from the "new" distribution of the field by use of /2/. The domain of calculation is one pole pitch. The grid is generated by two sets of coordinate lines $r = r_i$, $\varphi = \varphi_i$ of polar coordinates $r\varphi$. The number of grid points is in the range of 800. In φ-direction a condition of periodicity of the form

$$A(r,\varphi) = - A(r,\varphi + 90^{\circ}) \qquad /3/$$

is imposed.

The real current loading distribution $S(\gamma)$ is approximated by an idealized sinusoidal one:

$$S(\varphi) = J \cdot h\ N_c(\varphi) = \hat{S} \cdot \cos\varphi \qquad /4/$$

where J denotes the current density, h the conductor height and $N_c(\varphi)$ the position dependent number of conductors lying one upon the other. The MMF distribution is achieved by intergrating /4/

$$\Theta(\varphi) = \hat{\Theta} \cdot \sin\varphi \qquad /5/$$

$$\text{with } \hat{\Theta} = \frac{\tau p}{\pi} \cdot \hat{S} = \underline{\underline{\Theta}} \qquad /6/$$

The amplitude $\hat{\Theta}$ of the MMF wave equals the total d.c.MMF $\Theta_=$ per pole. The mechanical rotation is simulated by a two phase a.c. system in order to match the problem with the complex equations of the two dimensional field calculation

$$S(\varphi) = \hat{S} \cdot e^{j\varphi} = \hat{S}\,(\cos\omega t + j\sin\omega t) \qquad /7/$$

From the distribution of the vector potential the following secondary fields can be numerically calculated.
Magnetic flux density

$$B_r = \frac{1}{r}\frac{\partial A}{\partial \varphi}\,, \quad B_\varphi = -\frac{\partial A}{\partial r} \qquad /8/$$

Induced electric field strength

$$E = -\,j\omega \mathbf{A} \qquad /9/$$

Current density in the shield

$$J_R = \varkappa E = -\,j\omega\varkappa A \qquad /10/$$

Loss density

$$P = \mathrm{Re}\,(E\,J_S^*) \qquad /11/$$

The permeability is assumed to be constant within each elementary domain of the grid.

Table 1 Data of the shield systems

Shield diameter	7.0 m	Rotor diameter	1.20 resp. 1.80 m
Wall thickness	10 resp. 20 mm	Frequency	50 and 60 resp.
Material	Aluminum		25 and 30 r/s
Number of poles	2 resp. 4	D.c.MMF per pole	300 resp. 150 kA

RESULTS

The calculations are carried out for the parameters compiled in Table 1.

First the magnetostatic field valid for the unshielded rotor is numerically determined by means of the scalar potential. It was supposed that the extent of the calculation domain is of great influence on the results in case of undisturbed field spreading. However, it turns out that increasing the diameter of the calculation domain from 20 to 100 m enlarges the flux density on a circle line of 7 m in diameter by only about 10 %. When applying the shield a domain diameter of 20 m is sufficient from the beginning, because the shield reduces the outer flux density by about the factor 400 to 1000. The shielding effect increases with rising frequency and with enlarging thickness of the shield from 10 to 20 mm, as the penetration depth of aluminum at 50 Hz is about 12 mm.

Fig.1 demonstrates, that the radial field component continuously decreases towards the shield wall while the tangential field increases inside the cylindrical space compared with the unshielded field. Outside the shield both components are nearly equal.

The power losses diminish only little with increasing thickness of the

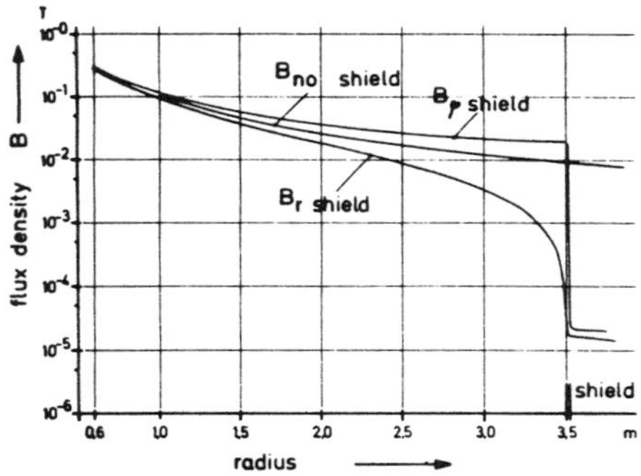

Fig.1 Maximum flux density components of the two-pole system, 50 Hz, 20 mm

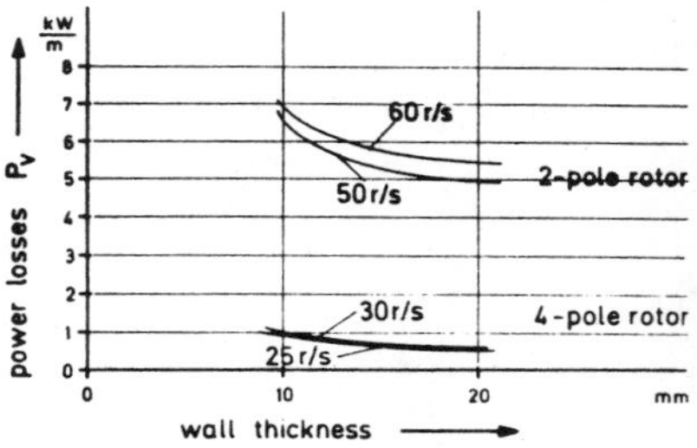

Fig.2 Power losses versus wall thickness

shield wall see Fig.2. Shields with a wall thinner than 10 mm were not investigated, as the losses increase about reverse proportional to the decreasing thickness. Fig.2 also shows that the four-pole rotor only generates a fraction of the power losses of the two-pole rotor. This is due to the bigger "air gap leakage flux" and to the smaller frequency.

The field plots in the following pictures Fig.3 to 6 make the physical effect of shielding visible. The flux lines of the undisturbed fields specially of the two-pole rotor seem to form curves like circles. Under the influence of a shield they are pressed together forming oval lines. Comparing Fig.3 with Fig.5 makes evident that the four-pole rotor indeed has a larger leakage flux than the two-pole rotor.

The nonlinear iterative field calculation of the unshielded steel tunnel gives the following power losses and shield MMFs for the two-pole rotor

$$P_V = 175 \text{ kW/m}$$

$$\Theta_S/\Theta_F = 7.8/300 = 0.026$$

The corresponding data of the aluminum shield with the same diameter write

$$P_V = 5 \text{ kW/m}$$

$$\Theta_S/\Theta_F = 135/300 = 0.45$$

The power losses are reduced to less than 3% of the losses in the steel walls by eddy currents 16 times larger than the currents in the steel walls.

ON THE SHIELD DESIGN

An important question of practice is the optimal design of the shield. This problem may be reduced here to the choice of the diameter of the cylindrical shield. The thickness of the material and its influence has already been treated.

Each shield diameter requires a calculation grid of its own and hence a certain amount of calculation work. An approximate determination only based on one shield diameter may be carried out as follows.

The shield is mainly linked with the tangential flux component B_φ, see Fig.1. The flux linkage Ψ_S of the shield also depends on the pole pitch

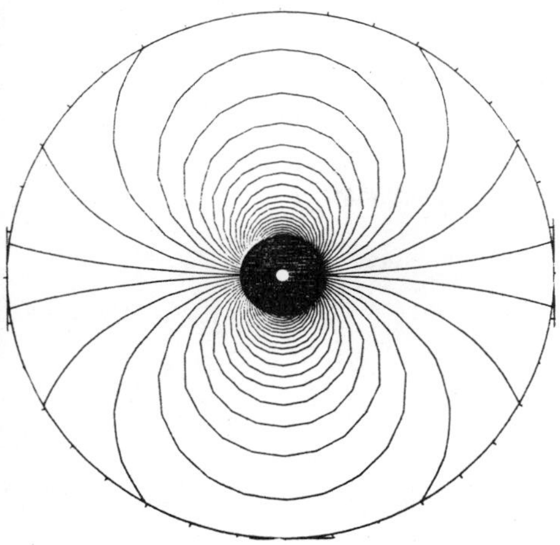

Fig.3 Field plot of the two-pole system without shield

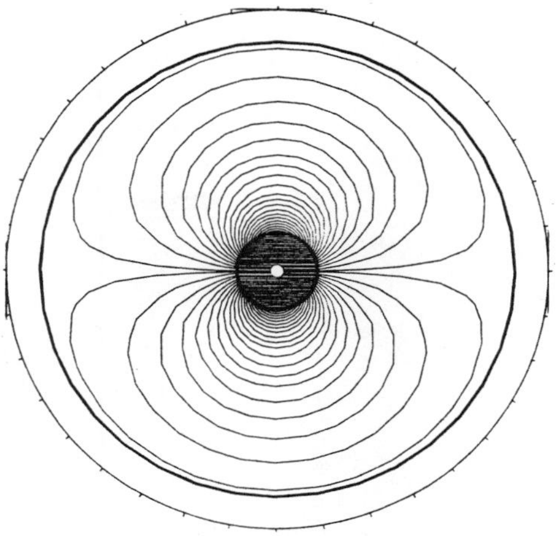

Fig.4 Like Fig.3 with shield 20 mm thick; 50 r/s

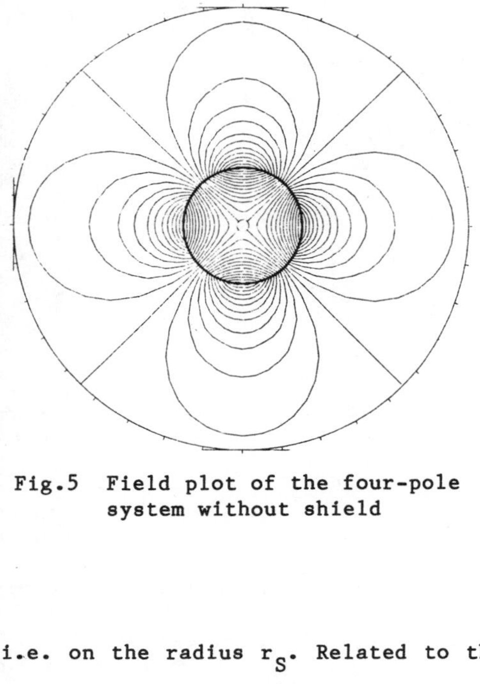

Fig.5 Field plot of the four-pole system without shield

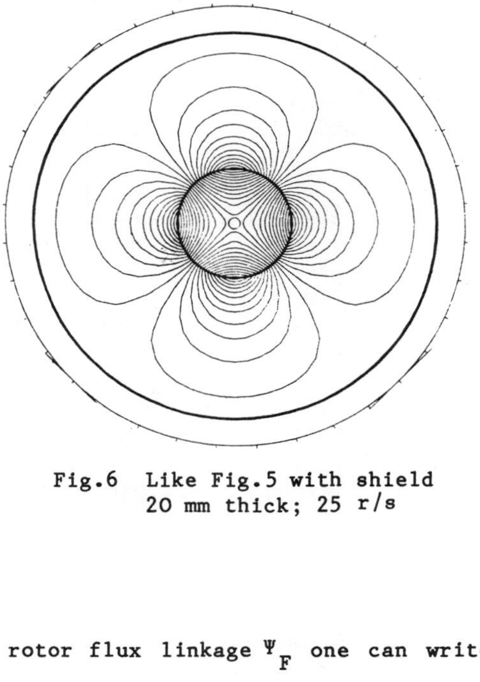

Fig.6 Like Fig.5 with shield 20 mm thick; 25 r/s

i.e. on the radius r_S. Related to the rotor flux linkage Ψ_F one can write

$$\Psi_S/\Psi_F = B_{\varphi S} \cdot r_S /(B_{\varphi F} \cdot r_F) \qquad /12/$$

This flux relation is identical with the relation between the induced shield MMF Θ_S and the exciting rotor MMF Θ_F. The numerical field calculation of the two-pole rotor at 50 Hz and 20 mm wall thickness yields

$$\Theta_S/\Theta_F = \Psi_S/\Psi_F = 0.45 \qquad /13/$$

Assuming that the tangential flux density B_φ of Fig.1 is nearly identical for shields with different radii - they will really be higher for smaller

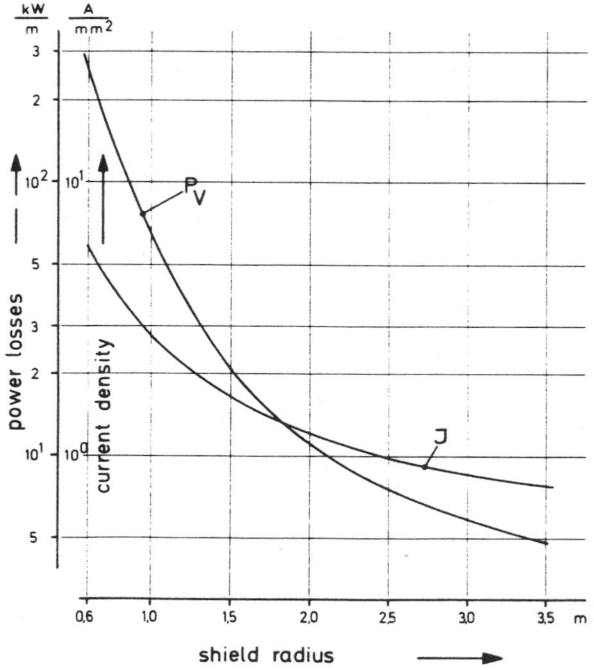

Fig.7 Approximate power losses and eddy current density of the two-pole 50 Hz 20 mm system versus shield radius

ones - the power losses may be estimated from one field calculation with the data $P_V(r_{So})$ and $B_\varphi\ (r_{So})$ by

$$P_V(r_S) = P_V(r_{So}) \cdot \left(\frac{B_{\varphi S}(r_S)}{B_{\varphi S}(r_{So})}\right)^2 \cdot \frac{r_S}{r_{So}} \qquad /14/$$

When introducing a resistance per pole on the basis of the current penetration depth it is also possible to evaluate the current density depending on the shield radius.

Fig.7 presents the losses and the current density versus shield radius. It is evident that the power losses increase significantly when the radius is reduced. It is caused by both increasing flux linkage and diminishing material volume.

A two-pole rotor with a length of 8 m generates power losses of about 500 kW in a shield of 2 m in diameter and only of about 40 kW in a shield of 7 m in diameter. The real value is increased by the losses in the end regions. As the power losses have to be supplied by the drive of the test facility the designer must compromise on the costs of the drive and on the costs of the shield. It has to be considered in addition that the small shield imposes heating and cooling problems. Moreover the maintenance is less comfortable as the small shield must be installed for any test run.

REFERENCES

1. W. Müller, "A new Iteration Technique for Solving Stationary Eddy-Current Problems Using the Method of Finite Differences", IEEE Transactions on Magnetics, vol. MAG-18, No.2 (March 1982).

AN EQUIVALENT CIRCUIT MODEL FOR INVERTER-FED SYNCHRONOUS MACHINES

M.J. Carpenter and D.C. Macdonald

Imperial College
London SW7 2BT, England

ABSTRACT

Commutation time of inverter fed synchronous motors is commonly expressed in terms of the commutation reactance X_c. A precise method of calculation is desirable especially for solid salient pole motors. A magnetic circuit method is used to model a machine cross section which is coupled to the stator windings making possible a direct coupling of the magnetic circuit with the inverter.

Commutation is effectively a line-to-line short circuit superimposed on a load condition and initial studies have been made on a line-to-line short circuit without rotor motion.

Calculated commutation on load agrees well with test results.

INVERTER-FED SYNCHRONOUS MOTORS

Inverter-fed synchronous motors are in use as variable speed drives as shown in Fig.1., thyristor inverters [1-3] being naturally commutated by machine voltage. To achieve commutation machine phase current has to be commutated before line-to-line voltage drops to zero that is the machine has to run over-excited. Power factor can be kept nearest unity (and machine size minimised) by keeping the firing angle, β, at which commutation is started before voltage zero, to a minimal value. The firing angle should be of sufficient value to allow for commutation time, μ, plus a safety angle, the safety angle being necessary to ensure the thyristor recovers its blocking capability.

Commutation time depends on:

i) load current magnitude
ii) voltage magnitude
iii) firing angle
iv) machine response to commutation usually expressed as commutating reactance, Xc.

The first three factors are known for specific operation. Commutating reactance has often been of rotor position at the start of commutation, likely to be a good approximation if there is a complete damper cage present and the subtransient reactances are approximately equal. There remains the question whether it is the correct reactance, although it is thought to be a good approximation.

Solid salient-pole machines present a substantially different picture as the subtransient reactances are less distinct and the axis reactances are noticeably different. It is thought that the absence of a high conductivity damper cage may give rise to high Xc and unacceptably long commutation times. This work is therefore concentrated in that area.

Fig.2 shows a section of a larger network use to model an actual motor.

MAGNETIC CIRCUIT MODEL

The approach used is based on the work of L. Haydock [4,5] and has been adapted to model a practical motor cross section interlinked with an electric circuit.

Haydock used network theory, magnetic terminals and linkages represented in the way described by Carpenter [6] to produce models, which are topologically accurate and representative of the actual magnetic condition in machines. Major flux paths are replaced directly with "magnetic capacitors" (capacitance being equivalent to permeance) [4,5]. And "magnetic resistors" are used to simulate damping in the solid rotor caused by eddy currents.

The methods developed in this work keep magnetic and electric circuits separate but properly linked using gyrators. The gyrator interlinks magnetic and electric circuits according to the familiar linkage equations (1) and (2) yielding a full representation of the magnetic and electric circuits.

$$\text{mmf} = I_e N \qquad (1)$$

$$\text{emf} = -I_m N = -\frac{d\phi}{dt} N \qquad (2)$$

where N = Number of turns
I_m = "Magnetic Current" - Rate of change of flux with time
I_e = Electric current

Commutation is like an unbalanced fault condition, necessitating a full representation of the three phase windings. This is accomplished with ease using gyrators having the correct linkages (for each individual stator tooth, L_1, L_2, L_3). Use of gyrators has the added advantage of removing the necessity to perform complex transferance and duality operations on the electric and magnetic circuits [5,7].

Air gap permeances and stator tooth leakage permeance are shown in Fig. 2 as C_g and C_s respectively. Calculation of circuit component values use standard equations adopted by Haydock. A layered structure of permeances regularly interspersed with appropriate magnetic

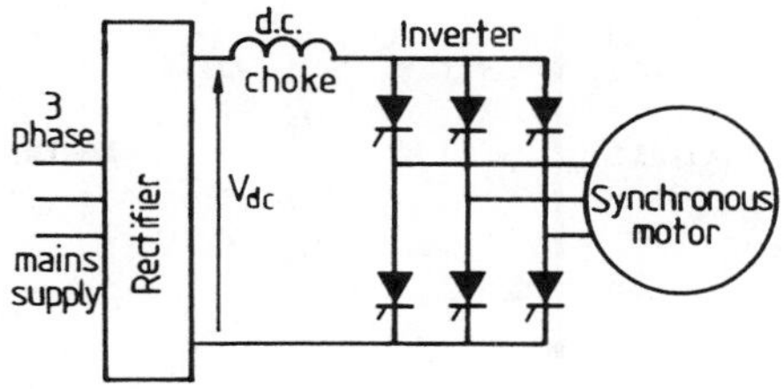

Figure 1

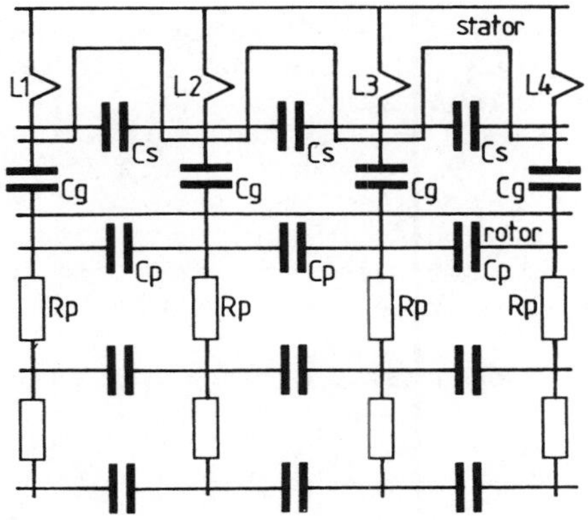

Figure 2

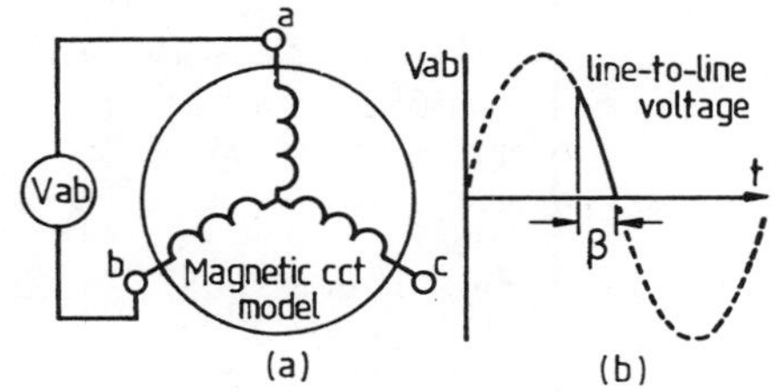

Figure 3

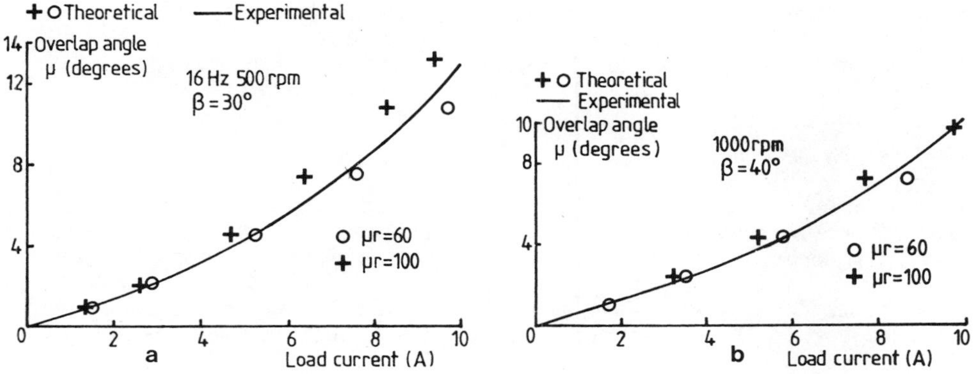

Figure 4

resistances, Cp and R_p respectively, models the solid rotor poles. Layers of permeance, C_p, and dampance, R_p, have been discretized into layers of thickness 0.5 mm at the top of the pole doubling for each layer down to the fifth of 8 mm thickness. In circuit terms the magnetic resistances can be seen as damping the magnetic current (rate of change of flux) and thereby resisting any sudden change of flux (charge movement between capacitors).

Incorporating the above ideas, a two pole-pitch model was used to simulate line-to-line short circuit tests.

LINE-TO-LINE SHORT CIRCUIT TEST

Treating commutation as a line-to-line short circuit, tests were performed on a motor magnetic circuit model for different motor speeds, firing angle, and excitation level, Fig. 3(a). The rotor was assumed stationary during commutation implying the motor back emf had to be an applied sine function, of typical shape in Fig.3(b), for a particular firing angle. At high motor speed the rate of fall of line-to-line voltage will be appreciable and this effect must be incorporated in any model since commutation is dependent on the forcing voltage level. The simulation also ignores the interlude periods between commutations.

The third phase was left open circuit since the current flowing in the winding was assumed to be unchanged during commutation. By adjusting gap capacitances accordingly the required rotor position was set for particular firing angles.

From test results commutation reactance has been found to be a function of time and rotor position. It is therefore unsatisfactory to think of Xc in the conventional manner as a function of the subtransient reactances. Gradual, although relatively fast, flux penetration of the rotor pole appears to create the variation of reactance, the effect being most dominant for short commutation times. To fully understand machine behaviour under these conditions it may be necessary to investigate and measure flux densities of the computer model, in the rotor pole model during commutation.

Calculations for commutation time versus load current have been made for several operating points. The general shape of resulting graphs compare well with practical measurements and absolute values compare favourably as shown in Fig. 4.

These initial tests make allowance for saturation with permeabilities set to a value higher (60 → 100) than the actual permeability of the machine. Curves at high field currents show larger discrepancies and this may be due to neglecting saturation. It should also be noted that no allowance has been made for motion or rotor load angle in these analyses.

COMMUTATION ON LOAD

A full representation of the machine magnetic circuit over two pole-pitches was assembled together with the inverter model. Rotational voltage generation was obtained by keeping the rotor stationary and by moving the stator conductors as suggested by Turner.[8] Iron relative permeability was taken as 100 throughout. The overall performance was obtained using SPICE and typical results are shown with measured values in Fig 5.

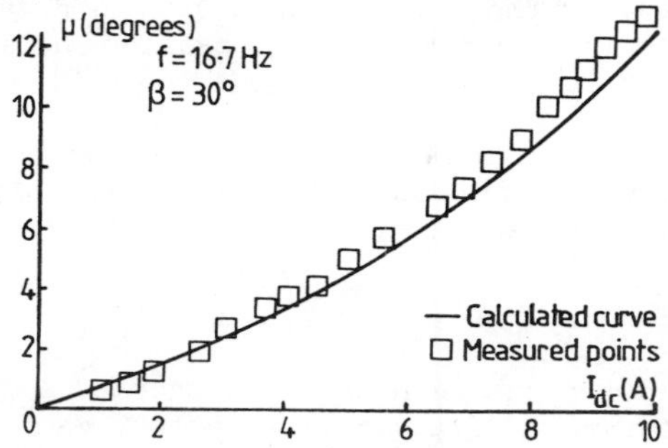

Fig. 6. Measured Commutation Time vs. Link Current for Solid and Laminated Rotor Machines.

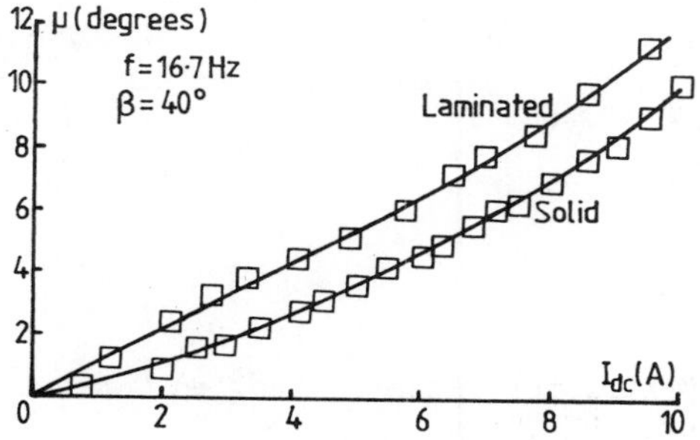

Fig. 5. Commutation Time vs. Link Current Calculated from a Full Synchro-Inverter Simulation, with Measured Results.

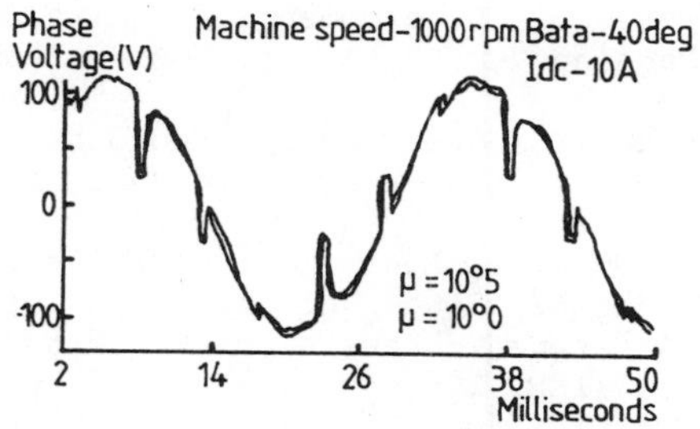

Fig. 7. Motor Waveforms for f=33.3 Hz.

Comparison with a laminated rotor machine fitted with a damper cage showed that at lower stator current loadings the solid pole machine commutated faster. This can be understood in that the outer surface of a laminated machine is unprotected by the damper cage and provides a low-reluctance path for commutating flux. In the solid pole machine eddy currents initially prevent flux entering the pole at all. As may be seen from Fig. 6., as the commutating currents rises the response is non-linear and at higher currents the position could reverse, being greater for the solid pole.

In spite of all the approximations, the method can produce a very good simulation of machine performance. Machine phase voltage is simulated virtually perfectly as shown in Fig. 7.

CONCLUSIONS

The method of simulation has been shown to provide a good approximation to commutation performance of a small laboratory machine.

The magnetic circuit representation is comparable with that of Ostovic [9]. For practical purposes the relatively crude approximation compared with finite element methods, gives sufficiently good simulation.

Surprisingly a solid rotor can have a smaller commutation time than a laminated machine with a damper cage.

Solution times were typically 10 - 13 minutes (cpu) for a two pole pitch model having 24 stator teeth,five capacitance permeance layers of rotor, yielding a 250 node network of 438 elements (capacitor,resistor, sources etc.).

REFERENCES

1. PELLY, B.R.: "Thyristor Phase Controlled Converters and Cycloconverters", Wiley, 1971.
2. GAYEK, H.W.: "Behaviour of Brushless Aircraft Generating Systems", IEEE Trans-Aerospace Support Conference Procedures, 1963, pp 594-621.
3. ADAMSON, C. and HINGORANI, N.G.: "High Voltage Direct Current Power Transmission", Garraway, 1960.
4. HAYDOCK, L.: "Systematic Development of Equivalent Circuits for Synchronous Machines", PhD thesis, Imperial College, University of London, 1986.
5. HAYDOCK, L.: "Magnetic Models for Synchronous Machines", 21st Universities Power Engineering Conference, April 1986, Imperial College, London.
6. CARPENTER, C.J.: "Magnetic Equivalent Circuits for Electrical Machines", Proc. IEE, 1968, Vol. 115, No. 10, pp 1303-1511.
7. LAITHWAITE, E.R.: "Magnetic Equivalent Circuits for Electrical Machines", Proc. IEE, 1967, Vol. 114 (11), pp 1805-1809.
8. TURNER, P.J.: "Finite Element Simulation of Turbogenerator Faults and Application to Machine Parameter Prediction IEEE Trans on Energy Conversion, 1987. Vol. EC2, No. 1, pp 122-131.
9. OSTOVIC, V.: "A Method for Evaluation Transient and Steady State Performance in Saturated Squirrel Cage Induction Machines", IEEE Trans on Energy Conversion, 1986. Vol. EC1, No. 3, pp 190-197.

MAGNETIC FIELD AND FORCE OF LINEAR INDUCTION MOTOR FED FROM NON-SINUSOIDAL SOURCE

Ewa Gierczak, Maciej Włodarczyk, and
Kazimierz Adamiak

Technical University of Kielce
Kielce, Poland

INTRODUCTION

Performances of rotating induction machines fed from a non-sinusoidal source have been investigated rather thoroughly whereas this same problem for linear machines has received much less attention [1,2,3]. The first published papers report an appreciable influence of the higher time harmonics on primary losses and their negligible influence on terminal performances. This paper describes computational and experimental results relating to the influence of a non-sinusoidal supply on the magnetic field and force of a small single-sided LIM controlled with variable-voltage-fixed-freguency /VVFF/ pattern. Theoretical analysis based on the Fourier series method, validated by experimental data, confirms the conclusion about a negligible influence of the higher time harmonics in the exciting current on a LIM performance.

COMPUTATIONAL MODEL OF LIM

A choice of computational technique should compromise between costs of computation and accuracy of results. Modern computational techniques can take into account most effects influencing the magnetic field in a LIM airgap. However, selecting the considered effects in a proper way, it is possible to obtain results with almost the same accuracy using much simpler approach. One of such techniques seems to be Fourier series method [4,5] applied in this paper. Apparently fictitious assumptions make it very simple, but idealizing of the model does not introduce big errors.
The magnetic field in the LIM has been analysed with the following simplifying assumptions:

1. The LIM is composed of four homogeneous, isotropic and linear layers /Fig.1/
2. The primary iron, made of an ideal magnetic material / $\mu = \infty$, $\sigma = 0$/, is infinitely long in x and z directions
3. The primary winding is replaced by the infinitely thin current sheet. Linear current density is defined as:

$$\underline{J}^s/x,t/=\sum_{k=1}^{\infty}\sum_{u=-1}^{\pm\infty} J^s_{uk}\exp\left[j/k\omega t+u\frac{\pi}{\tau}x+\psi_k/\right] \qquad /1/$$

where

J^s_{uk} - amplitude of the uk-th harmonic primary linear current density
u - space harmonic number/u=-1,5,-7,11,.../
k - time harmonic number/k=1,3,5,7,.../
τ - pole pitch
ψ_k - phase difference of k-th harmonic current density

The time harmonics that are a multiple of three occur only in this case if the circuit has a neutral connection.

4. The primary winding has a finite length equal to L, but it has been assumed that the primary currents create an infinite chain of identical current sheets. A distance between two adjacent sheets is equal to/τ_s-L/. A method of the distance τ_s evaluation has been described in [6].

The last assumption gives periodicity of the problem and the primary current can be developed in the Fourier series

$$\underline{J}^s/x,t/=\sum_{k=1}^{\infty}\sum_{u=-1}^{\pm\infty}\sum_{r=-1}^{-\infty} J^s_{ukr}\exp\left[j/k\omega t+\frac{u\pi}{\tau}x+\psi_k/\right]\cos\frac{r\pi}{\tau_s}x \qquad /2/$$

Application of the trigonometric transformation to Eq./2/ yields

$$\underline{J}^s/x,t/=0.5\sum_{l}\sum_{u}\sum_{k}\sum_{r}\underline{J}^s_{ukr}\exp\left[j/k\omega t+\frac{\pi}{\tau_{url}}x/\right] \qquad /3/$$

where

$$\underline{J}^s_{ukr}=\frac{4}{\pi r}\sin\left(\frac{r\pi L}{2\tau_s}\right)J^s_{ku}\exp\left[j/\psi_k/\right]; \qquad \tau_{url}=\frac{\tau\tau_s}{u\tau_s+r\tau l} \qquad l=1,-1$$

The current density described by the formula /3/is composed of two waves with the same amplitudes travelling along x axis with different velocities

$$v_{kurl}=-2fk\tau_{url} \qquad /4/$$

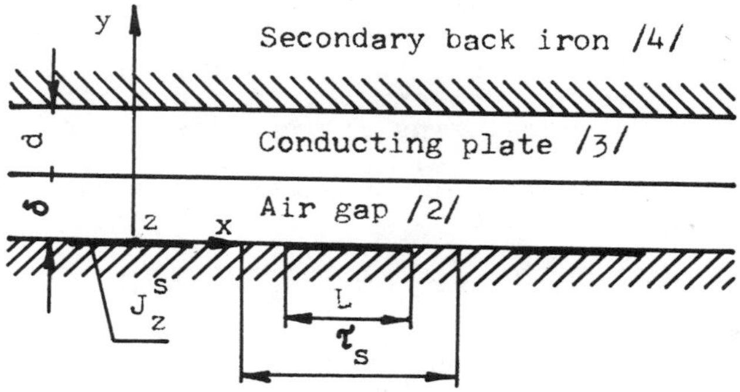

Fig. 1. Idealized model of a LIM with finite excitation length.

5. The airgap has been corrected by the Carter's coefficient k_C

$$\delta = \delta_a k_C \qquad /5/$$

6. Current symmetry has been assumed.

The magnetic field in a LIM can be easily analysed by the Fourier series method when the coordinate system is attached to the secondary. In this case the primary current density is described by the formula

$$\underline{J}^s/x,t/=0.5\sum_l \sum_u \sum_k \sum_r \underline{J}^s_{ukr} \exp\left[j/\omega t s_{kurl} + \frac{\pi}{\tau_{url}} x/\right] \qquad /6/$$

where

$s_{kurl} = 1 - \frac{v}{v_{kurl}}$ - slip referred to k-th time and ur-th space harmonic.

The two-dimensional magnetic field in the LIM can be calculated most conveniently by means of the magnetic vector potential $\underline{A}$. In the considered problem $\underline{A}$ has only one component along z axis and is governed by the equation

$$\nabla^2 \underline{A} - jk\omega\mu\sigma s_{kurl}\underline{A} = 0 \qquad /7/$$

As linearity of the problem has been assumed, the superposition principle can be used. Thus Eq. /7/ has been solved separately for each time and space harmonic, and next the results obtained have been summed up. Since the derivation of k-th time harmonic vector potential is similar to one made for the 1-st time harmonic, and published amongst others in [5] we write here the final expressions for the conducting plate /3/ of the LIM

$$\underline{A}^{/3/}_{kurl}/x,y/=\frac{-\mu_2}{\beta_2 \underline{M}}\underline{F}/x/\underline{J}^s_{kur}\left\{\cosh\left[\beta_3/y-d-\delta/\right]-\underline{D}\sinh\left[\beta_3/y-d-\delta/\right]\right\} \qquad /8/$$

where

$$\beta_p^2 = \left(\frac{\pi}{\tau_{url}}\right)^2 + jk\omega\mu_p\sigma_p s_{kurl} \quad \text{p-number of layer } /p=2,3,4/$$

$$\underline{F}/x/=\exp/j\frac{\pi}{\tau_{url}}x/; \quad \underline{B} = \frac{\mu_2\beta_3}{\mu_3\beta_2}; \quad \underline{D} = \frac{\mu_3\beta_4}{\mu_4\beta_3}$$

$$\underline{M}=/\underline{D}-1/\exp/-\beta_3 d/\left[\underline{B}\cosh/\beta_2\delta/-\sinh/\beta_2\delta/\right]+/\underline{D}+1/\exp/\beta_3 d/$$
$$\left[\underline{B}\cosh/\beta_2\delta/+\sinh/\beta_2\delta/\right]$$

The components of the magnetic flux density and the electric field intensity for each time harmonic at an arbitrary point is a sum of all space harmonics

$$\underline{B}_{xk}=\sum_l \sum_u \sum_r \frac{\partial \underline{A}_{kurl}}{\partial y} \quad /9a/ \qquad \underline{B}_{yk}=\sum_l \sum_u \sum_r -\frac{\partial \underline{A}_{kurl}}{\partial x} \quad /9b/$$

$$\underline{E}_{zk}=\sum_l \sum_u \sum_r -j\omega k s_{kurl}\underline{A}_{kurl} \qquad /9c/$$

The force density in the secondary for k-th time harmonic has been determined by virtue of the Lorentz formula

$$f_x = 0.5\sigma_3 \mathrm{Re}\left[-\underline{E}_{zk}\underline{B}^*_{yk}\right] \qquad /10/$$

The total magnetic and electromechanic quantities for all time harmonics can be found from equations

$$C = \sqrt{\sum_k |C_k|^2}\ ; \qquad C = B,E,J \qquad /11a/ \qquad\qquad f = \sum_k f_k \qquad /11b/$$

CALCULATION AND EXPERIMENTAL RESULTS

Calculation and experiments have been performed for a small linear machine, whose design data are given in Table 1.
In the first series of experiments the LIM was supplied with sinusoidal voltage waveform. In order to investigate the influence of the higher time harmonics in the primary excitation on the magnetic flux density the flux density distribution was measured when the LIM was supplied with the nonsinusoidal current due to VVFF control. The exciting current was recorded and next expanded in the Fourier series. The Perry's method which employs approximate formulae of numerical integration was used. Typical current time distribution is shown in Fig.2. In this current distribution most important are 11-th and 29-th harmonics which are about 8% of the first one.
Exciting the LIM with the current, distribution of which is shown in Fig.2, the normal component of the magnetic flux distribution on the conducting plate surface was measured. In Fig.3 are compared measured and calculated normal flux density distributions with sinusoidal and non-sinusoidal current of the same rms value and the same fundamental frequency.
The influence of the higher time harmonics on the electromagnetic field of the LIM has been evaluated analytically taking into account one space harmonic /u=-1/ and 20 harmonics resulting from assumption about periodicity of the primary current /r=-1,-3,...,-39/. Distributions of the normal flux density and the force density for only 1-st time harmonic /sinusoidal excitation/ current shown in Fig.2 and for all time harmonics from k=1 to k=37 have been compared in Fig.4.

Table 1.

primary length	$L = 0.2094$m
primary width	$a = 0.1$m
pole pitch	$\tau = 0.0501$m
tooth pitch	$\tau_t = 0.0167$m
slot number per pole per phase	$q = 1$
slot opening	$b = 0.008$m
number of wire per slot	$w = 210$
airgap	$\delta_a = 0.003$m
conducting plate thickness	$d = 0.006$m
conductivity of conducting plate	$\sigma_3 = 382\cdot10^5$S/m
conductivity of secondary back iron	$\sigma_4 = 591\cdot10^4$S/m

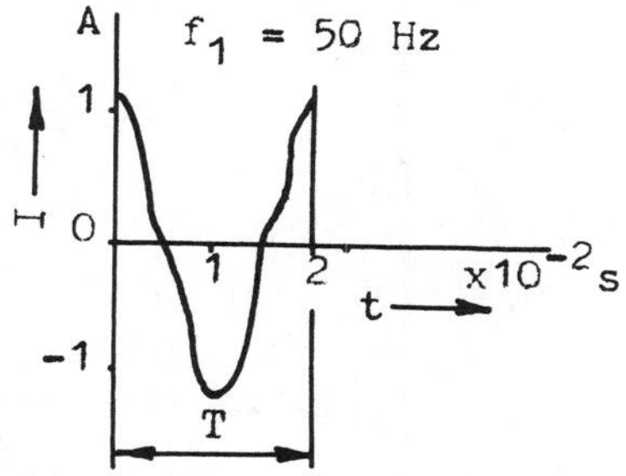

Fig. 2. Time distribution of exciting current in VVFF control of LIM.

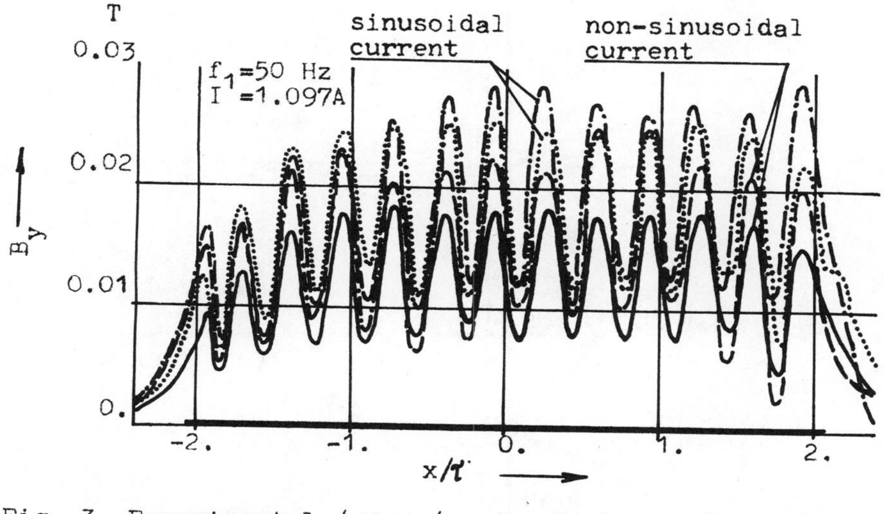

Fig. 3. Experimental /⋯/ and calculated /–·–·–/ results of the normal flux density on the secondary surface of the LIM.

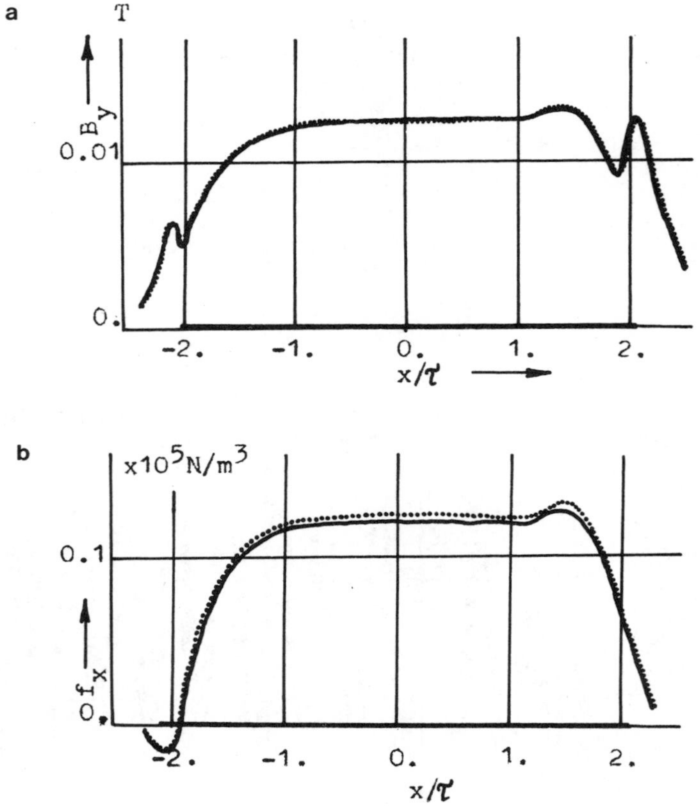

Fig. 4. Influence of the higher time harmonics on a/normal flux density, b/ force density on the secondary surface.

CONCLUSIONS

Results of the calculations indicate that the influence of the higher time harmonics on the magnetic field distribution and the force density of a LIM is negligible. This conclusion has been also confirmed by experimental data of the small LIM at standstill. The results presented in this paper have been obtained for not very high content of the higher time harmonics in the exciting current. However, a higher content is not very probable because a LIM terminal impedance has usually an inductive character and smooths the current time distribution. This is because even for a high content of the time harmonics in a voltage, a current is almost sinusoidal. This effect is more noticeable for linear motors rather than for rotary ones due to a large airgap which results in a low power factor.

REFERENCES

1. W. Paszek, B. Sliwa and T. Sztajer, Performance of a flat single-sided linear motor fed from non-sinusoidal voltage source, Problem Instalments-Electrical Machines EMA-KOMEL 24 (1976) /in Polish/.
2. W. F. Petrow and B. N. Slipliwyi, Influence of upper time harmonics of current load the linear induction motor characteristics, Izw. VUZ Elektromechanics 8 (1985) /in Russian/.
3. E. Masada, K. Fujisaki, M. Tamura, Influences of voltage harmonics on dynamic behaviour of the single-sided linear induction motor, Proc. of ICEM'86, vol.1, Munchen.
4. G. G. North, Harmonic analysis of a short stator linear induction machine using a transformation technique, IEEE Trans. PAS-92 (1973).
5. E. A. Mendrela, E. Gierczak, Two-dimensional analysis of linear induction motor using Fourier's series method, Arch. f. Elektr. 65 (1982).
6. E. A. Mendrela, J. Fleszar, E. Gierczak, A method of the distance between fictitious primaries in computational model of LIM used in Fourier series method, Arch. f. Elektr. 66 (1983).

ELECTROMAGNETIC PARAMETERS OF A TURBOGENERATOR DETERMINED BY THE FINITE ELEMENT CALCULATION

Wladyslaw Paszek, and Jan Staszak

Technical University of Gliwice, Poland
Technical University of Kielce, Poland

MACHINE MODEL AND ITS APPROXIMATION

Eddy currents induced in the solid rotor of large turbogenerators influence remarkably the electromagnetic properties of the machine being investigated. The magnetic linkage of the armature winding, of the field winding and the mutual field linkage between them results from the magnetic field distribution in the air gap evaluated for the case of a harmonic current impression. The eddy current reaction in the rotor was taken into account by the finite element method in the calculation of the field distribution. Constant permeability of the core was assumed and the end effects in the rotor were neglected but the complicated structures of the slotted solid rotor were considered and the aluminum alloy wedges located between the rotor teeth. The evaluated spectral transmittances (frequency plots of the fundamental transmittances) describe the electromagnetic properties of the machine and are suitable for the designation of the approximative operational transmittances containing the equivalent electromagnetic parameters of the machine model with lumped constants.

A typical two pole turboalternator with ratings : 200 MW, 15750 V, 8625 A, 50 Hz was investigated. The armature winding was represented by

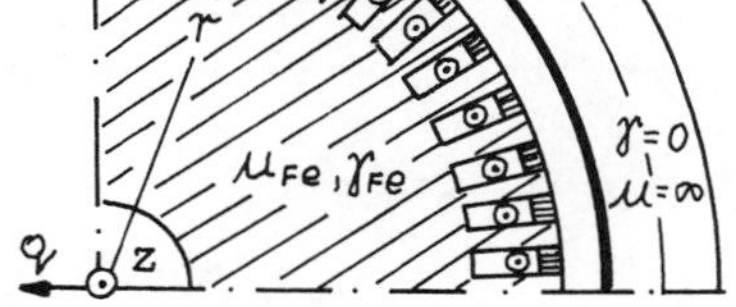

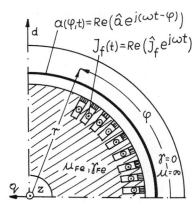

Fig. 1. Cross section of the machine model.

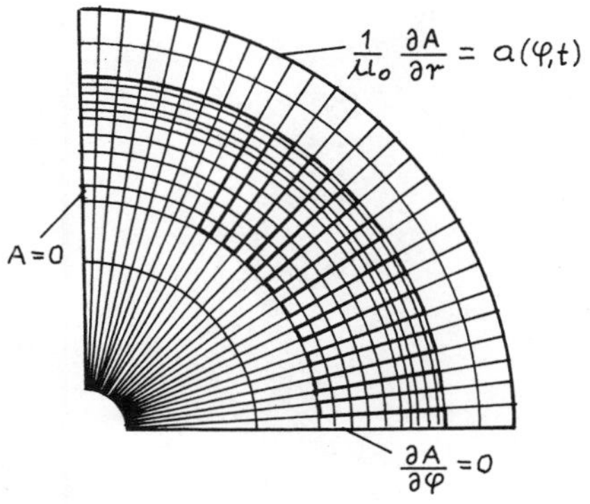

Fig. 2. Division of the area into finite elements.

an equivalent sinusoidally distributed current sheet with linear current density $a(\varphi,t)$ on the idealized smooth air gap sided stator surface. From the orthogonal transformation of the stator state variables into d, q frames attached to the rotor results the machine model represented in Fig. 1.

FIELD DISTRIBUTION AND FREQUENCY PLOTS

The field distribution was calculated at harmonic alternating current of angular frequency ω impressed into the equivalent armature resp. the fields winding. The quasi stationary distribution of the electromagnetic field $\mathbb{B}$, $\mathbb{E}$ was derived by the vector potential $\mathbb{A}$.

$$\mathbb{B}=\mathrm{rot}\mathbb{A}, \qquad \mathrm{div}\mathbb{A}=0, \tag{1}$$

$$\mathrm{rot}\left[\frac{1}{\mu}\,\mathrm{rot}\,\mathbb{A}\right] = -\gamma\frac{\partial\mathbb{A}}{\partial t} = \mathbb{J}, \tag{2}$$

$$\mathbb{J}=-j\omega\gamma\mathbb{A} \quad \text{- outside the windings} \tag{3}$$

Taking into account $\mathbb{J}=\mathbb{1}_z J$, $\mathbb{A}=\mathbb{1}_z A$ there results from (2)

$$\frac{1}{r}\,\frac{\partial}{\partial r}\left(\frac{1}{\mu}r\,\frac{\partial A}{\partial r}\right) + \frac{1}{r^2}\,\frac{\partial}{\partial\varphi}\left(\frac{1}{\mu}\,\frac{\partial A}{\partial\varphi}\right) = -J \tag{4}$$

Considering Dirichlet's boundary condition $(A=A_\beta)$ and Neumann's b. cond. $\frac{1}{\mu}\frac{\partial A}{\partial n} = -q$ with the known values A_β on the singular sectors of a contour limiting the area of the investigated field, respectively with the known function q on this contour,the solution of (4) appears as equivalent to the minimization of the functional

$$F(A)=\iint\limits_{\Omega}\left\{\frac{1}{\mu}\left(\frac{\partial A}{\partial r}\right)^2+\frac{1}{\mu}\,\frac{1}{r}\left(\frac{\partial A}{\partial\varphi}\right)^2-2JA\right\}r\,d\varphi\,dr + \int\limits_{C}2qA\,dC \tag{5}$$

In Eq. (5) Ω denotes the cross-section area of the machine model limited by the contour C with Neumann's b. cond. and $A_\beta=0$ on the singular sectors of C. The minimization of (5) was performed by the finite element

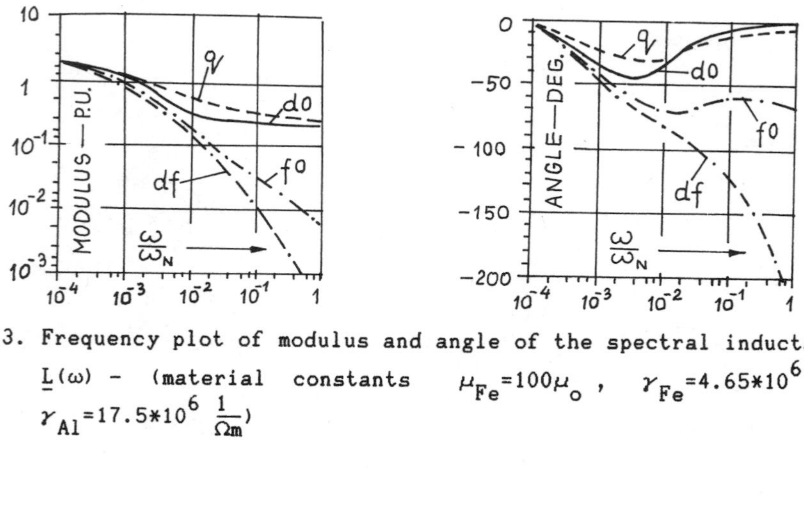

Fig. 3. Frequency plot of modulus and angle of the spectral inductances $\underline{L}(\omega)$ - (material constants $\mu_{Fe}=100\mu_o$, $\gamma_{Fe}=4.65*10^6\ \frac{1}{\Omega m}$, $\gamma_{Al}=17.5*10^6\ \frac{1}{\Omega m}$)

calculation. The area was divided into 378 segmental elements with 420 nodes (Fig. 2). For the distribution of A inside each element yields the polynomial

$$A=\alpha_1+\alpha_2 r+\alpha_3\varphi+\alpha_4 r\varphi \tag{6}$$

Solving the equation set with 840 unknown quantities, having a belt shaped coefficient matrix, we finally get the field distribution in the rotor and the air gap[1]. The flux linked with the armature resp. with the field winding determines the spectral inductances as functions of ω

$$\underline{L}_{d0}(\omega)=\underline{L}_{d0}=\left(\frac{\underline{\Psi}_d}{\underline{I}_d}\right)_{i_f=0} \quad ; \quad \underline{L}_{f0}(\omega)=\underline{L}_{f0}=\left(\frac{\underline{\Psi}_f}{\underline{i}_f}\right)_{I_d=0}$$

$$\underline{L}_{df}(\omega)=\underline{L}_{df}=\left(\frac{\underline{\Psi}_d}{\underline{i}_f}\right)_{I_d=0} \quad ; \quad \underline{L}_q(\omega)=\underline{L}_q=\frac{\underline{\Psi}_q}{\underline{I}_q} \tag{7}$$

shown in Fig. 3 as frequency plots of modulus and of phase angle.

From (7) result the spectral fundamental transmittances of the synchronous machine

$$\underline{L}_d=\left(\frac{\underline{\Psi}_d}{\underline{I}_d}\right)_{u_f=0}=\underline{L}_{d0}-\frac{j\omega\underline{L}_{df}^2}{j\omega\underline{L}_{f0}+R_f} \quad ; \quad \underline{G}=\omega_N\left(\frac{\underline{\Psi}_d}{\underline{u}_f}\right)_{I_d=0}=\frac{\omega_N\underline{L}_{df}}{j\omega\underline{L}_{f0}+R_f} \quad ;$$

$$\underline{H}=R_f\left(\frac{\underline{i}_f}{\underline{u}_f}\right)_{I_d=0}=\frac{R_f}{j\omega\underline{L}_{f0}+R_f} \tag{8}$$

presented in Fig. 4 as frequency plots.

The approximative operational transmittances of the machine to be evaluated have the form of rational polynomial fractions

$$L_d(p)=L_d\prod_{i=1}^{n_d}\left(\frac{1+p\tau_d^{(i)}}{1+p\tau_{d0}^{(i)}}\right) \quad ; \quad L_q(p)=L_q\prod_{i=1}^{n_q}\left(\frac{1+p\tau_q^{(i)}}{1+p\tau_{q0}^{(i)}}\right)$$

$$G(p)=\frac{\omega_N L_{df}}{R_f}\,\frac{\prod_{i=1}^{n_g}\left[1+p\tau_{g(i)}\right]}{\prod_{i=1}^{n_d}\left[1+p\tau_{d0}^{(i)}\right]} \quad ; \quad H(p)=\frac{\prod_{i=1}^{n_h}\left[1+p\tau_{h(i)}\right]}{\prod_{i=1}^{n_d}\left[1+p\tau_{d0}^{(i)}\right]} \tag{9}$$

The equivalent time constants in (9) were approximated by the minimization of the mean square errors in the modulus and in the phase angle of the frequency plots (8) and (9). Relatively good accuracy was obtained with $n_d=3$ and $n_q=4$, $n_g=2$. The operational approximative transmittances enable a simple calculation of the transients at constant rotational angular speed ω_N and at stationary conditions before the perturbation. When the equivalent time constants, being real values, obey some rules according to a R, L network (the phase minimum property corresponding to the condition of the positive time constants, interlacing property of zeroes and poles

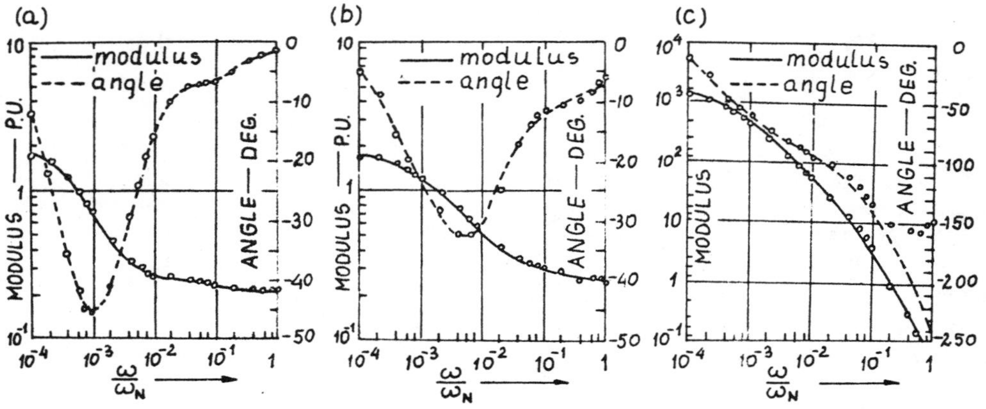

Fig. 4. Frequency plots of modulus, angle of the fundamental spectral transmittances and of the approximated operational transmittance.

(a) $\underline{L}_d(\omega)$, $L_d(p=j\omega)$, $L_d(p)=1.73\frac{(1+p1.81)(1+p0.74)(1+p0.0286)}{(1+p8.13)(1+p1.125)(1+p0.034)}$ p.u ;

(b) $\underline{L}_q(\omega)$, $L_q(p=j\omega)$,
$L_q(p)=1.67\frac{(1+p3.08)(1+p0.45)(1+p0.14)(1+p0.081)}{(1+p4.88)(1+p1.0)(1+p0.23)(1+p0.108)}$ p.u ;

(c) $\underline{G}(\omega)$, $G(p=j\omega)$, $G(p)=1279\frac{(1+p1.27)(1+p0.0016)}{(1+p8.13)(1+p1.125)(1+p0.034)}$.

(Time constants in seconds; o — approximated points)

in L(p) and H(p), satisfaction of the relationship $n_g \leqslant n_d$) it is possible to synthesize an approximative multiloop equivalent circuit of the machine composed of lumped R, L elements. The time constants present in this case the synthetic parameters of the multiloop network machine model. The really distributed constants involve discrepancies with regard to the approximative representation, which appear in some transients and other electromagnetic properties[2,3].

SYMMETRICAL SUDDEN SHORT CIRCUIT

The state equation set in operational form for the case of constant field winding voltage is

$$\begin{aligned}
\Delta U_d(p) &= R_a \Delta I_d(p) + p\Delta\Psi_d(p) - \omega_N \Delta\Psi_q(p) \\
\Delta U_q(p) &= R_a \Delta I_q(p) + p\Delta\Psi_q(p) + \omega_N \Delta\Psi_d(p) \\
\Delta\Psi_d(p) &= L_d(p)\ \Delta I_d(p) \ ; \qquad \Delta\Psi_q(p) = L_q(p)\ \Delta I_q(p) \\
\Delta i_f(p) &= -\frac{p}{\omega_N} G(p)\ \Delta I_d(p)
\end{aligned} \tag{10}$$

Let us consider in particuliar the field current during short circuit of the stator (after no load at rated voltage $U_{q0}=U_N$, and at field current I_{f0}) because of its great dependence on the accuracy of the model. We take into account the relative typically small value of the armature resistance.

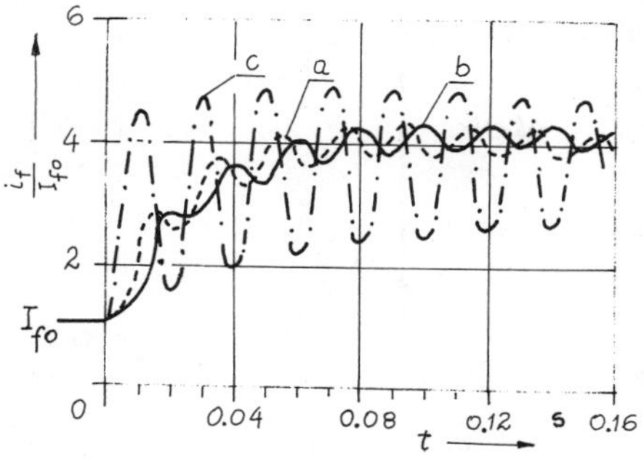

Fig. 5. Field current at sudden short circuit of the stator winding calculated taking into account R_a=0.0018p.u and τ_a=0.386 s.
(a) approximative machine model with lumped electromagnetic constants;
(b) machine model with distributed constants;
(c) simplified model with two equivalent electric circuits in the rotor.

$$\Delta i_f(p) = U_{q0}\,\frac{G(p)}{L_d(p)}\,\frac{1}{p^2 + \frac{2p}{\tau_a} + \omega_N^2} \quad ; \qquad \Delta i_f = \mathscr{L}^{-1}\left[\Delta i_f(p)\right]$$

$$\tau_a \approx \frac{1}{2R_a}\left[L_d(p=\infty) + L_q(p=\infty)\right] \tag{11}$$

On the other hand the field current was computed from the Fourier transform integral using frequency plots $\underline{G}(\omega)$ and $\underline{L}_d(\omega)$ evaluated before by the finite element calculation.

$$\Delta i_f = \frac{U_{q0}}{\pi}\,\mathrm{Re}\left\{\int_0^{\infty} \frac{\underline{G}(\omega)}{\underline{L}_d(\omega)}\,\frac{e^{j\omega t}}{(j\omega)^2 + j\frac{2\omega}{\tau_a} + \omega_N^2}\,d\omega\right\} \tag{12}$$

In Fig. 5 the results from (11) - curve (a) and from (12) - curve (b)

Table 1.

$\frac{i_{fp}}{I_{f0}}$	$\frac{i_{fp*}}{I_{f0}}$	φ_f	φ_{f*}
0.374	0.422	22.3°	57.7°

are presented. The curve (c) corresponds to the simplified machine model with $n_d=2$. Noticeably discrepancy appears in the periodic component. Its accurate shape was calculated for the special case of zero armature resistance. In Table 1 the amplitudes i_{fp} and the initial angles φ_f of the periodic component were compared . The discrepancies appear mainly in the initial angle (*-relates to the approximative model).

SUPPLEMENTARY REMARKS

Greater discrepancies appear in special transients of the machine, particularly when involved in the limits of the operational transmittances at $p \to \infty$. The exact analysis of the eddy current effects in a relatively simple solid iron core indicates that the operational transmittance depends rather more upon $\sqrt{p}$ than on p. This causes discrepancies in the frequency plots for large values of $p=j\omega$ and in the suitable initial shape of transients[4]. Eddy currents at higher frequencies occur mainly in the air gap sided skin of the solid rotor, that reduces the accuracy of the finite element calculation of the field distribution in the turboalternator furthermore.

SUMMARY

The distribution of the radial component of the magnetic flux density on the air gap sided internal stator surface at harmonic impression of the total current either in the armature or in the excitation winding - depends on the current frequency. From these field distribution set can be derived the frequency plot of the spectral transmittances describing the electromagnetic properties of the machine. For computation of the electromagnetic field distribution was applied the finite element method. The field calculations were carried out for a 200 MW generator at harmonic impression of the current either in the equivalent armature winding stationary with respect to the rotor, or in the excitation winding. The calculated spectral transmittances were approximated by the operational transmittances containing the approximative electromagnetic constants of the machine. The short circuit transients were calculated by the operational method and on other hand computed by the direct application of the spectral transmittances in the Fourier integral formula. The considerable dependence of the alternating component in the field current on the adequacy of the solid rotor representation was observed.

REFERENCES

1. J. Staszak, Analysis of the field distribution in the air gap of the synchronous machine by numerical method, (in Polish), Zesz. Nauk. Polit. Swiętokrzyskiej, Elektryka 13, (1983)
2. I. M. Canay, Ersatzschemata der Synchronmachine zur Berechnung von Polradgrossen bei nichtstationären Vorgängen sowie asynchronem Anlauf, BBC Mitteilungen 2, (1969)
3. A. Boboń and W. Paszek, Influence of the eddy current reaction in the solid rotor on the overvoltage in the open circuited field winding at asynchronous operation of the turboalternator, (in Polish), XX-Sympozjum Maszyn Elektrycznych, Kazimierz Dolny, (1984)
4. W. Paszek, Beitrag zur analytischen Erfassung der Ausgleichsvorgänge von Turbogeneratoren mit massiven Läufer, Archiv fur Elektrotechnik 61,(1979)

ANALYSIS OF THE IMPLICATIONS OF MAIN FIELD ENTITIES ON COMMUTATION IN FIELD REGULATED D.C. MOTORS

M. Rizzo[1], A. Savini[2], and C. Zimaglia[3]

[1]Dept. of Electrical Engineering, University, Palermo, Italy
[2]Dept. of Electrical Engineering, University, Pavia, Italy
[3]Dept. of Electrotechnics, Polytechnic, Torino, Italy

ABSTRACT

The paper applies a detailed field analysis by means of the finite element method to the study of the commutation in a d.c. motor speed-regulated by main field weakening; by comparing the resulting flux density diagrams in interpole airgaps to an ideal reference diagram, more reliable predictions about commutation are drawn than from usual evaluations. The method allows also some improved criteria for design purposes to be outlined.

INTRODUCTION

Although a.c. drives are enlarging their application areas, d.c. motors still are, and will be for a reasonable future, the most suitable devices for wide speed and torque regulations, provided one adopts all design and test ways in order to improve their static and dynamic operations. The capability of wide speed ranges at constant power operation is getting a rather central place for industrial d.c. motors, aiming at two purposes: a) to face variable speed and torque requirements within a single working process; b) to accomplish several different working processes, the "constant power" ranges actually enveloping several different "constant torque" ranges. Constant power ranges in speed ratios of 5 and more can be reached in spindle motors. Such a request involves static as well as dynamic consequences. The improvement of dynamical response is a major problem, for which solution criteria have been clearly individuated. From a static point of view, important field weakenings at rated armature voltage involve growing difficulties, as speeds and relevant electromagnetic strength levels grow; but a rather delicate aspect of the question is concerned with ratios more than with limit values of the speed. Precisely, the interpole adjustment (total gap entity), found as optimal in one extreme of the range, may not be suitable in the other; in worst cases, blackbands at extreme values of main flux can have no common points.

The problem is not new in itself, and a classical solution has been suggested in d.c. machine theory: to keep flux densities as low as possible in stator and rotor yokes, where main pole and interpole fluxes get common paths, so that saturation effects can be avoided. Today stronger technical requirements about speed ratios, as said above, as well as constraints about size limitations and application experiences prove that such a criterion is not completely satisfactory, and call for a deeper insight into the question.

On the obvious assumption that the design of rotor, pole face, interpole windings and brush-commutator system allows good separate settlements for commutation in separate operation conditions, the problem resides in matching such settlements when widely different conditions occur. Allowing a satisfactory match, reducing effects of discrepancies and getting optimal compromises are matter of design and adjustment cleverness, case by case. A general advancement on this ground could, however, occur if criteria for a more accurate prediction of possible discrepancies were available. To this purpose, attention should be focused not only on "global", or integral, or mean parameters, as commonly and differently defined and calculated "reactance voltages", but also on " local" detailed evaluations of the quantities to be compensated, as well as of the compensating ones, in different electromagnetic states of the machine. The knowledge of the resulting situations in interpole airgaps could make it possible to predict practical consequences of residual uncompensated effects in each relevant physical context. The present paper goes on along this way. It follows some former ones[1,2] which have applied the finite element method to such an analysis by: i) developing an overall field investigation of the machine, and setting up flux density diagrams in interpole airgaps; ii) taking on an ideal reference for the comparison of diagrams and for the evaluation of residual uncompensated effects. Moving from the provisonal conclusions[2], which pointed out the precision requirements necessary to draw sure indications, a further advancement in the resolution power of the method was pursued, so that discrepancies between analyzed and reference situations could be significantly evaluated and discussed. On the other hand, in order to keep computations within practical limits, a relatively small machine is now studied.

REFERENCE MODEL FOR COMMUTATION

A space model of "good interpole action" is needed, to which the resulting flux density diagram to be drawn in interpole airgap can be compared. The reference diagram should correspond as close as possible to the reactance voltages of armature coils as their commutations go on, so that a close compensation could be achieved; a model for good interpole action clearly supposes a model for commutation.

Let us consider a particular armature coil side, to be denoted by index n, at time t when commutating under a given brush; its instantaneous reactance voltage can be expressed as

$$e_{rn} = \mathcal{L}_n \, di_n/dt \qquad /1/$$

In the stady state operation of the machine (static commutation) t can be replaced by τ_n, a time variable whose definition is restricted only to the commutation interval T_n of coil side n. Inductance $\mathcal{L}_n$ has to be meant in general as a function of time τ_n and of a set $\{i_m\}$ of currents flowing in the coils which are commutating at time τ_n and coincide or are magnetically coupled with coil n:

$$e_{rn} = \mathcal{L}_n(\tau_n, \{i_m\}) \, di_n/d\tau_n \qquad /2/$$

The magnetic coupling concerns coil sides being in the slot where the considered one is located and (via paths through tooth heads, interpole face and airgap) also the preceding and subsequent ones so that the commutation should be properly examined on an "one slot" scale.

To allow external means of compensation to be used, any model for commutation must replace the Lagrangian (internal) viewpoint by the Eulerian (external) one, so that index n shall no longer appear in modelized formulations. The usual way of dealing with steady state commutation is to replace all equations as /2/ simply by:

$$e_r = \mathcal{L} \, \Delta I/T \qquad /3/$$

where:

a) the different commutating coil sides in a slot are equally identified with any one of them;

b) inductance $\mathcal{L}$, anyhow calculated, is constant;
c) current is assumed to vary linearly during the commutation.

The model assumed here, even though still involving simplifications, improves the approach, as:

a') the different commutating coil sides in a slot are replaced by a fictitious one, remaining in commutation as long as any actual coil side in the slot commutates, and being time by time subject to self and mutual reactance voltages as commutations of real coil sides go on; the external viewpoint takes on now a complete image of the set of coil sides in the slot;
b') inductance $\mathcal{L}$, still independent on currents, is a ladder function of a time τ, being defined on the whole commutation interval (or an angle Θ being defined on the whole "commutation zone") of the slot;
c') assumption c) is retained, and the mean speed of commutation is still evaluated according to the interval T of any real individual commutation.

The relevant formulation of the pattern is then:

$$e_r = \mathcal{L}(\Theta)\Delta I/T \qquad /4/$$

which allows local evaluations of reactance voltage, as well as of its differences from the function (as given by field analysis)

$$e(\Theta) \equiv B(\Theta) \qquad /5/$$

It could be remarked that the diagram $B(\Theta)$ actually depends on all present m.m.f.'s, then also on the whole set of commutations in the slot, which in turn depends on the diagram itself; so that, to be exact, an iterative process, starting from the assumption of a linear trend of armature m.m.f. on the commutation zone, should be developed. The pattern improvement brought in by a function $\mathcal{L}(\Theta)$ replacing a constant $\mathcal{L}$ is however sufficient to make even the first results practically valid.

THE MOTOR

The present analysis applies to a motor designed for a spindle-type utilization. Cutting motion drives on tool machines require: i) quite noiseless running at very low speeds; ii) wide ranges at constant rated power; iii) mechanical and electrical time constants as low as possible. Condition i) calls for proper choices of slot numbers and for slot skewing; ii) and iii) require, among other things, the pole face compensation even in small motors. Moreover, condition ii) clearly brings on the problem being dealt with here.

The motor under investigation has been chosen in order to make the analysis easier for some particulars, but it is quite similar to really built and tested spindle motors. A simplification has been introduced as concerns rated performances, with the aim of simplifying the ideal reference. For it, two assumptions help: i) a lap-type rotor winding; ii) a small number of coil sides per slot. Such conditions are allowed by a relatively low rated armature voltage.

The complete rated performance has then been adopted as follows:
58 kW - 240 V - 270 A- 1800/4000 rpm - independent excitation.
The motor has a polygonal,wholly laminated frame (Fig. 1); main design data are as follows: Pole number: 4. Length of stator and rotor cores: 270 mm.
Rotor core: diameters 225/75 mm; 54 open slot, each 6 mm wide, 21 mm high; 13 ventilation holes. Rotor winding: 6 conductors per slot.
Commutator: diameter 185 mm; 162 bars; brush width 12.5 mm.
Commutation zone: 6.98 commutator bar pitches.
Ideal flux density diagram: see Fig. 3 (dashed lines).
Main pole: arc length 105 mm; 6 face slots; total current per slot 540 A.
Interpole: core width 26/16 mm; total current per pole 1620 A.
Gap lengths: main pole 2.5 mm; interpole 4,5 mm (air) + 1.5 mm (nonmagnetic layers).

Excitation m.m.f.'s for rated voltage operation at 1800 and 4000 rpm: 3600 and 1000 A respectively.

FIELD COMPUTATION AND RESULTS

The cross section of the motor is considered, and variations in the axial direction are neglected. Due to fractional numbers of slots and ventilation holes per pole, the analysis of the magnetic field cannot be restricted to one forth of the section, corresponding to a pole pitch, but must be extended to the whole section. The latter is so covered by a fine mesh of first order triangular elements, amounting to 1821 in the whole. The mesh is especially refined in each interpole airgap, where as many as 49 elements are located, in the centers of which the magnetic flux density is evalued. Actual machine geometry and iron magnetic characteristics are carefully taken into account.

The analysis of the field is carried out using a finite element package which is based on the variational energy minimization technique, to get the solution in terms of magnetic vector potential.

The extreme different positions of the rotor teeth against the stator need to be taken into account. Being slots skewed by one tooth pitch on the rotor core length, all positions between the above mentioned ones actually occur for each pole in different planes orthogonal to the axis of the machine, so that conclusions being sought for could reasonably apply to an average of the different evaluated diagrams. As concerns the succession of slots and teeth, averaging has to be made on any two successive poles, whereas the presence of holes should be averaged on all four poles; differences between opposite poles are anyhow negligible. The diagrams obtained present asymmetries due to some residual influences, on the commutating zone of a slot, by the adjacent main poles; the asymmetry of the reference diagram is an inherent one, depending on the choice of a non diametral winding pitch. Figs. 2 give flux density diagrams at rated armature current, for rated main excitation values of 3600 and 1000 A respectively; diagrams a) and c) refer to a slot-centered interpole airgap, while diagrams b) and d) refer to a tooth-centered airgap.

A first rough evaluation concerns the mean values of flux density in single diagrams, namely: Fig. 2a): 0.074 T; Fig. 2b): 0.074 T; Fig. 2c): 0.084 T; Fig. 2d): 0.083 T. It can be remarked that a saturation level of the main flux, expressed by such flux densities as about 2.1 T in rotor teeth and 1.4 T in the stator yoke, together with an armature peripheral current density of about 310 A cm^{-1}, reduces the unsaturated mean values by about 12%.

A more thorough evaluation is pursued on average airgap diagrams, as given in Fig. 3, where the reference diagram is also reported. For it, the maximum and mean values are 0.131 and 0.081 T respectively.

i) (Saturated condition). The calculated mean value, as indicated above, is nearly 8.5% less than the mean ideal one, so that some decrease in interpole total gap might seem at first sight a proper measure. However, knowing $B(\Theta)$ on the whole communication zone, where pieces over the corresponding ideal ones already appear, allows local differences of both signs to be taken into account. Fig. 3a) indicates 0.035 T and 0.055 T as the largest positive and, respectively, negative discrepancy from the ideal reference; the correponding error voltages, at rated basis speed, are 0.23 V and 0.35 V, largely within the normal brush drops. Should an upward translation of the diagram be supposed, to make calculated and ideal maxima coincident, the maximum positive error voltage would become 0.42 V.

ii) (Unsaturated condition). The calculated mean value is now practically equal to the mean ideal one. The extreme errors, as shown by Fig. 3b), are now about ± 0.05 T, corresponding at top rated speed to about ± 1.3 V.

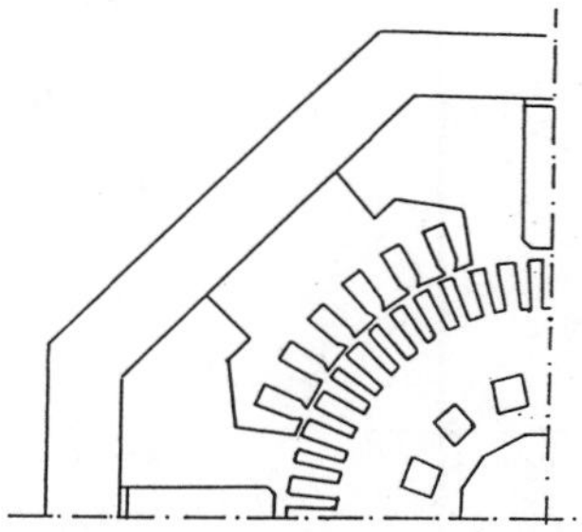

Fig. 1. Schematic cross section of one fourth of the machine.

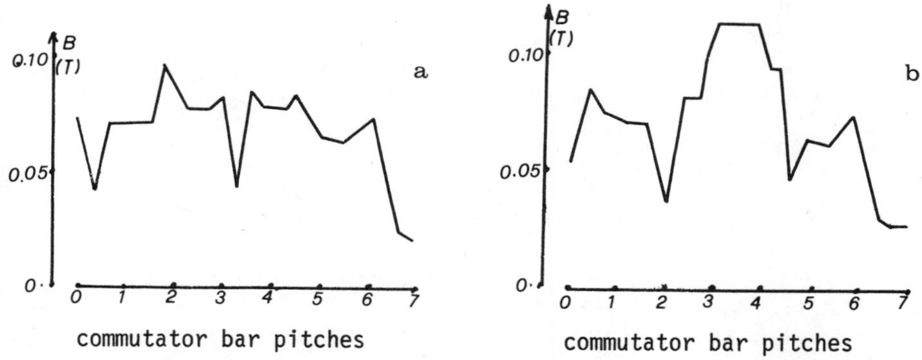

Fig. 2ab. Flux density diagram at rated armature current and total excitation current 3600 A for: a) slot-centered interpole airgap; b) tooth-centered interpole airgap.

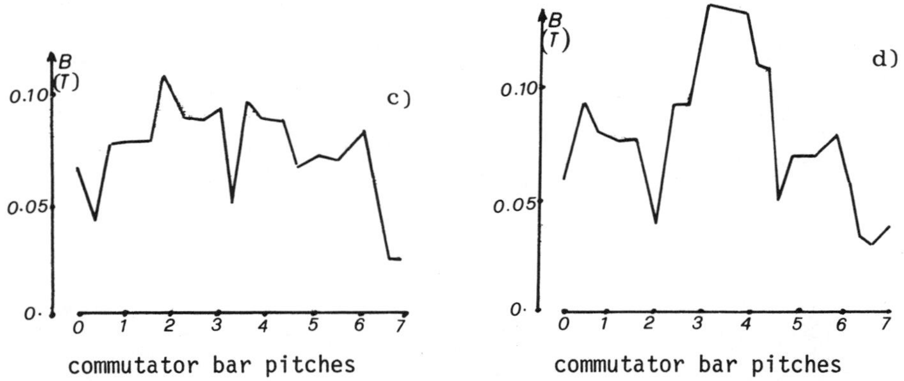

Fig. 2cd. Flux density diagram at rated armature current and total excitation current 1000 A for: c) slot-centered interpole airgap; d) tooth-centered interpole airgap.

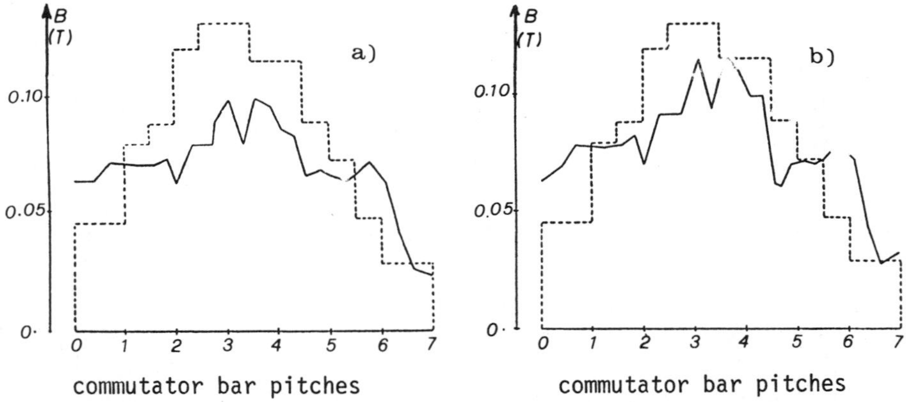

Fig. 3ab. Average flux density diagram in the interpole airgap at rated armature current and: a)total excitation current 3600 A; b) total excitation current 1000 A.

It is worthwhile to note that the field analysis, while able to quantify the mean weakening of interpoles as saturation goes on, also reveals and quantifies local discrepancies opposite to the mean trend (for example, regions of the commutation zone where the action of the interpole is lacking even in unsaturated condition, or excessive even in saturated one). In the present case, error voltages of both signs at top speed, although can be still accepted, are at limits of brush absorption possibilities, while a gap variation in either sense would worsen the situation. The decision of not modifying the design interpole gap could then get rationally justified.

CONCLUSIONS

General remarks about some design aspects are suggested by investigations of the above type.

Lateral discrepancies have been found, having in general opposite sign compared to the central region of the diagrams, so that mutual obstructions to reductions by overall gap modifications rise up.

It should also be considered that the skewing of slots varies the interval between the theoretical position of the brush and the real one; the averaging process should, in a strict sense, be more complex than above, involving also, for a given pole, the cross sections of the machine at both extremities of rotor core. The mutual positions of analyzed and reference diagrams, as have been considered here, can be thought with a sufficient approximation to represent a mean cross section of the machine, from which the other sections as less differ, as the rotor tooth pitch is small.

Practicable tendencies could then concern:

i) numbers of rotor slots as high as possible, to get commutation zones narrower, then more separate from main pole arcs, and also to get small tooth pitches;

ii) narrower brushes, in spite of larger partial reactance voltages, again to get narrower commutation zones;

iii) sharper interpole bevels, with larger total interpole gaps and correspondingly higher compensation ratios along the polar pitch;

iiii) narrower main pole arcs, lower main pole gaps, proper shaping of main pole profiles, in order to reduce leakages towards interpoles.

Of course, such general tendencies should, case by case, be compared to all other specifications being imposed to the motor design.

In any case, a validity should be acknowledged to the method, as helping in predictions at design stage, and in test interpretation for getting an optimal final settlement of the machine. Compared to usual ones, it seems to give a not negligible contribution to consolidate the operation of widely regulated d.c. motors, and even to reasonably approach some range extensions, as applications more and more call for.

REFERENCES

1. M. Rizzo, A. Savini and C. Zimaglia, "Field investigation on a compensated d.c. traction motor by finite element method", Symposium on Modelling and Simulation of Electrical Machines and Converters, Liège, Belgium (1984).
2. M. Rizzo, A. Savini and C. Zimaglia, "Analysis of the commutating field in variable speed d.c. motors", Symposium on Modelling and Simulation of Electrical machines, Converters and Power Systems, Quebec City, Canada (1987).
3. C. Zimaglia, "Rinnovamento dei concetti di impostazione di una moderna serie di motori a c.c.", L'Elettrotecnica, vol. LXX, n. 9 (1983).
4. S. Loutzky, "Calcul pratique des machines electriques à courant continu", Editions Eyrolles, Paris (1963).

4. **ELECTRICAL MACHINES - Induction Machines**

Introductory remarks

E. Andresen

Institut fur Elektrische Energiewandlung
Technicsche Hochschule Darmastadt
Schlossgraben 1 - 6100 Darmstadt, West Germany

In the last decade many efforts have been made to precalculate the characteristic data of induction machines and to predetermine the performance more exactly by means of computer calculation, simulation and modelling. The aim is a better knowledge of what happens in induction motors at steady state and transient performance, when supplied by the mains and by frequency inverters as well. The aim is furthermore to improve the design methods and to increase the efficiency of the machines for the benefit of energy savings.

The geometrical structure of induction machines is rather unfavorable for field calculation. A very small air gap width and different slot numbers of stator and rotor cause big field gradients and thus demand a very fine discretization varying with changing rotor position. The field diffusion problem of the rotor cage winding could be solved under the condition of sinusoidal field quantities. The slots and the nonlinear magnetization characteristic cause current harmonics. The solution requires time discretization of the currents. The end windings and the rotor bar skewing may be of big influence on the performance but require a threedimensional calculation method. It is therefore impossible to develop a true model of the induction machine valid for all problems. It is rather necessary to work out a model appropriate for any problem or any group of problems.

The contributions on electromagnetic fields in induction machines of ISEF'87 included in this chapter can be understood on this background.

An attempt to master the field calculation with one and the same discretization for arbitry rotor positions is made by K. Bill, K. Pawluk and W. Perzanowski. The linear air gap region is treated by boundary elements and thus needs not be discretized by area elements. The nonlinear regions of stator and rotor are calculated by finite elements. The fields of the three regions are determined successively using the results of the preceding calculation as corrected boundary values for calculating the field of the adjacent region. So there are two cycles of iteration, one for the nonlinear process of stator and rotor field and one for coupling the two fields by the air gap field. The paper shows that only the small number of about 15 iterations is necessary to bring about a good convergence. The method is applied on the magnetostatic field at no-load.

A. Demenko presents another numerical method. He follows the aim to consider the third dimension of a squirrel cage rotor by modelling the end parts of the cage through a network with lumped parameters and by coupling the ordinary differential equations of the cage meshes with the partial differential equations of the two-dimensional field. A special parallel matrix processor is used for solving the field equations and the network relations in parallel. So the current density distribution in the cross section of the rotors bars and the bar current time functions of transient processes, e.g. switching on at standstill, can be determined and observed simultaneously.

M.I. Dabrowski has investigated the effect of saturation of teeth and yokes on the air gap flux density distribution in induction machines. This is an analytical approach based on the method of Arnold which determines the effect of tooth saturation on the air gap field. The yoke saturation influence has been neglected in the analytical solutions until now. The author shows the interdependency of the saturation in yoke and teeth and finally gives a graphical solution for determining the pole flux.

An analytical solution of the solid iron rotor with deep bar cage is presented by W. Paszek and A. Kaplon. The aim is to find out the steady state and transient performance, specially the starting up behaviour. The authors replace the slot tooth structure of the rotor by a continuous multilayer model with average permeability and conductivity in the tangential and radial direction. The permeability is assumed independent of the field strength. The cylindrical layers are approached by linear ones. The frequency dependent rotor flux linkage is considered as operational admittance and expanded as infinite series of partial fractions. The equivalent circuit of the machine is thus enlarged by additional secondary meshes. The authors investigate the influence of the number of meshes on the admittance locus and on the transient torque slip curve of an 8 pole 400 kW motor. Moreover the effect of solid steel teeth compared with ideally laminated teeth can be seen.

T. Sliwinski investigates the electromagnetic field in slots of electrical machines generally. By means of the describing partial equations he finds out that the field lines in slots have the same shape regardless where the excited current winding is located in the slot. It is therefore possible to reduce the field problem from two to one dimension, using coordinates running parallel with the flux lines. Dividing the slot in small strips a formula with double series is achieved to determine the permeance of a current carrying slot of arbitrary shape. As a preconditon the permeance of the separate strips must be calculated, e.g. by conformal transformation. The author recommends to approximate the flux lines by circular arcs. Two examples, a trapezoidal slot part and a circular slot, are considered and the formula for approximate calculation are given.

COUPLING OF THE FINITE ELEMENT AND BOUNDARY ELEMENT METHODS BY ITERATIVE TECHNIQUE

Krzysztof Bill, Krystyn Pawluk, and Witold Perzanowski

Department of Fundamental Research
Institute of Electrotechnics
Warsaw, Poland

INTRODUCTION

The magnetic field distribution in the cross-section of an induction motor at no-load is computed. An iterative assembling of the Finite Element Method with the Boundary Element Method is achieved. The field quantities of interest are found as a result of successive separated calculations of stator and of rotor fields. These fields are linked by means of the BEM input and output quantities. An algorithm of the procedure and conclusive estimation of the iteration process are shown in detail.

GENERAL REMARKS

When an electric machine or apparatus cross-section is to be dealt with, subregions of very different material parameters (the air + ferromagnetic, the air + conductor) are encountered. Applying the FEM to the field calculation in the regions that consist of different homogeneous zones the uniform methodical approach is preserved, but often results in a series of problems that must be solved. This fact can easily be accounted for. The magnetic field distribution in an electric machine can be found using the FEM by itself, and the accuracy of field computation in ferromagnetic parts can be recognized as sufficient. It does not refer to the machine air gap. The air gap might be covered with finite elements, or gap elements could be resorted to. It is shown in earlier paper[1] that in this case the core magnetic field is correctly computed. Such a model is absolutely insufficient, when the field of the very air gap is of interest, e.g. its harmonics distribution. The subdivision of the air gap region into elements introduces non-existing harmonics resulting from the discretization. Thus, the necessity of the air gap special treatment arises.

In tackling the problem of investigation of the induction motor magnetic field, the authors intended to introduce the model not distorting the air gap field and taking into account the B-H curve of the stator and rotor core. This task can be solved combining the FEM applied to the stator and rotor with the BEM used in the region of the air gap /as in Fig.1/. A mixed approximation of the boundary quantities is the distinctive feature of the presented method. This approximation, unconventional in BEM, is forced here by the assumption of linear approximation of the magnetic potential over an element in the FEM. That is why, the BEM involves both linear approximation of the potential and constant-valued approximation of its normal derivative. Due to these assumptions the continuity of physical boundary conditions is preserved.

The FEM applied separately to the stator and rotor regions creates no serious difficulties. On the contrary, the field computation of such a very, very slim region as the air gap, when resorting to the BEM involving this unconventional approximation, brings about some problems. Some comprehensive research has had to be done to deal with these problems.

When coupling the FEM with BEM, a direct or iterative technique could be employed. The direct algorithms suggested in paper[2] were to be exploited, but some computational obstacles occured. For this reason 'the double iterative algorithm' has been applied.

ALGORITHM OF ITERATIVE METHOD

In general, the considered algorithm can be referred to as a system of two 'nonlinear' regions with 'a linear' one inserted between them. The algorithm for the purpose of calculation of the induction motor at no-load has been developed.

The iterative algorithm of the field computation is accomplished on the basis of its simpler form pertaining to the circuit model of the motor. Each iteration, based on the field model, has its counterpart based on the circuit one.

Circuit Model

The idea of iterative approach to the field which is due to be computed by two different methods in three subregions (stator, air gap, rotor) may be clearly illustrated on the circuit model given on Fig.2. The iterative calculation process for Φ with $R_{ms}=R_{ms}(\phi)$ and $R_{mr}=R_{mr}(\phi)$ is presented in tab. 1.

Field Model

Fig.1 shows the whole field region under consideration. The circuit quantities: Θ, Φ, V_m, R_m have been succesively replaced by the field quantities: J - current density in the stator windings, A - magnetic

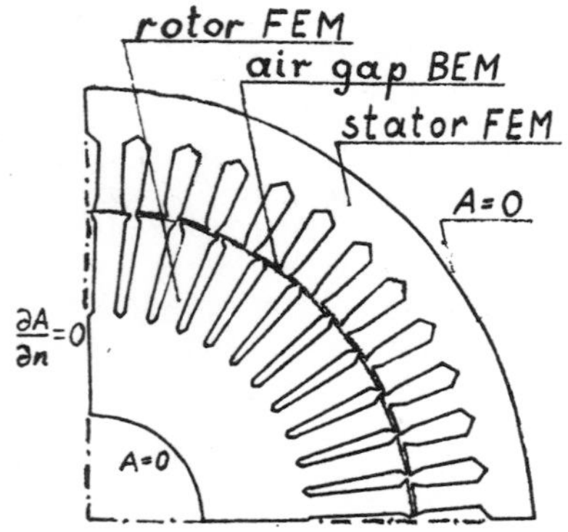

Fig.1. Cross-section of the motor under investigation.

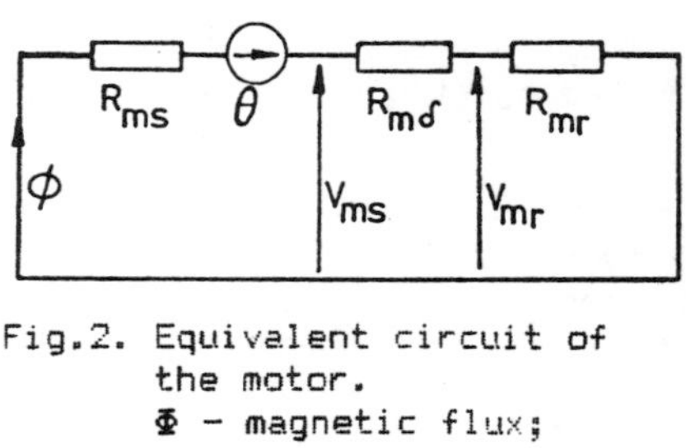

Fig.2. Equivalent circuit of the motor.
Φ - magnetic flux;
Θ - current linkage of the stator windings;
V_{ms}, V_{mr} - stator and rotor magnetic potential differences;
$R_{ms}, R_{m\delta}, R_{mr}$ - stator, air gap and rotor reluctances.

potential, $q=\frac{\delta A}{\delta n}$ the tangential component of the flux density, ν – reluctivity. The iterative calculation process for A with ν depending on the magnetic iron characteristic is presented in tab.2, where B – subscript related to the quantities treated as input or output data in the BEM, F – subscript denoting field quantities obtained from the FEM.

Tab. 1. Iterative process employing the circuit model

The first iteration

step 1	step 2	step 3
$V_{mr}^{(1)} = 0$ since infinite permeability of the rotor core is assumed at the first iteration	$[R_{ms}(\Phi^{(1)})+R_{m\delta}]*$ $*\Phi^{(I)} = \theta$ $\Phi^{(I)}$ – has to be found from non-linear equation	$V_{ms}^{(1)} = \theta - R_{ms}(\Phi^{(1)})\Phi^{(1)}$

The next iterations

step 1	step 2	step 3
$V'^{(I)}_{mr} = R_{mr}(\Phi^{(I-1)})*$ $*\Phi^{(I-1)} \Longrightarrow$ $\Rightarrow V_{mr}^{(I)} = V_{mr}^{(I-1)} + URQ*$ $*(V'^{(I)}_{mr} - V_{mr}^{(I-1)})$	$\Phi^{(I)} = (V_{ms}^{(I-1)} +$ $-V_{mr}^{(I)})/R_{m\delta}$	$V'^{(I)}_{ms} = \theta - R_{ms}(\Phi^{(I)})\Phi^{(I)} \Rightarrow$ $\Rightarrow V_{ms}^{(I)} = V_{ms}^{(I-1)} + URQ(V'^{(I)}_{ms} -$ $+V_{ms}^{(I-1)})$

Tab. 2. Iterative process employing the field model

The first iteration

step 1	step 2	step 3
$q_r^{(1)} = 0$ since $\nu_r = 0$ is assumed in the first iteration	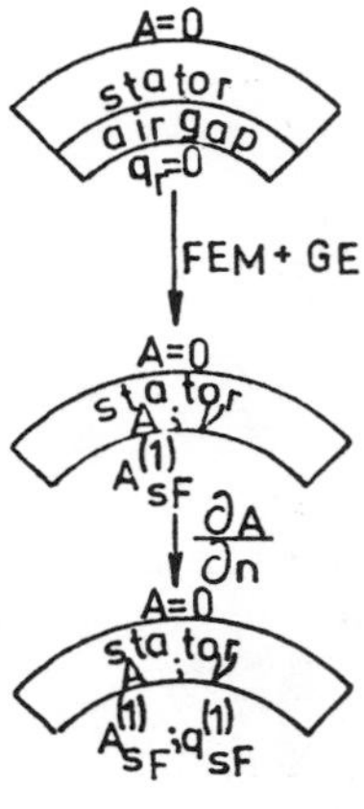	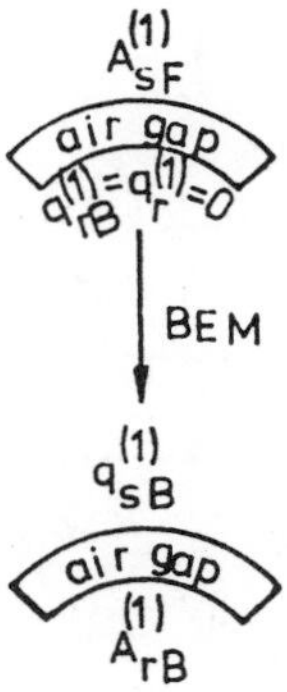

The next iterations

step 1	step 2	step 3
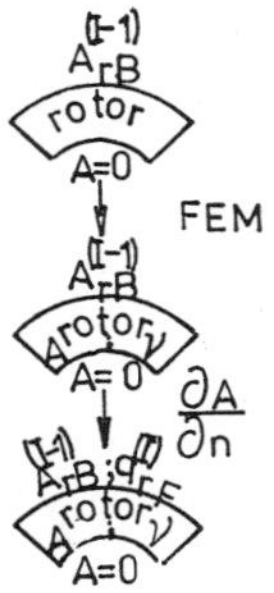	$q_{sB}^{(I)} = q_{sB}^{(I-1)} + URQ * \left(\frac{\nu_F}{\nu_B} q_{sF}^{(I-1)} - q_{sB}^{(I-1)}\right)$ $q_{rB}^{(I)} = q_{rB}^{(I-1)} + URQ * \left(\frac{\nu_F}{\nu_B} q_{rF}^{(I)} - q_{rB}^{(I-1)}\right)$ 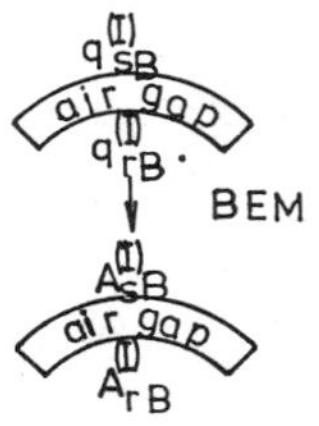	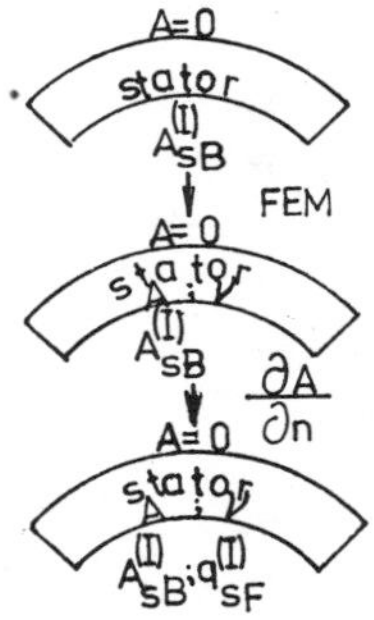

GENERAL IDEA OF THE COMPUTER PROGRAMME

Calculation by FEM

As has already been mentioned, the stator and rotor field was obtained using the FEM. The reluctivity variations $\nu=\nu(B)$ were taken into account at internal loop iterations (Fig.3). The underrelaxation factor of reluctivity was assumed URF=0.25. The iterative process of solving nonlinear set of equations describing the stator and rotor field stops, when the difference C(i) between quantities referring to the error of field computation in two successive iterations was $C(i) \leq 0.03$

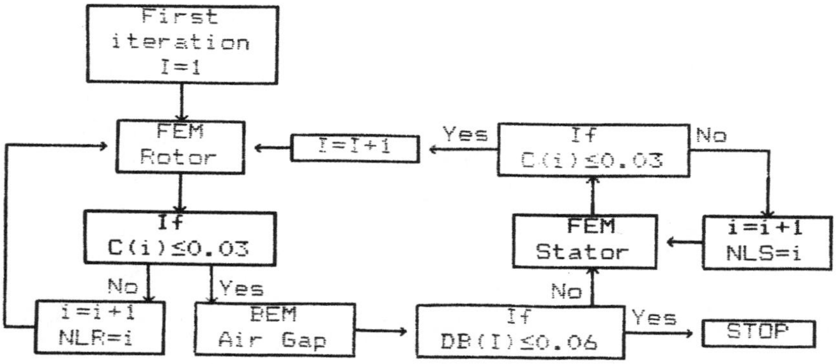

Fig. 3. Flow chart of the double iteration process

Calculation by BEM

The field quantities along the air gap boundary line are computed using the BEM. When the approximation of these quantities is not a constant-valued one, the boundary conditions cannot easily be imposed, in a unique way. The shape of the air gap is, as we know well, very slim. Both these circumstances require special treatment to be involved. Therefore, the least squares method and scaling of the matrix of the BEM equations are employed. To verify both the correctness of the software package and application range of the BEM, numerous tests were carried out.

Coupling the FEM with BEM

The coupling consists in modification of the values of the normal derivative $q_{KB}^{(I)}$ (pertaining to the BEM) by the normal derivative $q_{KF}^{(I)}$ values obtained using the FEM in the rotor and stator region according to formula (1)

$$q_{KB}^{(I+1)} = q_{KB}^{(I)} + URQ * \left[\frac{\nu_{KF}^{(I)}}{\nu_{KB}^{(I)}} q_{KF}^{(I)} - q_{KB}^{(I)} \right] \qquad (1)$$

Where:

I - the number of external loop iteration,

q_{KF}, q_{KB} - the normal derivatives at node K along the air gap boundary,

ν_{KF}, ν_{KB} - the reluctivities at node K set up when approaching towards the boundary line from inside the stator /rotor/ or air gap region respectively,

URQ - the underrelaxation factor of the normal derivatives.

The value of URQ was fixed through tests and equal to 0.17. The values of A_{sB} and A_{rB} along the air gap boundary resulting from the BEM were used during successive iterations I as the boundary conditions for the FEM stator and rotor field computation. The iterative process stops, when condition (2) was satisfied

$$DB(I) = \max_{K} \left[\frac{\nu_{KF}^{(I)}}{\nu_{KB}^{(I)}} q_{KF}^{(I)} - q_{KB}^{(I)} \right] \leq 0.06 \qquad (2)$$

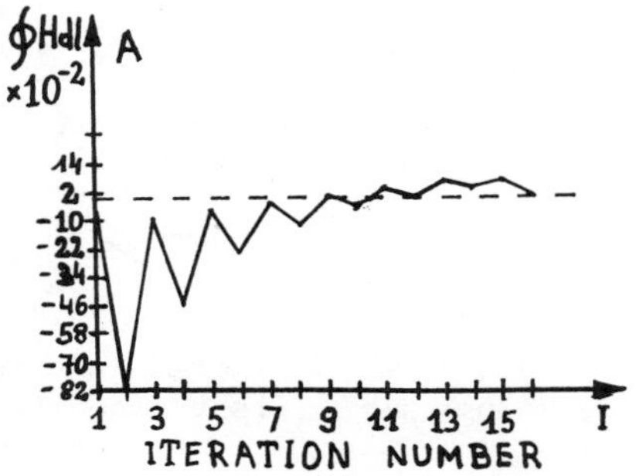

Fig. 4. ∮Hdl along the air gap boundary versus iteration number (outer loop)

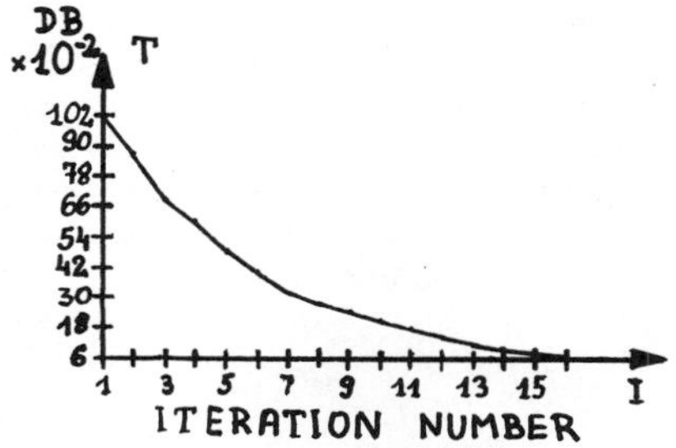

Fig. 5. The greatest difference between input value of q_B in the BEM and the values of q_F calculated from the FEM during iterative process

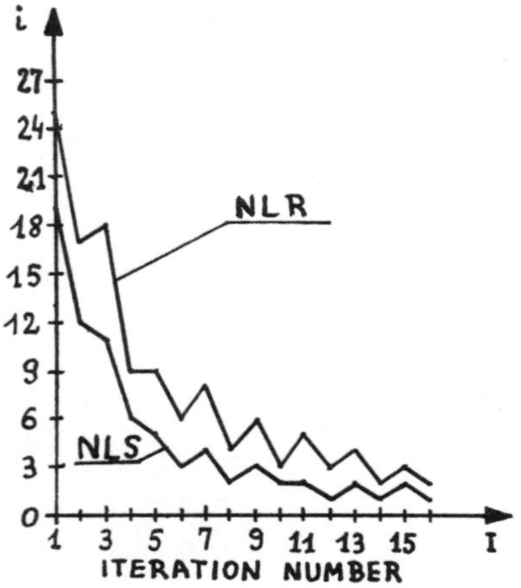

Fig. 6. The numbers of iterations in inner loops pertaining to the stator - NLS and rotor - NLR during the iterative process

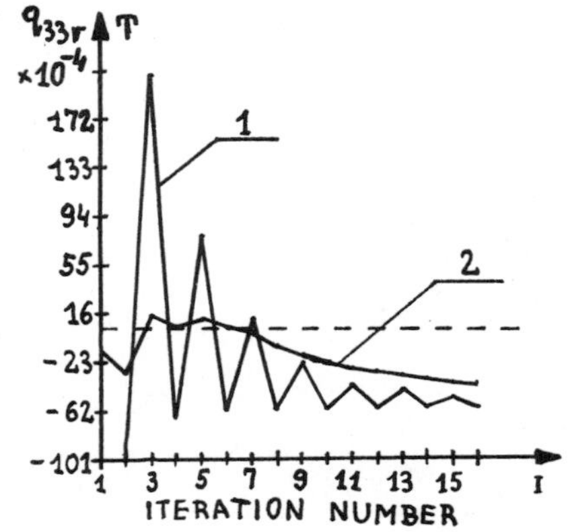

Fig. 7. Tangential component of the flux density at the point adjacent to the tooth tip and the air gap boundary. Obtained from the field-1; input data of the BEM - 2

ESTIMATION OF THE ITERATIVE PROCESS

The value of $\oint Hdl$ oscillating tends to zero (Fig.4.) during the iterative process.

The number of iterations, required to reach the assumed accuracy of the stator and rotor field computations, diminishes in the oscillating manner (Fig.5.), when iterative process in external loop proceeds.

All the figures show that iterative process in both inner loops (Fig.6.), and in the outer one (Figs.4,5) is stable, and the whole double iterative process converges properly.

CONCLUSIONS

The iterative method of coupling the FEM with BEM is developed for the field calculation of the induction motor at no-load. Assuming the set of the stator winding currents, the stator and rotor field are found using the FEM. The nonlinear Poissons (stator) and Laplace's (rotor) equations are solved separately. Mutual influence of the fields is taken into account, by introducing the BEM applied, in turn, to the air gap region. The use of the BEM allows us to avoid discretization of the air gap into elements resulting in the generation of field harmonics, and enables simple modelling of rotor shifting (if required). Having as input data the potentials and their normal derivatives along the air gap boundary, more information could be elicited, namely the field quantities at internal air gap points, which are of great practical interest. The displayed results corroborate fully the general and detailed correctness of the presented iterative technique.

REFERENCES

1. K.Bill, Field Analysis of an Induction Motor at No-Load by Means of the Finite Element Method with Gap Elements, Archiv für Elektrotechnik, 69(1986), pp. 379-384.
2. K.Bill, K.Pawluk, Hybrid Finite/Boundary Element Method Approach to Induction Motor Magnetic Field Analysis, Proceedings of International Symposium on Electromagnetic Fields, Sept.26-28(1985), Warsaw,Poland.

EFFECT OF MAGNETIC SATURATION ON THE AIR-GAP FLUX DENSITY WAVE IN POLYPHASE INDUCTION MACHINES

Mirosław I. Dąbrowski

Electrical Engineering Department
Technical University of Poznań, Poland
ul. Piotrowo 3A, PL-60965 Poznań

INTRODUCTION

Both the performance analysis and the design of electric machines is based more and more often upon the field methods instead of circut theory. However, on the basis of the electromagnetic field theory, it is not possible to obtain analytical relations between functional parameters and structural parameters of a machine. Thus, for implementation of the field methods numerical algorithms are necessary and therefore only approximate results may be obtained. This means that the solution is of the implicit form and not of the usually desired explicit form. Indeed, the solution in the implicit form is not very useful for the design of optimal machines, because in this case we are looking, among other things, for appropriate dimensions of the machine's magnetic circuit. That is why analytical methods for computation of magnetic circuits are still very important, even though they are worked out with some simplifying assumptions.

EFFECT OF THE CORE SATURATION ON THE FLUX DENSITY WAVE

Computations of the flux density distribution in the air gap are based on the magnetomotive force (mmf) equation, which in the case of an induction machine, for the line L_x passing the air gap in a point with coordinate x, may be written as follows[4]:

$$V_\delta(x) + V_{t1}(x) + V_{t2}(x) + 0,5\ V_{y1}(x) + 0,5\ V_{y2}(x) = \theta(x), \qquad (1)$$

where $\theta(x)$ is the magnetic excitation enclosed in the line L_x; other symbols are explained in Fig.1.

When considering the effect of the mmf on the magnetic field distribution it may be assumed that the air gap is uniform on the whole rotor periphery and equal to:

$$\delta' = k_C \delta\ , \qquad (2)$$

where k_C is the Carter coefficient and δ is the real air gap length. Then:

$$V_\delta(x) = \frac{k_C \delta}{\mu_0} B(x) = c_\delta(x). \qquad (3)$$

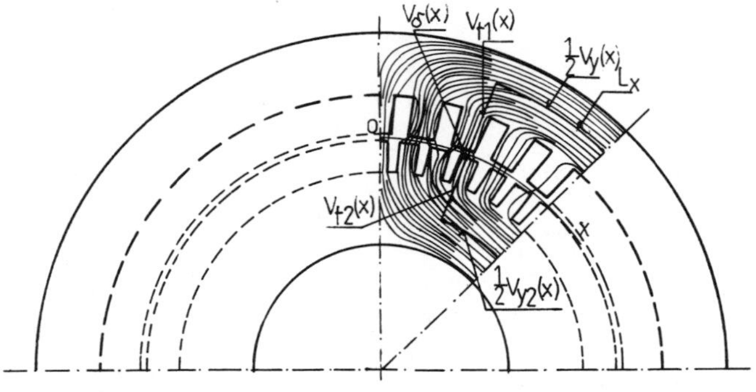

Fig. 1. Magnetic field distribution in polyphase induction machine.

Denoting:

$$V_t(x) = V_{t1}(x) + V_{t2}(x), \tag{4a}$$

$$V_y(x) = V_{y1}(x) + V_{y2}(x), \tag{4b}$$

we obtain the flux density waveform:

$$B(x) = \frac{1}{c_\delta}\,[\theta(x) - V_t(x) - 0{,}5\ V_y(x)]. \tag{5}$$

Graphical interpretation of the equation (5) for the cosinusoidal mmf is shown in Fig.2. The curve $V_t(x)$ of mmf for the teeth is by reason of saturation always somewhat sharper-pointed than sinusoid, whereas the curve $V_y(x)$ of mmf for the yokes is always flatter than sinusoid[1,2]. From these observations the following conclusions may be derived:

- for a given distribution of mmf $V_y(x)$ for the yokes, the sharper the curve $V_t(x)$ is the flatter the line $V_\delta(x)$ is and thus, the flatter also the flux density distribution $B(x)$ is;
- for a given distribution of mmf $V_t(x)$ for the teeth, the greater the participation of the yoke mmf in the whole magnetic excitation curve $\theta(x)$, the less flat the flux density distribution $B(x)$ will be;

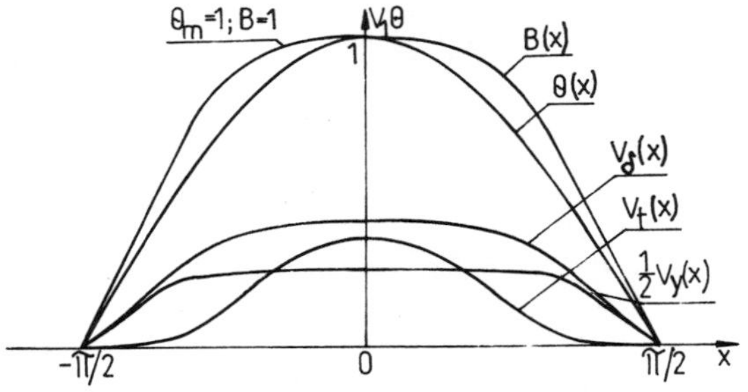

Fig. 2. Distribution of the excitation curve $\theta(x)$, the mmf: $V_\delta(x)$, $V_t(x)$, $V_y(x)$ and the air-gap flux density $B(x)$.

- mmf for the yokes and mmf for the teeth effect the magnetic field distribution just in an opposite manner.

In order to exactly compute the space distribution of the air-gap flux density, and especially - to determine the maximum flux density, an investigation of the distribution of mmf for yokes is necessary.

MAGNETOMOTIVE FORCE ALONG THE LINE L_x IN THE YOKE

Using the following expressions, e.g., for the stator yoke:

$$0,5\ V_{y1}(x) = \int_x^{\pi/2} H_{y1}(x)\,dx, \tag{6}$$

$$H_{y1}(x) = \varphi[B_{y1}(x)], \tag{7}$$

$$B_{y1}(x) = B_{ym1}\frac{\int_0^x B(x)\,dx}{\int_0^{\pi/2} B(x)\,dx}, \tag{8}$$

the distribution of mmf for the yoke may be computed for a given flux density B(x) in the air gap.

From equations (5) ÷ (8) it follows that B(x) can only be computed by the method of successive approximations. In order to investigate the effect of B(x) on the yoke mmf, computer calculations of the curve $V_y(x)$ have been made with assumption that the air-gap flux density wave is: triangular, sinusoidal, rectangular. In Fig.3 one can find functions $H_y(x)$ determined by (7) and (8) for the maximum flux density B_{ym} = 1,2 ÷ 2,05 T in the yoke; the line 1 is the magnetization curve of the chosen magnetic material. Then from (6) the functions 0,5 $V_{y1}(x)$ have been computed. The results in relative values are shown in Fig.4 (by way of example - only according to the sinusoidal air-gap flux density wave). As a basic unit the mmf:

$$0,5\ V_y(0) = \int_0^{\pi/2} H_y(x)\,dx \tag{9}$$

along the line L_x passing through the air in the point 0, has been chosen.

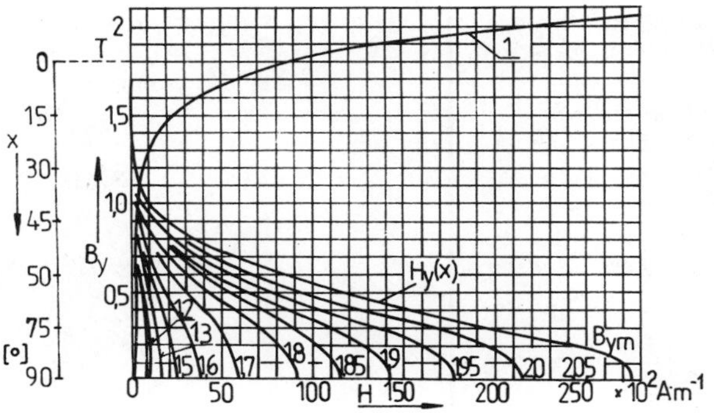

Fig. 3. Magnetic field intensity $H_y(x)$; 1 is the magnetization curve.

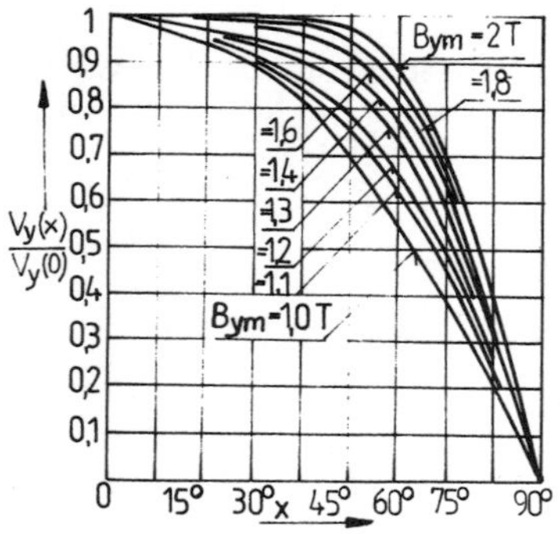

Fig. 4. Relative values of mmf for the yoke - the air gap flux density wave is sinusoidal.

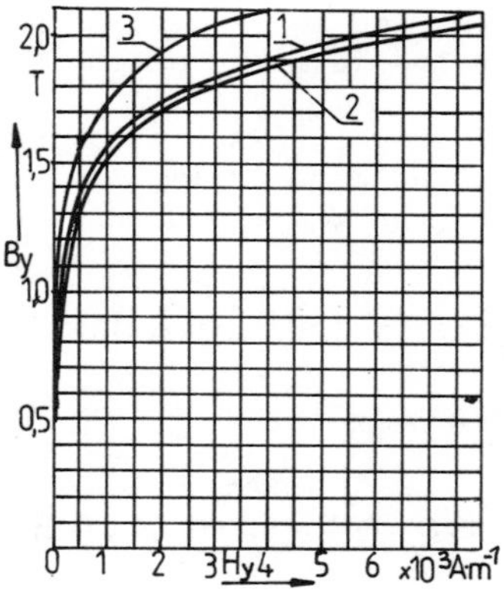

Fig. 5. The effective magnetic field intensity $H_{ye}(B_{ym})$ - the air gap flux density wave is: 1 sinusoidal, 2 triangular, 3 rectangular.

The obtained results clearly show that the greater the maximal flux density B_{ym}, the flatter the mmf curve in the yoke, even if the flux density wave in the air gap varies in a wide range (from triangular to rectangular). The effect of B(x) on mmf curve in the yoke my be seen also from Fig.5 which shows results of computation of the effective magnetic field intensity H_{ye} in relation to the flux density B_{ym}:

$$H_{ye}(B_{ym}) = \frac{2}{\pi} \int_0^{\pi/2} H_y(x,\ B_{ym})\ dx \ . \qquad (10)$$

Above it is shown that the mmf in the yoke sharpens the air-gap magnetic field distribution; it follows from Fig.5 that the sharper the curve B(x) and the total magnetic flux Φ constant, the greater the field intensity H_{ye} in the yoke, and thus, the greater the mmf $0{,}5\ V_y$. These two phenomena are related by an undesirably positive feed-back.

COMPUTATION OF THE MAXIMAL AIR-GAP FLUX DENSITY

In order to compute the maximal flux density B_m in the air gap the author had proposed a method[4] based on a modification of Arnold's approach[6]. In the method given by Arnold the effect of the mmf in the yokes on the flux density B_m was omitted. The idea of the new method is illustrated in Fig.6. The curves in Fig.6 describe: (1) magnetic excitation, (2) mmf in the air gap, (3) sum of the mmfs in the air gap and in the teeth, (4) mmf in the yokes, (5) sum of the mmfs along the line L_x. Curve (5) has been plotted by the method explained in[2].

According to Arnold's approach the saturation coefficient is:

$$k_{sA} = \frac{V_\delta + V_t}{V_t} = \frac{\overline{hk}}{\overline{hi}} \ , \qquad (11)$$

while, in accordance to the method presented in this paper, it is:

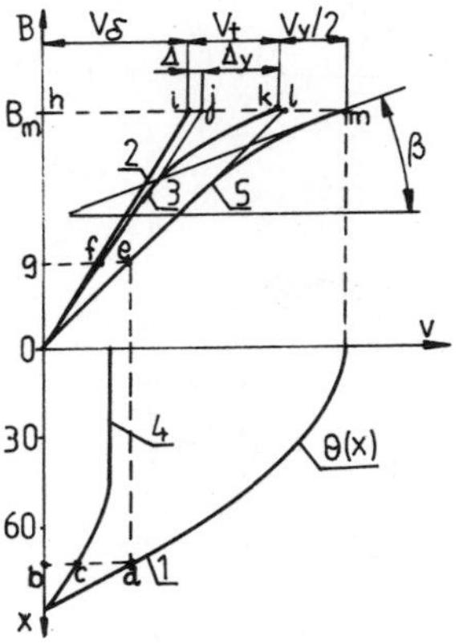

Fig. 6. Diagram to formula (12).

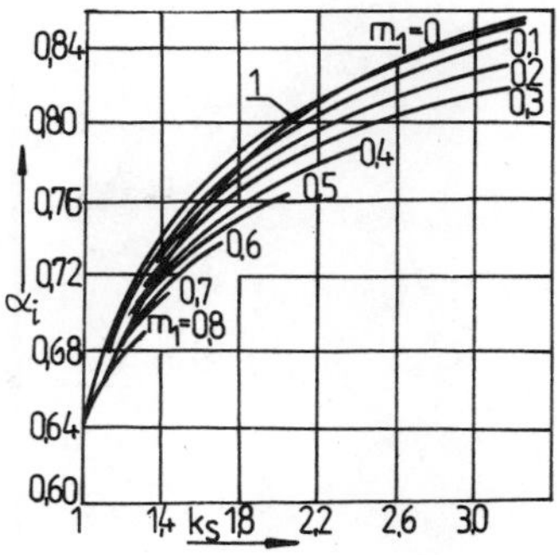

Fig. 7. Coefficient $\alpha = f(k_s, m_1)$, 1 curve given by Arnold.

$$k_s = \frac{V_\delta + V_t + 0{,}5\, V_y}{V_\delta + \Delta + \Delta_y} = \frac{\overline{hm}}{\overline{hl}} \tag{12}$$

In paper[4] it has been proved that if flux density in the yoke is in the range of (1 ÷ 1,6) T, formula (12) my be replaced by:

$$k_s = \frac{2(V_\delta + V_{t1} + V_{t2}) - V_{y1}(1{,}04B_{ym1} - 0{,}74) - V_{y2}(1{,}04B_{ym2} - 0{,}74)}{2(V_\delta + \Delta)}, \tag{13a}$$

where Δ is the linear part of mmf in the teeth (Fig.6), given by:

$$\Delta = 3(B_{ta1}h_{t1} + B_{ta2}h_{t2}) \cdot 10^2, \tag{13b}$$

and: B_{ta1}, B_{ta2} are the average flux densities in the atator tooth and in the rotor tooth, respectively (in T); h_{t1}, h_{t2} are lengths of the magnetic field line in the stator tooth and in the rotor tooth, respectively (in m).

The maximal flux density in the air gap is:

$$B_m = \frac{\Phi}{\alpha_i \tau L_i}, \tag{14}$$

where τ is the pole-pitch, L_i is the ideal core-length and the coefficient:

$$\alpha_i = \frac{2 \int_0^{\pi/2} B(x)\, dx}{\pi B_m} \tag{15}$$

We have computed coefficient α_i as a function of the coefficients k_s and m_1, where $m_1 = \mathrm{tg}\beta$ is slope of the tangent line to the curve (5) in Fig.6 at point m. The curve (5) has been approximated by the function:

$$B = \frac{aV + bV^2}{c + V} \tag{16}$$

expressed in relative values. The coefficients in (16) are[3]:

$$\left.\begin{array}{ll} a = \dfrac{k_s m_1(2k_s - 1)}{d} ; & b = \dfrac{2k_s^2(1 - m_1) - k_s(2 - m_1) + 1}{d} ; \\ c = \dfrac{k_s - m_1(2k_s - 1)}{d} ; & d = 2k_s(k_s - 2) + m_1(2k_s - 1) - 1. \end{array}\right\} \quad (17)$$

Assuming that the mmf expressed in relative values is $V = \cos x$ we obtain:

$$\alpha_i = \frac{2}{\pi}\int_0^{\pi/2} \frac{a\cos^2 x + b\cos x}{c + \cos x}\,dx = \frac{2}{\pi}\left[a + (ac - b)(cA_0 - \frac{\pi}{2})\right], \quad (18a)$$

but for $c^2 > 1$ is:

$$A_0 = \frac{2}{\sqrt{c^2 - 1}} \operatorname{arc\,tg} \frac{\sqrt{c^2 - 1}}{c + 1}, \quad (18b)$$

and for $c^2 < 1$ is:

$$A_0 = \frac{1}{2\sqrt{1 - c^2}} \ln \frac{1 + \sqrt{1 - c^2}}{1 - \sqrt{1 - c^2}}. \quad (18c)$$

Results of computations are shown in Fig.7, where (1) is the curve given by Arnold.

CONCLUSIONS

A method of analysis of the magnetic field distribution and also of the computation of the maximal air-gap flux density as well as of the mmfs in yokes has been presented. Complex interactions between mmfs in elements of the magnetic circuit have been taken into account. The analysis have shown that the magnetic saturation of yokes and the magnetic saturation of teeth have opposite effect on the distribution of the magnetic field. The method is particularly useful for design and iterative optimization of electric machines, i.e., for many computations repeated in a loop with appropriately changed dimensions of elements of the magnetic circuit. Indeed, using the given expressions we can easily deduce how the tooth-slot zone dimensions must be changed in order to minimize the sum of mmfs in the rotor core and in the stator core.

REFERENCES

1. M. Dąbrowski, The effect of the magnetomotive force in a core on the air gap flux density distribution in an induction motor, ZN Politechniki Łódziej. 8:23 (1961). (In Polish).
2. M. Dąbrowski, The effect of magnetic potential in the teeth and yokes on magnetic field in a tree-phase asynchronous machine, Elek. čas. SAV. 2:92 (1964). (In Czech.).
3. M. Dąbrowski, The effect of magnetomotive force in the teeth on magnetic field distribution in the air gap of an asynchronous machine, Roz. Elektrot. PAN. 1:207 (1964). (In Polish).
4. M. Dąbrowski, "Magnetic fields and circuits of the electrical machines", WNT, Warszawa (1971). (In Polish).
5. T. Śliwiński, Berechnung des Magnetisierungsstromes von Asynchronmotoren, A.f.E. 53:299 (1970).
6. R. Richter, "Elektrische Maschinen. Bd. IV", Birkhäuser Ver., Basel (1954).

HYBRID SIMULATION OF ELECTROMAGNETIC FIELD IN SQUIRREL-CAGE WINDING

Andrzej Demenko

Electrical Engineering Department
Technical University of Poznań, Poland
ul. Piotrowo 3a, PL-60965 Poznań

INTRODUCTION

The complete mathematical model of the phenomena in the electromagnetic field of electrical machines should include not only the field equations but also the Kirchhoff's equations which describe the scheme of windings connections. The methods of the formation of such a model and of the electromagnetic field simulation in the windings composed of a series of connected coils have been given in works[1,2]. However, in the presented paper the author describes the method of direct simulation of the electromagnetic field in a region with squirrel-cage windings. The specific hybridism of the method consists in connection of the field and Kirchhoff's equations, and also in application of a hybrid computer system composed with a parallel matrix processor for the field simulation and a conventional digital computer. The systems with a two-dimensional magnetic field have been considered. The sectors of end rings with parts of bars which protrude from the ferromagnetic core have been treated as concentrated resistance and inductance elements.

In order to determine the problem, a mathematical model with partial and ordinary differential equations has been used. This model consists of:

- equation which describes the distribution of the magnetic vector potential A

$$\frac{\partial}{\partial x}(\nu \frac{\partial A}{\partial x}) + \frac{\partial}{\partial y}(\nu \frac{\partial A}{\partial y}) = \sigma(\frac{\partial A}{\partial t} - \frac{U_{pj}}{l}) \quad , \qquad (1)$$

- integral expressions which describe the currents i_{pj} in the bars

$$i_{pj} = R_{pj}^{-1} U_{pj} - \iint_{S_{pj}} \sigma \frac{\partial A}{\partial t} \, dxdy \qquad j=1,2,\dots,z \quad , \qquad (2)$$

- Kirchhoff's equations which define the scheme of bars and end rings connections in the squirrel-cage winding

$$(R_d + pL_d)[k]^T[i_{pj}] + (R + pL)[i_j] + [k]^T[U_{pj}] = [0] \qquad (3a)$$

$$[i_{pj}] = [k][i_j] \tag{3b}$$

Here ν is the reluctivity, σ is the conductivity, l is the length of the magnetic core, S_{pj} is the j-th bar cross-section area, and R_{pj} is the resistance of j-th bar in core. $[k]$ is a matrix which transforms currents $[i_j]$ in the end rings into currents $[i_{pj}]$ in the bars and z is the number of bars. Others symbols are shown in Fig. 1.

To achieve the univocal character of these equations solution, the boundary conditions must be given. In a region with the squirrel-cage winding the boundary conditions are produced by the flux wave travelling around the winding. Usually, this wave is generated under the time-varying voltage constraint or current constraint in the primary windings. Due to this the equations (1), (2), (3) should be solved simultaneously with the field and Kirchhoff's equations for the region of primary windings. The methods of electromagnetic field simulation in this region have already been presented[1,2] and therefore, to simplify the description, only the region with the squirrel-cage winding is going to be considered further on.

METHOD OF SIMULATION

The equations presented above have been solved numerically. The region with the squirrel-cage winding is subdivided into meshes and the partial differential equation (1) is approximated by a system of ordinary differential equations. Then time is segmented and by approximating the differentiation d/dt this system together with equations (2), (3) are transformed into a system of nonlineary algebraic equations which describe the distribution of magnetic flux lA in the nodes of the discretization net and the distribution of currents in bars and end rings for successive time steps t_n[3]. Taking into accound that

$$(R_d + pL_d)[i_{pj}] + [U_{pj}] = [U_j]$$

these nonlineary algebraic equations can be expressed in the following matrix form

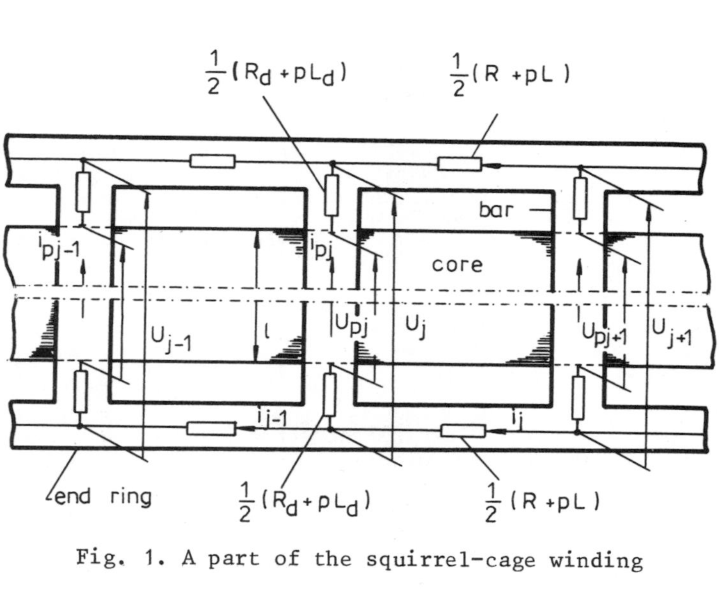

Fig. 1. A part of the squirrel-cage winding

$$\begin{bmatrix} \frac{\Delta t}{2}[R_{un}] + [D] & -[D][k_1]^T & [0] \\ -[k_1][D] & [Y_d] + [R_{pj}]^{-1} & -[Y_d] \\ [0] & -[Y_d] & [Y_d] + Y[k][k]^T \end{bmatrix} \begin{bmatrix} [\Delta\Phi_n] \\ [U_{pn}] \\ [U_n] \end{bmatrix} = \begin{bmatrix} -[\theta_{zn-1}] \\ x_d[i_{pj}(t_{n-1})] \\ x_w[i_{pj}(t_{n-1})] \end{bmatrix} \quad (4)$$

where

$$[\Delta\Phi_n] = \frac{1}{\Delta t}\{[\Phi_n] - [\Phi_{n-1}]\}, \qquad [\theta_{zn-1}] = \frac{1}{2}\{[R_{un}] + [R_{un-1}]\}[\Phi_{n-1}],$$

$$[Y_d] = (R_d + \frac{2}{\Delta t}L_d)^{-1}[1], \qquad [U_{pn}] = \frac{1}{2}\{[U_{pj}(t_n)] + [U_{pj}(t_{n-1})]\},$$

$$Y = (R + \frac{2}{\Delta t}L)^{-1}, \qquad [U_n] = \frac{1}{2}\{[U_j(t_n)] + [U_j(t_{n-1})]\}, \qquad x_w = \frac{2L}{\Delta t}Y - x_d.$$

In the above relations, $[\Phi_n]$ is the vector of values 1A in the nodes of the discretization net for $t=t_n$, $[R_{un}] = [R_u([\Phi_n])]$ is the matrix of reluctances, $[D]$ is the matrix of elementary conductances associated with nodes, $x_d = 2L_d(2L_d + \Delta t R_d)^{-1}$, and $\Delta t = t_n - t_{n-1}$. Matrix $[k_1]$ transforms currents (in the subregions surrounding nodes) into the total currents in the bars.

The specialized parallel matrix processor which is a multiplane network of an analog or a digital type modules (resistance or a microprocessor network[3,4]) has been successfully used for the solution of the equations (4). Part of the node modules of the network parallely solved the field equations and the others simulate the connections of bars and end rings. The portion of the resistance network which can be used for inversing the matrix of coefficients in equations (4) is shown in Fig. 2. The resistors in surface S represent elements of matrix of reluctances. Between surfaces S , S_u there are resistors that imitate the elements of conductances matrix [D]. The conductance of resistors in surface S_c is proportional to the "difference admittance" of end ring sectors.

The system of equations given by (4) is, in general, nonlinear due the

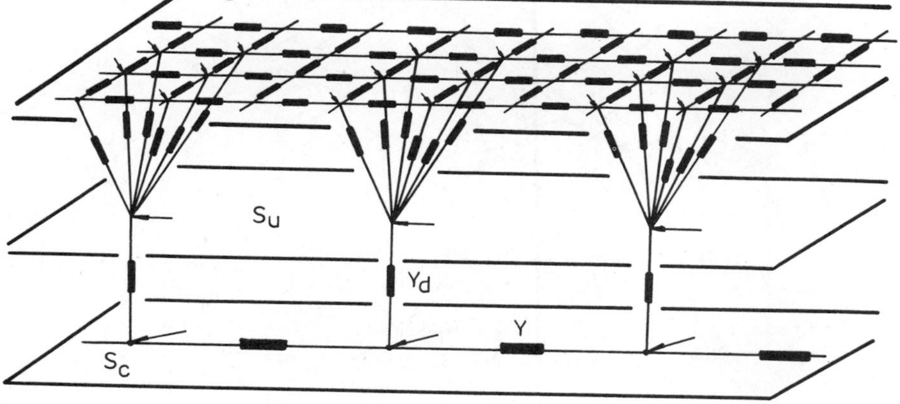

Fig. 2. A portion of the resistance network

saturable ferromagnetic materials. That is why iterative computations are necessary. Two iterative methods are practicable: (a) PS method with solution obtained in the parallel processor (resistance network), (b) PM method with solution obtained in the master processor which controls the parallel network. In the PS method, at first the reluctivity in elements of matrix $[R_{un}]$ is modified by the so called "chord method"[5] and then $[\Delta\phi_n]$, $[U_{pn}]$, $[U_n]$ are solved by the parallel processor. In the PM method, a parallel processor is used for the calculation of amendments $[\delta\Delta\Phi_n^{(k)}]$, $[\delta U_{pn}^{(k)}]$, $[\delta U_n^{(k)}]$ whereas the solution

$$[\Delta\Phi_n^{(k+1)}] = [\Delta\Phi_n^{(k)}] + [\delta\Delta\Phi_n^{(k)}]$$

is computed by the master processor. To assure the maximum rate of convergence of the PM method, the matrix inverted by the network must be similar to the matrix of coefficients in equations (4), but the static reluctivity in elements of the submatrix $[R_{un}]$ should be replaced by a dynamic reluctivity computed for the previous iteration step. In addition, the vector on the right hand side of equations (4) must be replaced by the vector of remains[3]. The decision about the end of the iterative process is taken by node elements in the PS method and by the master processor in the PM method.

EXAMPLE

In order to illustrate the presented method an example of electromagnetic field computation has been given. The transient states produced by the application of supply voltages in the stator windings of small power squirrel-cage motor with motionless rotor have been investigated. The two-phase, four-pole motor with 8 slots in the stator and 11 slots in the rotor has been considered - see Fig. 3. The stator windings consist of 2000 turns per phase and its resistance is equal to 160 Ω. The resistance of rings, bars and the length of the magnetic core are: $R=4\cdot10^{-6}\Omega$, $R_{pj}=0,16\cdot10^{-3}\ \Omega$, $R_d=0,01\cdot10^{-3}\Omega$, $l=75$ mm. The region considered is divided by a network of 738 nodes. The electromagnetic field in the region with the stator windings

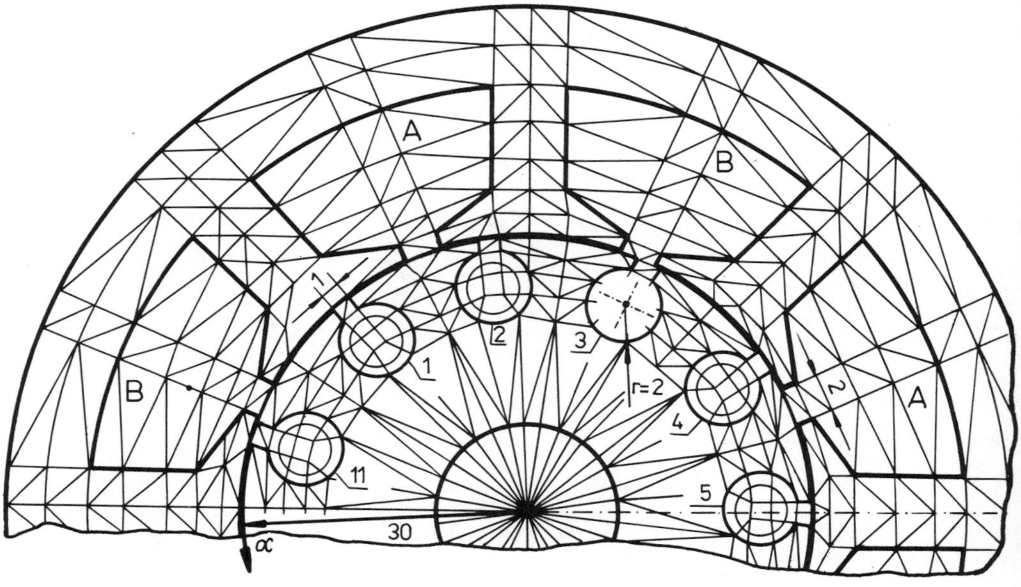

Fig. 3. Geometry of test machine.

is calculated as it is described in the paper[1]. However, the region with the squirrel-cage winding is modelled by the method presented above. Both the regions are simulated simultaneously. The calculations have been realized for sinusoidal voltage switching on stator windings:

$$U_A = U_m \cos(\omega t + \varphi), \qquad U_B = U_m \sin(\omega t + \varphi)$$

where $f=\omega/2\pi=400$ Hz, $\varphi=4{,}05^{\circ}$, $U_m=220$ V.

The calculated currents in rotor bars have been plotted in Fig. 4. The currents can be seen to consist of steady sinusoidal components and transient parts. The steady sinusoidal components are nonsymetrical in consequence of the nonlinearity and differential leakage. In Fig. 5 the calculated distributions of current density j in the p-th bar of the rotor are shown. The additional displacement of the current density in the direction of the stator slot with the winding of the phase in which the current is maximum can be seen. This phenomenon does not appear in the classical analysis of the skin effect in the bars of the squirrel-cage winding.

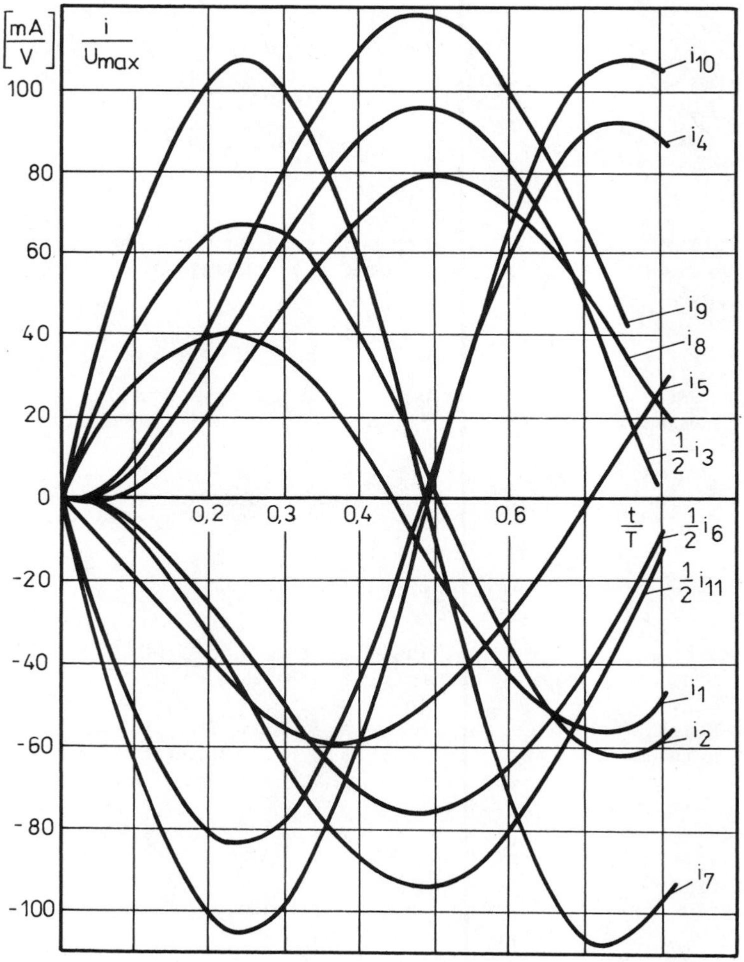

Fig. 4. Currents i_p in the rotor bars. (p is number as in Fig. 3.)

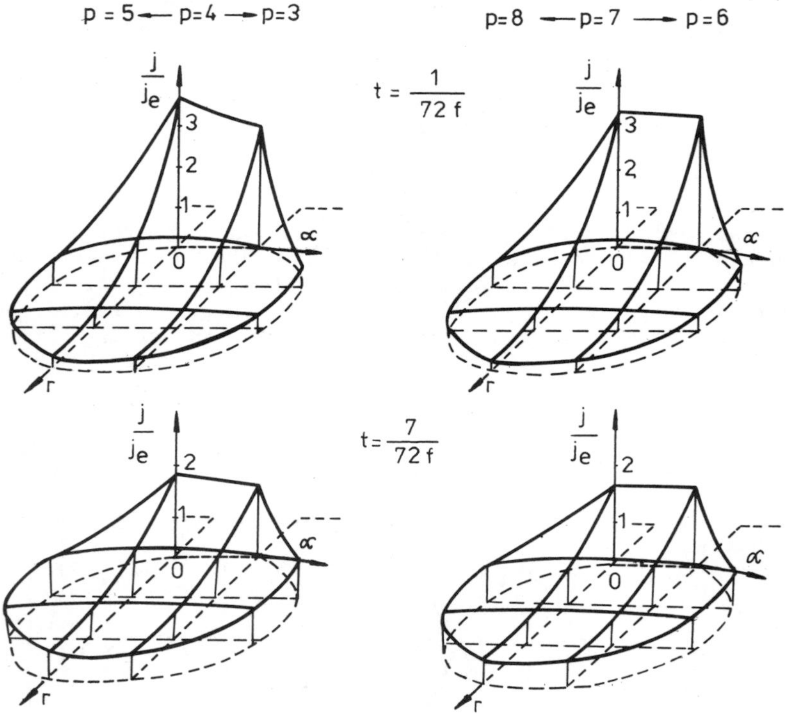

Fig. 5. Distribution of current density j in p-th bar for $t=\frac{1}{72f}$ and $t=\frac{7}{72f}$. (j_e is the average density in the bar.)

CONCLUSION

The presented method can be considered as a performing tool in the magnetic circuits analysis. It can be successfully applied for the calculation of the electromagnetic field in electrical machines with squirrel-cage windings. Due to the use of the network parallel processor the computation time can be reduced, the method of simulation becomes simpler and more visual.

REFERENCES

1. M. Dąbrowski and A. Demenko, The electromagnetic field simulation of multiturn windings by specialized hybrid method, in: "Proceedings of ISEF'85," Electrotechnical Institute, ed., Warszawa (1985).
2. T. Nakata, N. Takahashi, and Y. Kawase, Numerical analysis of magnetic field in loaded electrical machinery connected to a constant voltage sourse, in: "Proceedings of ISEF'85," Electrotechnical Institute, ed., Warszawa (1985).
3. A. Demenko, "Modelling of Electromagnetic Field Diffusion in Electromechanical Converters," WPP Rozprawy 162 (in Polish), Poznań (1985).
4. W. R. Cyre, C. J. Davis, A. A. Frank, L. Jedynak, and V. C. Rideout, WISPAC: a parallel array computer for large-scale system simulation, Simulation, 25, Nov., 11:165-172 (1977).
5. S. C. Tandon, A. F. Armor, and M. V. K. Chari, Nonlinear transient finite element field computation for electrical machines and devices, IEEE Trans. PAS-102, 2:1089-1095 (1983).

INDUCTION MACHINE WITH ANISOTROPIC MULTILAYER ROTOR MODELLING THE ELECTROMAGNETIC AND THE ELECTRODYNAMIC STATES OF A SYMMETRICAL MACHINE WITH DEEP BAR CAGE IN SOLID IRON ROTOR CORE

Wladyslaw Paszek, and Andrzej Kaplon

Technical University of Gliwice, Poland
Technical University of Kielce, Poland

DIFFERENTIAL EQUATIONS OF TRANSIENT STATES

Eddy current phenomena in the deep bar cage situated in the slots of the solid iron rotor complicate immensely the analysis of transients of such symmetrical polyphase induction machines. When the end effects in the rotor including the influence of end rings are neglected, one obtains two dimensional electromagnetic field distributions. In this case the copper bars and the solid iron teeth can be substituted accurately with an anisotropic two-layer continuous secondary structure having different electromagnetic constants (permeability, conductivity) in the tangential, radial and axial direction[1]. The successive layers in the multilayer machine model are: isotropic air gap, the first magnetically anisotropic rotor layer substituting the slot openings and tooth-top space, the second anisotropic layer substituting the deep bars with adjacent solid iron teeth and the isotropic solid iron layer under the slots. The sinusoidally distributed symmetrical m_1-phase winding was assumed in the stator with p pole pairs, $\xi_1 w_1$ effective number of turns per phase winding and an ideal sheeted magnetic nonsatureted core. The orthogonal transformation of the stator phase current in a two phase current $\underline{I}_1(t)$ expressed in new frames attached to the rotor, enables the solution of the electromagnetic field in the rotor, excited by the primary equivalent current sheet fixed relative to the rotor body $a_1(x,t) = -2\sqrt{\frac{m_1}{2}}\,\frac{w_1\xi_1}{\bar{p}\tau_p} I_1(t)\sin(\frac{\pi}{\tau_p}x)$. In the case of cylinder shaped layers, cylindrical functions in the solution of the field distribution occur. Remarkable simplification is obtained substituting the cylindrical layer with a flat one. In the field distribution better behaved hyperbolic functions occur creating a slightly less accurate result in the field solution. The magnetic field-linkage $\Psi_{1\delta}(p)$ of the stator and consequently at a given stator current the operational stator inductance $L_{1\delta}(p) = \frac{\Psi_{1\delta}(p)}{I_1(p)} = \frac{N(p)}{M(p)}$ result from the electromagnetic field distribution on the air gap sided surface. Separating the magnetizing inductance $L_\mu = L_{1\delta}(p=0)$, one obtains[2] the rotor admittance

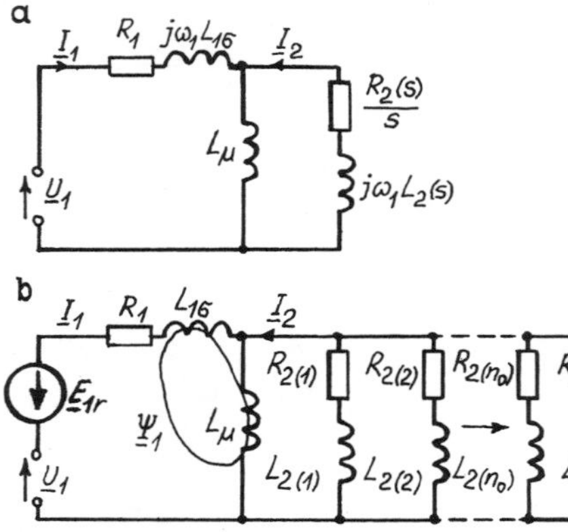

Fig. 1. Equivalent circuit of the induction machine. (a) for steady state behaviour; (b) for transient behaviour.

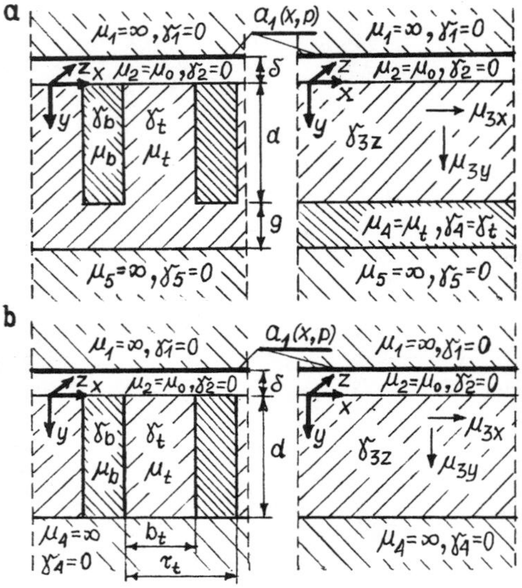

Fig. 2. Substitution of the secondary by anisotropic conducting layers. Deep bars: (a) inside a solid iron layer; (b) between solid iron teeth.

$$\frac{1}{Z_2(p)} = \frac{1}{pL_{1\delta}(p)} - \frac{1}{pL_\mu} \quad (1); \qquad \frac{Z_2(p=js\omega_1)}{s} = \frac{R_2(s)}{s} + j\omega_1 L_2(s) \quad (2)$$

The equivalent circuit (Fig. 1a) of the machine suitable for the determination of the steady state behaviour (at given slip s and impressed symmetrical stator voltage of angular frequency ω_1) results from (1). The operational inductance is a transcendent function of the differential operator p and a meromorphic function with an infinite number of negative real zeroes p_n but is not directly applicable for the calculation of the dynamic transients at variable rotational speed. By the expansion of the operational rotor admittance into the infinite sum of partial fractions we consequently get a structually transformed equivalent circuit of the machine composed of a network of infinite loops.

$$\frac{1}{Z_2(p)} = \sum_{n=1}^{\infty} \frac{1}{R_{2(n)} + pL_{2(n)}} \quad ; \qquad \frac{1}{R_{2(n)}} = - \frac{M(p_n)}{p_n^2 \left[\frac{dN(p)}{dp}\right]_{p_n}} \quad ;$$

$$p_n = - \frac{1}{\tau_{e2(n)}} \quad ; \qquad L_{2(n)} = \tau_{e2(n)} R_{2(n)} \qquad (3)$$

The infinite sum in (3) is quickly convergent, especially for small values of p, hence we get a suitable approximation breaking down the sum after n_o and using an additional two-pole component R_o, L_o for the rejected rest. The resistance R_o results from the superimposed equality of the steady state static conductance of the secondary, as well as in the accurate equivalent circuit as in the approximative representation; the inductance

L_o results[3,4] from the equal equivalent electromagnetic time constant τ_e

$$\frac{\tau_e}{R_2} = -\frac{d}{dp}\left[\frac{1}{Z_2(p)}\right]_{p=0} = \sum_{n=1}^{\infty}\frac{\tau_{e2(n)}}{R_{2(n)}} \; ; \qquad \frac{1}{R_2} = \frac{1}{Z_2(p=0)} = \sum_{n=1}^{\infty}\frac{1}{R_{2(n)}} \tag{4}$$

At a limited number of the two-poles taken into account we get the approximative equivalent circuit composed of lumped R,L elements (Fig. 1b). From this approximative circuit and from the obvious electrical and mechanical constraints we simply derive the ordinary differential equation set in a canonic form describing the electrodynamic system, well adapted for digital integration. Using synchronous rotating frames for the electromagnetic state variables[1] we have

$$\frac{d\underline{\Psi}_1}{dt} = \underline{U}_1(0) - j\omega_1\underline{\Psi}_1 - R_1\underline{I}_1 \; ; \qquad \frac{d\underline{\Psi}_{2(i)}}{dt} = -R_{2(i)}\underline{I}_{2(i)} - j(\omega_1-\omega)\underline{\Psi}_{2(i)}$$

$$\left[\underline{\Psi}\right] = \left[L\right]\left[\underline{I}\right] \qquad i=0,\dots,n_o \tag{5a}$$

$$\frac{d\omega}{dt} = -\frac{\bar{p}}{J}\left[T_e - T_1(\omega)\right] \; ; \qquad T_e = \bar{p}\,\mathrm{Im}(\underline{I}_1\underline{\Psi}_1^*) \tag{5b}$$

EVALUATION OF THE PARAMETERS

Two variants of deep bar cage situation a) inside a solid iron layer - Fig. 2a, b) between solid iron teeth, situated on an ideally sheeted rotor - Fig. 2b, c) inside an ideally sheeted rotor core as results for singular case of iron material constants - Fig. 2b , were taken into account.

The equivalent electromagnetic constants of each anisotropic rotor layer μ_{nx}, μ_{ny}, γ_{nz} result from the parallel paths of bar and the tooth for the current in z direction, the parallel paths of the magnetic permeance in the y direction and the series paths of permeance in the x direction[1]

$$\gamma_{nz} = \frac{b_t}{\tau_t}\gamma_t + \frac{\tau_t-b_t}{\tau_t}\gamma_b \; ; \quad \mu_{nx} = \frac{\tau_t}{\dfrac{b_t}{\mu_t} + \dfrac{\tau_t-b_t}{\mu_b}} \; ; \quad \mu_{ny} = \frac{b_t}{\tau_t}(\mu_t-\mu_b) + \mu_b \tag{6}$$

The field distribution in each layer is described by Maxwell's equations in an operational form (displacement current being neglected)[1] which were solved by means of vector potential

$$\mathrm{rot}\,\mathbb{H}(p) = \vec{n}_z\mathbb{J}(p) \; ; \quad \mathrm{rot}\,\mathbb{E}(p) = -p\mathbb{B}(p) \; ; \quad \mathbb{J}(p) = \vec{n}_z\gamma_z\mathbb{E}(p)$$

$$\mathbb{B}(p) = \mathrm{rot}\,\mathbb{A}(p) \; ; \quad \mathrm{div}\,\mathbb{A}(p) = 0 \; ; \quad \mathbb{A}(p) = \vec{n}_z\mathbb{A}(p) \tag{7}$$

$$A_n(x,y,p) = \left[C_{1n}(p)\exp(-\varepsilon_n y) + C_{2n}(p)\exp(\varepsilon_n y)\right]\cos\left(\frac{\pi}{\tau_p}x\right)$$

where:

$\varepsilon_2^2 = \left(\frac{\pi}{\tau_p}\right)^2$ - yields for the air gap and

$\varepsilon_n^2 = \frac{\mu_{nx}}{\mu_{ny}}\left(\frac{\pi}{\tau_p}\right)^2 + \gamma_{nz}\mu_{nx}p$ - for the each anisotropic layer

$C_{1n}(p)$, $C_{2n}(p)$ are integrating constants which result from the boundaries

$$A_n(p) = A_{n+1}(p) \tag{8}$$

$$\vec{n}_y\left[\frac{1}{\mu_{(n+1)x}}\mathrm{rot}_x A_{n+1}(p) - \frac{1}{\mu_{nx}}\mathrm{rot}_x A_n(p)\right] = \begin{cases} a_1(x,p) \text{ -between stator and air gap} \\ 0 \text{ - elsewhere} \end{cases}$$

Further

$$H_{2x}(x,y,p) = \frac{1}{\mu_2}\frac{\partial A_2}{\partial y} \; ; \quad H_{2y}(x,y,p) = -\frac{1}{\mu_2}\frac{\partial A_2}{\partial x} \quad \text{yields for the air gap} \tag{9a}$$

and for the anisotropic layers

$$H_{nx}(x,y,p) = \frac{1}{\mu_{nx}}\frac{\partial A_n}{\partial y} \; ; \quad H_{ny}(x,y,p) = -\frac{1}{\mu_{ny}}\frac{\partial A_n}{\partial x} \tag{9b}$$

$$\left[B_{nx},\, B_{ny}\right]^T = \mathrm{diag}\left[\mu_{nx},\, \mu_{ny}\right]\left[H_{nx},\, H_{ny}\right]^T .$$

Taking into account (8) in the equations (7,9) we get the flux linkage of the stator winding

$$\Psi_{1\delta}(p) = \sqrt{\frac{3}{2}}\, w_1 \xi_1 l_i \int_{-\frac{\tau_p}{2}}^{\frac{\tau_p}{2}} B_y(x, y=-\delta, p)\, dx$$

and subsequently the operational stator inductance for rotor variant a) at negligence of a small slot opening layer (for simplification of the resulting formula) - Fig. 2a:

$$L_{1\delta}(p) = \frac{C\mu_2}{\varepsilon_2} \frac{\frac{\varepsilon_4}{\mu_{4x}}\mathrm{sh}(\varepsilon_4 g)\left[\frac{\varepsilon_3}{\mu_{3x}}\mathrm{ch}(\varepsilon_3 d)\mathrm{sh}(\varepsilon_2\delta) + \frac{\varepsilon_2}{\mu_2}\mathrm{sh}(\varepsilon_3 d)\mathrm{ch}(\varepsilon_2\delta)\right] + \frac{\varepsilon_3}{\mu_{3x}}\mathrm{ch}(\varepsilon_4 g)\left[\frac{\varepsilon_3}{\mu_{3x}}\mathrm{sh}(\varepsilon_3 d)\mathrm{sh}(\varepsilon_2\delta) + \frac{\varepsilon_2}{\mu_2}\mathrm{ch}(\varepsilon_3 d)\mathrm{ch}(\varepsilon_2\delta)\right]}{\frac{\varepsilon_4}{\mu_{4x}}\mathrm{sh}(\varepsilon_4 g)\left[\frac{\varepsilon_3}{\mu_{3x}}\mathrm{ch}(\varepsilon_3 d)\mathrm{ch}(\varepsilon_2\delta) + \frac{\varepsilon_2}{\mu_2}\mathrm{sh}(\varepsilon_3 d)\mathrm{sh}(\varepsilon_2\delta)\right] + \frac{\varepsilon_3}{\mu_{3x}}\mathrm{ch}(\varepsilon_4 g)\left[\frac{\varepsilon_3}{\mu_{3x}}\mathrm{sh}(\varepsilon_3 d)\mathrm{ch}(\varepsilon_2\delta) + \frac{\varepsilon_2}{\mu_2}\mathrm{ch}(\varepsilon_3 d)\mathrm{sh}(\varepsilon_2\delta)\right]} \tag{10}$$

For rotor variant b) and c) - Fig. 2b:

$$L_{1\delta}(p) = \frac{C\mu_2}{\varepsilon_2} \frac{\frac{\varepsilon_3}{\mu_{3x}}\mathrm{sh}(\varepsilon_3 d)\mathrm{sh}(\varepsilon_2\delta) + \frac{\varepsilon_2}{\mu_2}\mathrm{ch}(\varepsilon_3 d)\mathrm{ch}(\varepsilon_2\delta)}{\frac{\varepsilon_3}{\mu_{3x}}\mathrm{sh}(\varepsilon_3 d)\mathrm{ch}(\varepsilon_2\delta) + \frac{\varepsilon_2}{\mu_2}\mathrm{ch}(\varepsilon_3 d)\mathrm{sh}(\varepsilon_2\delta)} \tag{11}$$

$$C = 6\,\frac{l_i (w_1 \xi_1)^2}{\bar{p}\tau_p}$$

Substituting (10), (11) into (1), (2), (3) we get the R,L parameters in the equivalent circuit of the rotor.

STEADY STATE AND DYNAMIC PROPERTIES

In Fig. 3 the slip dependent stator admittances of a 8-poles, 3-phases induction motor (6 kV, 400 kW ratings) with rectangular deep bar copper rotor cage (4x53 mm^2) having the following constructive date: ideal

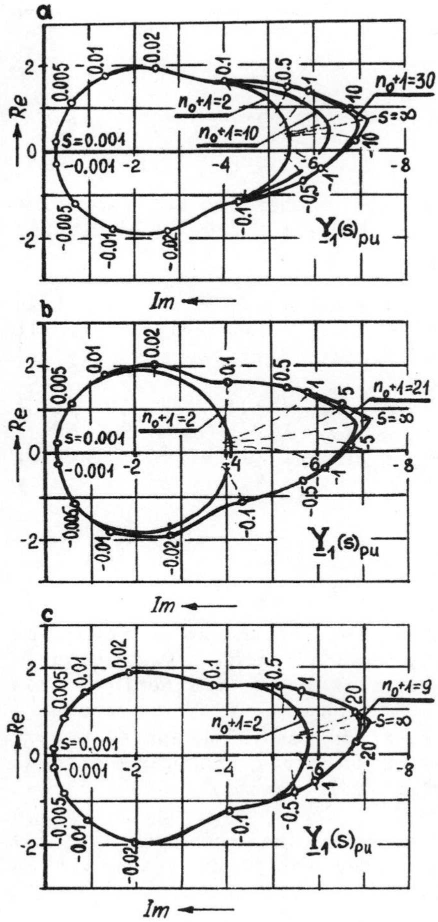

Fig. 3. Slip dependent stator admittance of the induction machine with deep bar cage.
(a) inside the iron rotor;
(b) between solid iron teeth;
(c) in sheeted iron rotor.

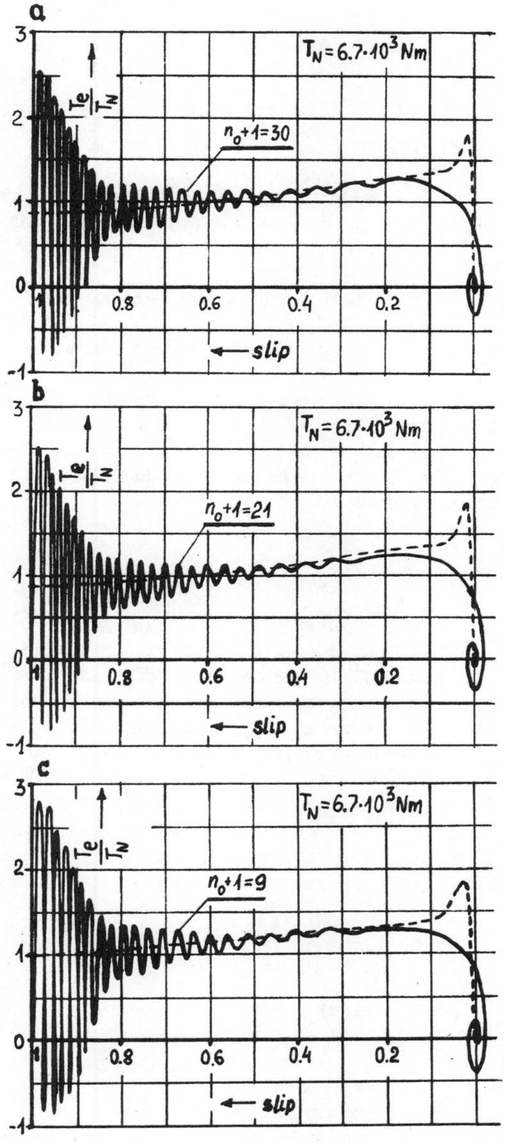

Fig. 4. The trajectory of the working point (the electromagnetic torque T_e – slip) at running up of the motor with no mechanical load after switching on rating supply voltage (cases of rotor structure as in Fig. 3).

length l_i=35 cm, δ=1.5 mm, pole pitch τ_p =27.8cm, slot pitch τ_t =2.53 cm, deep bar height d=53 mm and thickness of the solid iron layer g=8.4 mm are presented.

Fig. 3a corresponds to $\mu_t=100\mu_o$ (equivalent saturated value due to leakage flux), $\gamma_t=4.9*10^6 \frac{1}{\Omega m}$ in Fig. 2a.

Fig. 3b corresponds to $\mu_t=100\mu_o$, $\gamma_t=4.9*10^6 \frac{1}{\Omega m}$ but $\mu_4=\infty$, $\gamma_4=0$ (solid iron teeth) in Fig. 2b.

Fig. 3c corresponds to $\mu_t=\infty$, $\gamma_t=0$ (ideally sheeted rotor core) in Fig. 2b.

The solid iron in the rotor increases the number of R,L two-poles necessary for good approximation of the equivalent circuit by lumped electromagnetic parameters.

Increasing the thickness of the solid layer under the slots we get an increased number of necessary R,L two-poles in the equivalent circuit while the shape of the stator admittance undergoes slight changes only.

Fig. 4 presents the trajectory of an electromagnetic working point (torque-slip) of the machine running up with no mechanical load after switching the stator winding on rating supply voltage (moment of inertia J = 80 kgm^2). The dotted line points out the steady state: torque - slip characteristic.

SUMMARY

The copper deep bar cage situated in the slots of the solid rotor was substituted by an anisotropic multilayer continuous secondary structure. The operational stator inductance of the machine model was derived from the electromagnetic field distribution in the multilayer structure calculated from the solution of Maxwell's equation set. The approximative equivalent circuit results from the expansion of the primary operational admittance into partial fractions. The ordinary differential equation set describing the electrodynamic states results from this approximative circuit and from the obvious electrical and mechanical constraints. The accurate and approximative slip dependent stator admittance as well as the electromagnetic torque in transient states were presented.

REFERENCES

1. W. Paszek and A. Kaplon, Induktionsmaschine mit mehrschichtiger Läuferstruktur als Modell zur Abbildung transienter Vorgänge, 31. Internationales Wiss. Kolloq., Vortragsreihe A3, Heft 1, Ilmenau (1986)
2. W. Paszek and A. Kaplon, Equivalent circuit of the induction machine with a two-layer secondary part, which reproduces equations of electromagnetic states, (in Polish), Zesz. Nauk. WSI, Opole (1985)
3. W. Paszek, Transientes Verhalten der Induktionsmaschine mit Hochstabläufer, Archiv f. Elektrotechnik, 63 (1981)
4. W. Paszek and Z. Pawelec, Ersatzschaltung für transiente Vorgänge der Induktionsmaschine mit Keilstabläufer, Archiv f. Elektrotechnik, 67 (1984)

PECULIARITIES OF ELECTROMAGNETIC FIELD IN SLOTS

Tadeusz Śliwiński

Department of Fundamental Research
Instytut Elektrotechniki, Warsaw, Poland

INTRODUCTION

A lot of papers have been published on the electromagnetic field in slots of electrical machines and on the slot leakage. In many books and papers the slot leakage is calculated in a very primitive way neglecting physical principles. Simplified boundary conditions at the slot opening are often assumed thus causing remarkable inaccuracies. Analytical methods are applicable for calculation of slot leakage in the case of negligible skin effect and of bar impedance with the influence of the skin effect only for simple slot shapes (e.g. circular, rectangular, trapezoidal). For often used more complicated slot forms analytical solutions do not exist or are inaccurate. Recently, numerical methods based on variational principles have been introduced for the slot leakage calculation[1,2]. The practical use of the afore-mentioned methods is until now rather limited especially in the case when the skin effect exists.

In this paper some general properties of the electromagnetic field in slots are deduced. It is shown that these properties are useful in practical calculating of the slot leakage inductance in the case when the skin effect may be neglected as well as of the impedance of bar located in a slot taking skin effect into account.

SHAPE OF FLUX LINES

Let us assume that the electromagnetic field in the slot is two-dimentional and the iron is non-saturated. In this case the field equations in the slot are (Fig. 1):

$$\left.\begin{aligned} &\frac{\partial B_y}{\partial x} - \frac{\partial B_x}{\partial y} = \mu_o J_z(x,y) \; ; \qquad J_z(x,y) = [E - E_z(x,y)]\gamma \\ &\frac{\partial E_z}{\partial y} = - \frac{\partial B_x}{\partial t} \; ; \qquad\qquad \frac{\partial E_z}{\partial x} = + \frac{\partial B_y}{\partial t} \end{aligned}\right\} \quad (1)$$

where $J_z(x,y)$ is the current density and $E_z(x,y)$ the electric field strength of the slot field at the point (x,y). E is the outer electric field strength acting on the conductors being constant on their whole section area in the slot and γ is the material conductivity.

Three idealized typical cases of the electromagnetic field in slots described by these equations may be distinguished:

(1) slot part with a single bar lying close to its walls - in this case E is constant in the whole section area of this slot part and J_z is variable due to skin effect.

(2) slot part filled uniformly with thin insulated wires connected in series - in this case E as well as J_z are constant in sections of all the conductors; no significant error is committed if the wire and slot insulation and free space between wires are neglected and constant E and J_z assumed in this slot part.

(3) slot part without winding - magnetic field is produced by conductors located between this part and the slot bottom.

The shape of flux lines is described by the equation

$$\frac{dy}{dx} = \frac{B_y}{B_x} \tag{2}$$

From equations(1) total differential of electric field strength is

$$dE_z = \frac{\partial}{\partial t}(B_y dx - B_x dy) \tag{3}$$

It follows from (1),(2) and (3) that the electric field strength and the current density along specific flux line of magnetic field are constant.

Let us analyse now the shape of a flux line in a slot outside the conductors. A narrow strip of conductor embraced by two close flux lines (Fig.1) produces the z-component of magnetic vector potential at point P(x,y)

$$A_{zs}(x,y) = \frac{\mu_o}{4\pi}\int_S \frac{J_z dS}{r} \tag{4}$$

where r is the distance between the point P and the element dS of the strip. The integration shall be performed not only for the whole real conductor strip S but also for all the mirror images of this strip in relation to the slot walls. If the strip contains a homogeneous conductor, the current density J_z is constant in the strip itself and in all its mirror images. Formula (4) may be then written in the form

$$A_{zs}(x,y) = J_z F_s(x,y) \tag{5}$$

where F_s is depending only on the shape and dimensions of the slot. Taking into account that

$$B_x = \frac{\partial A_z}{\partial y} \quad \text{and} \quad B_y = -\frac{\partial A_z}{\partial x} \tag{6}$$

the equation of the flux line produced by all of the n conductor strips is

$$\frac{dy}{dx} = \frac{\sum_{i=1}^{n} J_{zi}\frac{\partial F_{si}}{\partial x}}{\sum_{i=1}^{n} J_{zi}\frac{\partial F_{si}}{\partial y}} \tag{7}$$

In the slot part without conductors the slot walls are equipotential surfaces and the field in this part depends only on the total current in the slot part situated between the slot bottom and the part in question. It follows from (7) that the necessary condition for it is

$$\frac{\frac{\partial F_{S1}}{\partial x}}{\frac{\partial F_{S1}}{\partial y}} = \frac{\frac{\partial F_{S2}}{\partial x}}{\frac{\partial F_{S2}}{\partial y}} = \ \ldots \tag{8}$$

This means that the constituents of the magnetic field produced at a specific point by separate strips go in the same direction. It is to be seen from (7) that condition (8) is satisfied also in the case of the slot part containing conductors with current.

The final conclusion is that the shape of flux lines is identical in a slot part without conductors as in the same slot part filled with conductors at constant current density. The shape of flux lines does not change even if the skin effect appears and the current density is different in separate strips but constant within each strip. This conclusion is of great importance for practical calculation methods. The field picture of a slot part without current may be received exactly e.g. by means of conformal transformation. The same picture is valid also for the same slot part with current in the case when conductivity of the conductor between two close flux lines is constant.

STRIP ELEMENTS

On the basis of the deduced properties of the electromagnetic field in slots the two-dimensional problem described in (1) can be reduced to one dimension. This is done by introducing "strip elements" which are defined as the area contained between two neighbouring flux lines. These elements are similar to the three-dimensional Maxwell's tubes developed by Hammond & others[1,2] into the dual "tubes and slices system". Anyhow, owing to the pecularities of the electromagnetic field in slots these strip elements may be applied also to the conductors with eddy currents (skin effect).

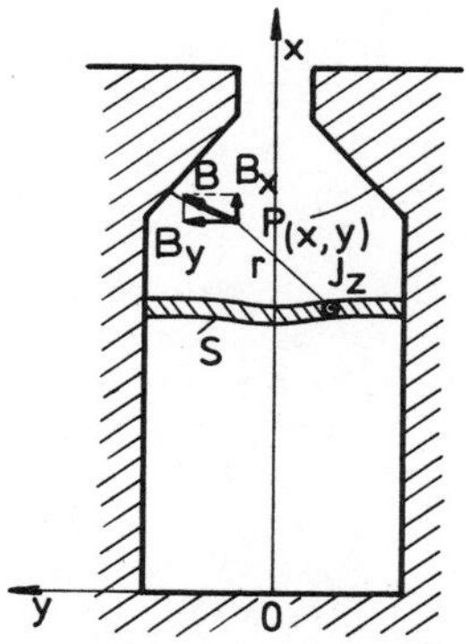

Fig. 1 Electromagnetic field produced in slot by conductor element dS

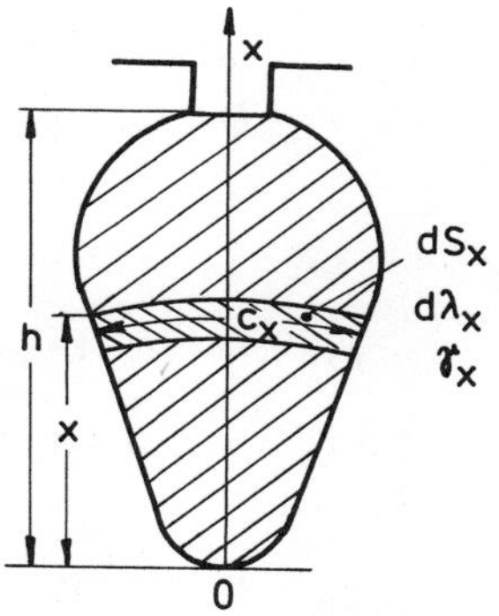

Fig. 2 Conception of strip element

Let us assume at the beginning that the strip element is filled up with a continous conductor or with air. The quantities defining the strip element are (Fig. 2): the surface dS_x, material conductivity γ_x and permeance $d\lambda_x$ of the strip element between both the slot walls. The position of the element is defined by the coordinate x. As permeance (or magnetic conductance) is understood here the ratio of the resultant breadth of the strip to its length i.e. to the middle length c_x of the flux line between the slot walls. In the simplest case of a rectangular strip element its permeance is

$$d\lambda = \frac{dx}{c_x} = \frac{dS_x}{c_x^2} \tag{9}$$

Let us assume now that the source of the electromagnetic field in the slot with a single bar (case 1) is a sinusoidal electric field strength acting on the conductor (i.e. voltage per unit of conductor length). Then the field in the slot is described by following equations

$$\left.\begin{aligned}
&\underline{\Theta}_x = \int_0^x \underline{I}_x \, dS_x \quad ; \quad \underline{J}_x = (\underline{E} - \underline{E}_x)\,\gamma_x \\
&d\underline{\Phi}_x = \mu_o \, \underline{\Theta}_x \, d\lambda_x \quad ; \quad \underline{\Phi}_x = \mu_o \int_x^h \underline{\Theta}_x \, d\lambda_x \\
&\underline{E}_x = j\,\omega\,\underline{\Phi}_x
\end{aligned}\right] \tag{10}$$

where underlined letter symbols are used to designate complex quantities.

It follows from (10) the equilibrium equation for the slot field

$$\underline{E} = \frac{\underline{J}_x}{\gamma_x} + j\,\omega\,\mu_o \int_x^h \left[\int_0^x \underline{J}_x \circ dS_x \right] d\lambda_x \tag{11}$$

The total bar current is $I = \int_0^h \underline{I}_x \, dS_x$ and the bar impedance per unit length is $\underline{Z} = \underline{E}/\underline{I}$.

To enable practical calculations, the discretization of the slot area into n strip elements should be done. Equation (11) is transformed into an equation for the i-th element with the current $\underline{I}_i$

$$\underline{E} = \frac{\underline{I}_i}{\gamma_i S_i} + j\,\omega\,\mu_o \sum_{j=i+1}^{n} \left(\sum_{k=1}^{j} \underline{I}_k \right) \lambda_j \tag{12}$$

A system of n linear equations is received which may be solved in the usual way. But it is more convenient to follow the way shown by Klokov[3] and to calculate the currents I_i successively begining from I_1. The suitable formula is received after a mutual substracting of equations (12) for elements with numbers i and $i+1$

$$\underline{I}_{i+1} = \underline{I}_i \frac{\gamma_{i+1} S_{i+1}}{\gamma_i S_i} + j\,\omega\,\mu_o (\gamma_{i+1} S_{i+1}) \lambda_i \sum_{k=1}^{i} \underline{I}_k \tag{13}$$

In case 2 (wire winding, J_z = const) the resultant permeance of the slot part with winding is

$$\lambda = \frac{1}{\left[\int_0^h dS_x\right]^2} \int_0^h \left[\int_0^x dS_x\right]^2 d\lambda_x \tag{14}$$

and discretization yields

$$\lambda = \frac{1}{\sum_{i=1}^{n} S_i^2} \sum_{i=1}^{n} \lambda_i \left(\sum_{k=1}^{i} S_k\right)^2 \tag{15}$$

Is to be stated clearly that it would be incorrect to calculate the values of λ by (14) or (15) for successive parts of the slot and simply to add them. The resulting slot permeance must be computed for the whole slot part occupied by the winding.

In case 3 (the slot part without winding) the followingformulae hold $\lambda = \int d\lambda_x$ and $\lambda = \sum \lambda_i$. It means that in this case the permeances of separate strip elements may just be added.

ALGORITHMS FOR PRACTICAL CALCULATIONS

The main difficulty in using the strip element method consists in evaluating the permeances of separate strips. A good approximation may be achieved on the base of magnetic field analysis in slot parts of different shapes made by conformal transformation. The results of such an analysis were put together by the author[4,5] In an interesting case (Fig. 3) of a trapezoidal slot part "2" situated between the rectangular parts "1" and "3" with conductors beneath part "3" the resultant permeance is

$$\lambda = \underbrace{\frac{h_1}{b_1} - \Delta\lambda_1}_{\text{"1"}} + \underbrace{\frac{1}{2\beta} \ln \frac{b_3}{b_1}}_{\text{"2"}} + \underbrace{\frac{h_3}{b_3} + \Delta\lambda_3}_{\text{"3"}} \tag{16}$$

The permeance of rectangular slot parts "1" and "3" (confined by the flux lines starting at the slot apices) consists of components h_1/b_1 or h_3/b_3 exactly corresponding to the assumption of straight and parallel flux lines and small corrections: $\Delta\lambda_1$ and $\Delta\lambda_3$. These corrections take into

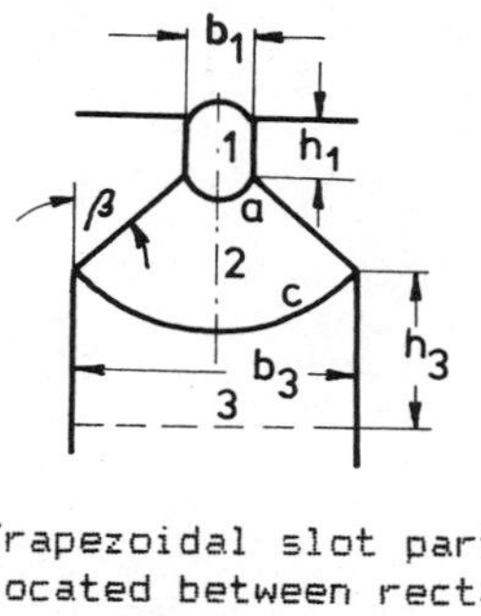

Fig.3 Trapezoidal slot part "2" located between rectangular parts "1" and "3"

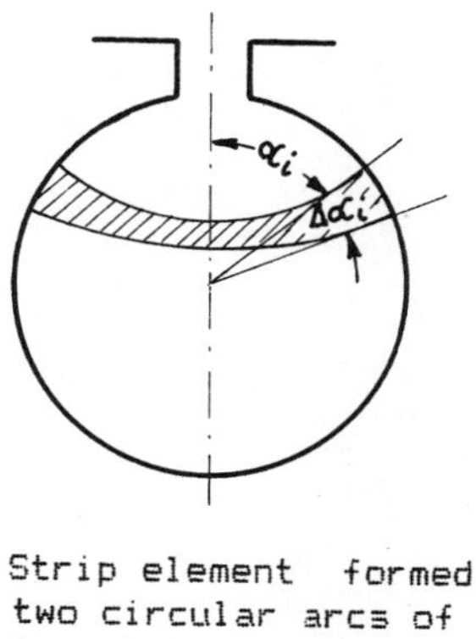

Fig.4 Strip element formed by two circular arcs of different radii

account the flux line deformation in the vicinity of the trapezoidal parts and their exact values have been evaluated by means of the conformal transformation for some values of angle β (Fig.3). For computer calculations formulae have been worked out[5] giving approximated values of $\Delta\lambda_1$ and $\Delta\lambda_2$ as functions of β and b_1/b_3. An amazing conclusion is that the real permeance of slot part "2" (confined by the flux lines "a" and "c") has exactly the same value as the calculated one with the assumption that the flux lines are circular arcs being orthogonal to the trapezoidal slot walls. Basing on these observations, a practical calculation method has been elaborated. In this method it is assumed that the flux lines are circular arcs orthogonal to the slot walls. In the case of rectangular slot parts the flux lines convert into straight lines.

Due to technological reasons the outline of the slots in electrical machines is made of straight lines and circular arcs only. The possible shapes of strip element are therefore limited to three types: 1) rectangular strips, 2) strips of constant breadth formed by arcs of the same radii (in trapezoidal slot parts), 3) strips formed by two arcs of different radii (in circular slot parts). For the two first types of strip elements the permeance is $\lambda_i = S_i/c_i^2$. For the third type the permeancecalculated by the conformal transformation (Fig.4) is given by the formula $\lambda_i = \Delta\alpha_i/\pi \sin \alpha_i$.

In exact calculations the deformation of flux lines in the boundary region between several slot parts of different shapes should be taken into account. This may be done by means of corrections $\Delta\lambda$ similar to (16). In the case of slot parts containing conductors with current the corrections should be made not only on permeances but also on the surfaces in the boundary region. These last corrections (with + or - sign) must be introduced in such a way that the sum of strip surfaces will be exactly the same as the real total slot part surface.

CONCLUSIONS

Basing on the demonstrated pucularities of the electromagnetic field in slots the conception of strip elements confined by flux lines has been introduced and formulae for calculating impedance of a bar located in slot and slot permeance have been derived. Assuming the simplified shape of the flux lines as circular arcs being orthogonal to the slot walls it is possible to use this method to all complicated slot forms including double and triple cages. Numerical results obtained with the use of corrections derived by conformal transformation are quite exact in the case of slot parts without winding and give a good approximation in the case of slot parts with winding.

REFERENCES

1. P. Hammond, M. C. Romero-Fuster, S. A. Robertson, Fast numerical method for calculation of electric and magnetic fields based on potential-flux duality, IEE Proc.A. 132:84 (1985)
2. A. Krawczyk, P. Hammond, Comparison between the boundary-element method and the method of tubes and slices for the calculation of electrical machine parameters, in the Proc.of ICEM'86,Munich
3. B. Klokov, Calculation of skin effect in bars of arbitrary shape, Elektrotechnika. 9:25 (1969) (in Russian)
4. T. Sliwinski, Nutenstreuung elektrischer Maschinen, Wiss.Z.der El. 16:35 (1970)
5. T. Śliwiński, A.Głowacki, Starting performance of induction motors PWN, Warsaw, (1982) (in Polish)

5. MECHANICAL AND THERMAL EFFECTS

Introductory remarks

A. Viviani

Department of Electrical Engineering
University of Genova
via all'Opera Pia 11 - 16145 Genova, Italy

The topic of coupled fields has recently gained increasing interest. It is well known that, on one hand, heat dissipation due to applied and induced currenty accompanies the operation of practically all electromagnetic devices so that heat transfer plays a considerable role in their design; on the other hand, the electromagnetic field gives origin to mechanical stresses and, when applicable, movement in electromechanical devices. This interaction is, actually speaking, reciprocal in that temperature distribution and movement caused by mechanical stress influence, in turn, the electromagnetic field. As a consequence, the investigation of coupled fields, including electric, magnetic, thermal and mechanical fields, is complex and troublesome. Despite the availability of big computers and new methodologies, the solution of the resulting non linear time varying equations in three dimensions for any device is not yet feasible. So, at the moment either simplified structures are considered for solving the whole problem or complicated structures are analyzed, making however approximations on the formulation of the problem. Five contributions are included in the chapter.

The work by S. Iskierka deals with mechanical effects on ferromagnetic conducting media placed in a static magnetic field. A cross section of a magnetic bearing is considered and the force and torque on the moving shaft is evaluated by means of a finite element analysis of the magnetic field. The effects of air gap width, permeability, speed and current distribution are examined.

G. Krusz investigates the only thermal problem in the cross section of a linear motor. Heat generated by conductors in slots is assumed and the resulting temperature distribution obtained by the application of the finite element method to the non linear heat conduction problem.

The goal of the work by Y. Lefevre et al. is the evaluation of local forces of magnetic origin on a device. It is shown how the integration of Maxwell's tensor on a surface partially enclosing a device part allows the calculation of magnetic forces acting on such a part. Experimental results are also reported.

The paper by K. Pienkowski presents the results of the magnetic field analysis in a simplified model of two-sided linear induction motor during d.c. breaking. The equations are solved by Fourier transformations taking into account end effects.

Finally a paper deals with heat produced by eddy currents induced by a time vaying magnetic field. The close coupling of Maxwell equations governing the magnetic field and Fourier equation for heat transfer makes the problem very difficult to solve. The authors, A. Stochniol and V.S. Nemkov, investigate the problem in a long ferromagnetic slab of rectangular cross section in a axially directed magnetic field which is supposed to vary sinusoidally with time. The finite difference method is applied taking into account the effects due to non linear magnetic characteristics and the variations of electric and thermal properties with temperature.

NUMERICAL ANALYSIS OF EFFECT OF MATERIAL AND STRUCTURAL PARAMETERS ON MAGNETIC BEARING PERFORMANCE

Sławomir Iskierka

Technical University of Częstochowa
ul. A. Deglera 35
42-200 Częstochowa, Poland

INTRODUCTION

In recent years the phenomenon of levitation has attracted a lot of attention from engineers. It is applied in many branches of technology, e.g. crucibleless melting of conductors and semiconductors, high-speed ground transportation and frictionless bearings. Although there are many ways of suspension of bodies the means based on utilisation of energy of the electromagnetic or the magnetic field are most commonly applied. This paper presents the analysis of magnetic bearing with a rotating ferromagnetic shaft.

A body is in equilibrium if the sum of forces acting upon the body is equal to zero. Moreover, the state of equlibrium may be stable, unstable or neutral, depending on the fact whether a slight displacement of the body causes its: return to the positon of equilibrium, further movement or motionlessness. In static field of forces $\mathbb{F}(x,y,z)$ the condition of stable equlibrium in the point (x_o,y_o,z_o) is expressed by means of the relations[1].

$$\mathbb{F}(x_o,y_o,z_o) = 0 \tag{1}$$

$$\nabla\, \mathbb{F}(x_o,y_o,z_o) < 0 \tag{2}$$

The condition (1) is the equlibrium condition and (2) is the stability condition. If the electromagnetic field of forces $\mathbb{F}(x,y,z)$ is irrotional then there is a potential $\Psi(x,y,z)$ and the force $\mathbb{F}(x,y,z)$ can be expressed by the relation

$$\mathbb{F}(x,y,z) = -\nabla\, \Psi(x,y,z) \tag{3}$$

The condition of stable equilibrium in the point (x_o,y_o,z_o) assume the forms.

$$\Psi(x_o,y_o,z_o) = 0 \tag{4}$$

$$\nabla^2\Psi(x_o,y_o,z_o) = 0 \tag{5}$$

In order to achieve the state of stable equilibrium of a body suspended in the magnetic field it is necessary to apply a special converter regulating the force of attraction.

ANALYSIS OF THE MAGNETIC BEARING

Basic relations

In the analysis presented below it has been assumed that:
- the magnetic permeability of the stator is infinitely large,
- the magnetic induction vector has only two components,
- the shaft rotates with uniform motion,
- the conductivity γ and the magnetic permeability μ are uniform within the whole area of the shaft's cross-section,
- the forcing is in the form of exciting current linear density distributed on the surface of the stator.

The cross-section of the system under consideration has been presented in Fig.1. The electromagnetic field in a conducting medium moving in a static magnetic field is described by Eq.(6) which results from Maxwell's equations

$$\nabla^2 \mathbb{A} + \mu \gamma \, (\mathbb{V} \times \text{rot}\, \mathbb{A}) = 0 \qquad (6)$$

For the analysis the Cartesian system of coordinates has been assumed. Taking into consideration the assumptions the vector potential $\mathbb{A}$ has only one component A_z and Eq.(6) becomes

$$\nabla^2 A - \mu \gamma \left(v_x \frac{\partial A}{\partial x} + v_y \frac{\partial A}{\partial y} \right) = 0 \qquad (7)$$

For an arbitrary point of the shaft's cross-section $P\ (x_o, y_o)$ the linear velocity may be determined by means of the equations

$$\begin{aligned} v_x &= -2\pi n y_o \\ v_y &= 2\pi n x_o \end{aligned} \qquad (8)$$

Exciting current distribution

In order to control eddy currents accompanied by a force decreasing the force of the shaft suspension in the considerations a forcing with direct current has been assumed. The distribution of the exciting current linear density around the circumference of the stator has been presented in Fig.2.

By expansion of the exciting current linear density into the Fourier series we obtain

$$J_L(\varphi) = \sum_{k=1}^{\infty} \frac{J_L}{k\pi} \left[\cos(k\alpha) - \cos(k\beta)\right] \sin(k\varphi) \qquad (9)$$

Multiplying the exciting current linear density by the winding span we obtain the current which flows in the winding

$$w\, I' = J_L\, R\, (\beta - \alpha)$$

It has been assumed that $w\, I' = I$.

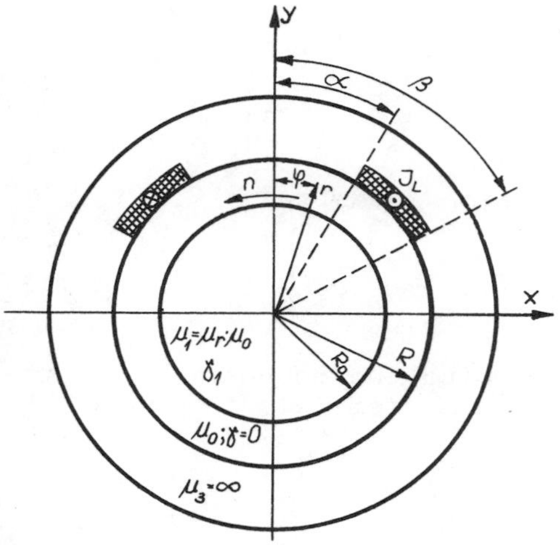

Fig.1 Cross-section of a magnetic bearing

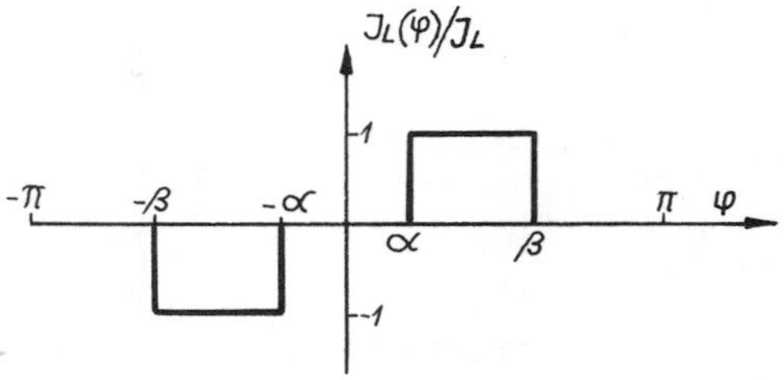

Fig.2 Distribution of the exciting current linear density

<u>Description of the applied method</u>

By application of the Galerkin method to Eq.(7) we obtain the system of equations:

$$\iint \text{grad } A \text{ grad } \psi_m + \mu \gamma \left(v_x \frac{\partial A}{\partial x} + v_y \frac{\partial A}{\partial y}\right) \psi_m \, d\Omega = \oint \frac{\partial A}{\partial n} \psi_m dS \qquad m = 1,2,3,\ldots,N \tag{10}$$

Assuming the base functions in the form

$$\omega_m = \frac{1}{2\Delta}(a_m + b_m x + c_m y) \qquad m = i,j,k$$

$$a_i = x_j y_k - x_k y_j \qquad b_i = y_j - y_k \qquad c_i = x_k - x_j \qquad \Delta = \frac{1}{2}\begin{vmatrix} 1 & x_i & y_i \\ 1 & x_j & y_j \\ 1 & x_k & y_k \end{vmatrix}$$

and interpolating the function A inside the element Ω^E

$$A^E = [\, \omega_i \quad \omega_j \quad \omega_k \,] \begin{bmatrix} A_i \\ A_j \\ A_k \end{bmatrix}$$

for a single element we obtain the system of equations

$$\iint \text{grad } A \text{ grad } \omega_m + \mu \gamma \left(v_x \frac{\partial A}{\partial x} + v_y \frac{\partial A}{\partial y}\right) \omega_m \, d\Omega^E = \oint \frac{\partial A}{\partial n} \omega_m dS \qquad m = i,j,k \tag{11}$$

From the flow law we infer that:

$$\left.\frac{\partial A}{\partial n}\right|_{S=R} = \mu_o \, J_L(\varphi) \tag{12}$$

Interpolation of the boundary condition leads to:

$$\left.\frac{\partial A}{\partial n}\right|_{S=R} = \mu_o \, J_L \left(1 - \frac{s}{d} \right) + \mu_o \, J_L \, \frac{s}{d} \tag{13}$$

For elements having no common nodes with the boundary S of the area Ω the right side of the equation system (11) is equal to zero.

NUMERICAL CALCULATIONS

The force acting on the ferromagnetic shaft can be determined by integration of the forces arising from the exciting field action with an equivalent specific loading on the shaft surface. However, it is more convenient to make use of the principle of reciprocation and calculate the interaction between the exciting current and the field corresponding to the potential $A - A_o$ as has been shown in paper[2]. This field is determined as the difference of the total electromagnetic field, after the introduction of a rotating ferromagnetic shaft, and the exciting field A_o (after the removal of the shaft). The infinitesimal force acting upon the stator may be calculated using the equation

$$d\,\mathbb{F} = \mathbb{J} \times \mathbb{B} \, dV \tag{14}$$

The components of the force $\mathbb{F}$ acting upon the stator are calculated by integration of Eq.(14) around the circumference of the stator and multiplication by the stator length in the Oz axis direction

$$F_x = - l \oint J_L \, B_y \, dS \tag{14a}$$

$$F_y = l \oint J_L \, B_x \, dS \tag{14b}$$

where: $B_x = \dfrac{\partial (A-A_o)}{\partial y}$; $B_y = - \dfrac{\partial (A-A_o)}{\partial x}$

The electromagnetic torque acting upon the stator is calculated by means of Eq.(15) by: vector multiplication of the radius by the infinitesimal force and integration around the stator circumference

$$M = \oint x \, dF_y - \oint y \, dF_x = l \oint x \, J_L \, B_x \, dS + l \oint y \, dF_y \, dS \tag{15}$$

or using the approximate Eq.(16)

$$M = \sum_{i=1}^{E} (x_o' F_{yi} - y_o' F_{xi}) \quad (16)$$

where: F_{xi}, F_{yi} denote the components of the force $\mathbb{F}$ acting on an element of the stator circumference,
x_o', y_o' denote the coordinates of the stator circumference element centre.

Eq.(16) can be obtained assuming that the force acting upon the stator circumference element is concentrated in the element centre. The shaft is affected by: forces and electromagnetic torque equal in value but imposed in opposite directions. The total of power losses connected with eddy currents and magnetization of the shaft may be determined by calculating the power resulting from the electromagnetic braking torque which acts on the shaft.

$$P = 2 \pi n M \quad (17)$$

The power losses caused by the currents induced in the shaft are calculated by means of the equation

$$P = l \gamma \iint (v_x \frac{\partial A}{\partial x} + v_y \frac{\partial A}{\partial y})^2 \, d\Omega \quad (18)$$

The exemplary calculations have been performed for the following parameters: R = 0.02 m, γ_1 = 6 MS/m. The obtained results have been presented in Fig. 3 and Fig. 4.

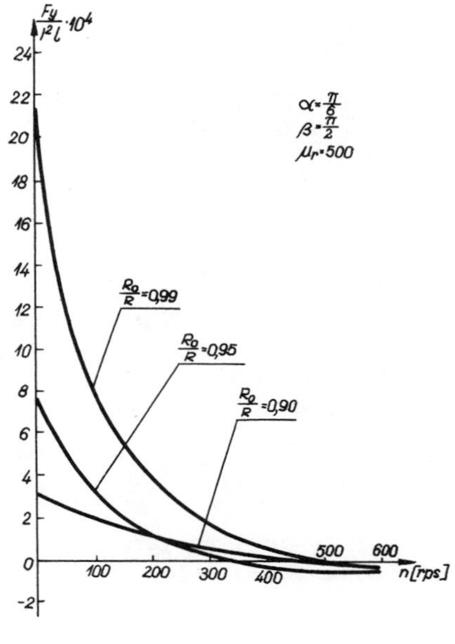

a

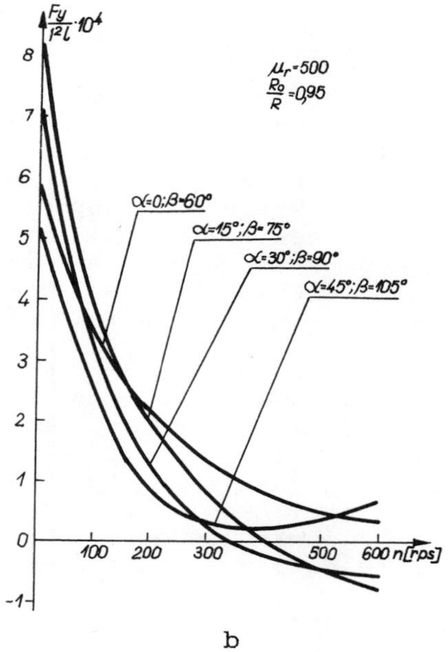

b

Fig.3 The shaft suspension force:
a) effect of the shaft air gap width,
b) effect of the excitation winding distribution.

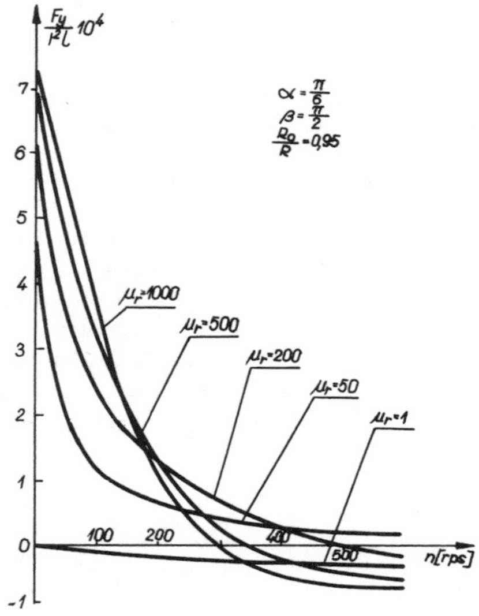

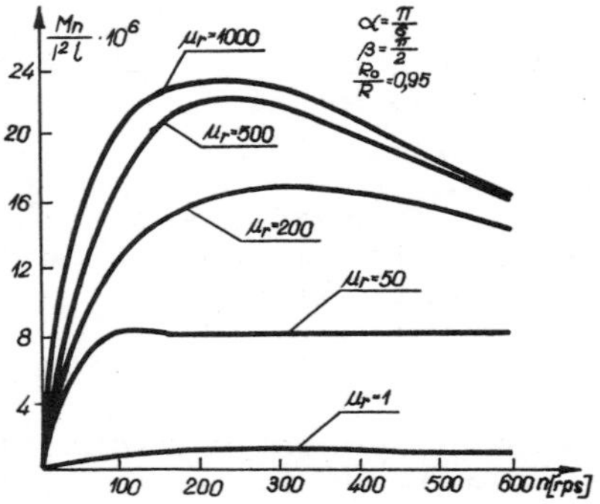

Fig.4 Effect of the shaft magnetic permeability on:
a) the suspension force,
b) the braking torque acting upon the shaft.

CONCLUSIONS

On the basis of the obtained results of calculations and graphs we may conclude that:

- the increase of the shaft magnetic permeability at low-speed rotation of the shaft (up to 150 r.p.s.) is followed by the increase of the force suspending the shaft, whereas at higher rotational speeds the force decreases with the growth of the shaft magnetic permeability, or even it is reversed (the shaft is repelled),
- the rotating shaft is affected by a relatively large force in Ox-axis direction proportional to the rotational speed and the magnetic permeability of the shaft,
- a displacement of the excitation winding is followed by considerable changes of the force suspending the shaft, especially for small rotational speeds of the shaft (up to 200 r.p.s.),
- the shaft suspension force is largely affected by air gap width between the shaft and the stator, especially at low rotational speeds (up to 200 r.p.s.),
- the rotating shaft is affected by a braking torque proportional to the rotational speed and the magnetic permeability of the shaft,
- total power losses in the shaft are proportional to the rotational speed and the magnetic permeability of the shaft.

REFERENCES

1. B. V. Jayawant, Electromagnetic suspension and levitation, IEE Proc., Pt. A, 8:549 (1982).
2. W. Lipiński, R. Sikora, K. M. Gawrylczyk, Magnetische Lagerung rotierender Körper, Z.elektr. Inform.-u Energietechnik, 4:354 (1979).

THE FINITE ELEMENT METHOD SOLUTION
FOR THE STATIONARY TWO-DIMENSIONAL THERMAL PROBLEM

Grażyna Krusz

Institute of Electrical Machines and Transformers

Technical University of Łódź, Poland

INTRODUCTION

The problems of the heat flow and the electro-magnetic field are strongly coupled in all electrical equipments.
Heating of inside of electrical machines and other low voltage electrical devices is caused by Joule effect of current flowing in the windings, eddy-currents and hysteresis in the magnetic core.
The temperature rise modifies the resistivity, resulting in a change of a loss distribution and a heat generation.
The heat transfer coefficients depend on the surface temperature too.
In this way the problem of the temperature distribution in electrical devices can be highly non-linear.
On the other hand the thermal properties of materials and loss density throughout the volume of the object under investigation are non-homogeneous and show the directionality.
The traditional ways of solving this type of problems are analytical techniques with many simplifying assumptions or resistance analog method[1].
The well known finite difference method and the finite element method[2,3,4,5] have proven in recent years to be powerful techniques for the analysis of thermal fields in various electrical equipments.

STATEMENT OF PROBLEM

The stationary temperature distribution in the primary of a linear induction motor and elements of other electrical devices is governed by the non-linear heat conduction equation:

$$\nabla\cdot(\lambda \Delta t) = - p_v \qquad (1)$$

where t is the temperature,
λ is the thermal conductivity,
p_v is the thermal power density.
In cartesian coordinates equation (1) for cross section of the machine can be expressed as

$$\frac{\partial}{\partial x}\left(\lambda_x \frac{\partial t}{\partial x}\right) + \frac{\partial}{\partial y}\left(\lambda_y \frac{\partial t}{\partial y}\right) = -p_v, \qquad (x,y) \in \Omega \qquad (2)$$

where x, y are the coordinates in two dimensional space Ω,
λ_x, λ_y are the conductivities in the x and y directions.
Finally equation (2) is elliptic type, non-linear equation.
On the part of external surface of the motor, i.e. on the part $\partial_1\Omega$ of a boundary temperature t_0 is given as a set:

$$t = t_0, \qquad (x,y) \in \partial_1\Omega$$

On the part $\partial_2\Omega$ of the boundary specific rate of heat flow q is given as

$$\lambda_x \frac{\partial t}{\partial x}\cos(\nu,x) + \lambda_y \frac{\partial t}{\partial y}\cos(\nu,y) = -q, \qquad (x,y) \in \partial_2\Omega$$

where ν represents the external-normal versor.
From the last part $\partial_3\Omega$ of the boundary heat is transferred to ambient air and may be written as

$$\lambda_x \frac{\partial t}{\partial x}\cos(\nu,x) + \lambda_y \frac{\partial t}{\partial y}\cos(\nu,y) = \alpha_k\,(t - t_{uk}) + \alpha_r\,(t - t_{ur})$$

$$(x,y) \in \partial_3\Omega$$

where α_k is the film heat-transfer coefficient of the natural convection,
α_r is the heat-transfer coefficient for radiated heat,
t_{uk} is the temperature of ambient air for convection,
t_{ur} is the ambient temperature for radiation.
The convection heat-transfer coefficient is dependent on Nusselt number Nu_m, according to the equation:

$$\alpha_k = \frac{Nu_m\,\lambda_m}{\delta}$$

where λ_m is the thermal conductivity of the ambient air,
δ is the charakteristic dimension of the body.
The heat-transfer coefficient for the radiated heat:

$$\alpha_r = \sigma_0\,(T_{ur}^{\;2} + T^2)\,(T_{ur} + T)\,\varepsilon$$

where σ_0 is Boltzmann constant,
T_{ur}, T are the temperatures t_{ur}, t in Kelvin temperature scale,
ε is the thermal emissivity of the body under consideration.

ANALYSIS OF THE PROBLEM

A week form of equation (2) and all types of boundary conditions is

$$\int_\Omega \left(\lambda_x \frac{\partial t}{\partial x}\frac{\partial v}{\partial x} + \lambda_y \frac{\partial t}{\partial y}\frac{\partial v}{\partial y}\right) d\Omega = \int_\Omega p_v\, v\, d\Omega + $$
$$- \int_{\partial_1\Omega+\partial_2\Omega+\partial_3\Omega} \left[\lambda_x \frac{\partial t}{\partial x}\cos(\nu,x) + \lambda_y \frac{\partial t}{\partial y}\cos(\nu,y)\right] v\, dl$$

where $(t - t_0) \in H_0^1(\Omega)$, $\forall\, v \in H_0^1(\Omega)$.

It is easy to see that a Galerkin finite element formulation is

$$\int_{\Omega} (\lambda_x \frac{\partial t_h}{\partial x}\frac{\partial v}{\partial x} + \lambda_y \frac{\partial t_h}{\partial y}\frac{\partial v}{\partial y})\, d\Omega + \int_{\partial_3\Omega} (\alpha_k + \alpha_r)\, t_h\, v\, dl =$$

$$= \int_{\Omega} p_v\, v\, d\Omega + \int_{\partial_2\Omega} q\, v\, dl + \int_{\partial_3\Omega} (\alpha_k\, t_{uk} + \alpha_r\, t_{ur})\, v\, dl$$

where $(t - t_0) \in H^r_{FEM}$, $\forall\, v \in H^r_{FEM}$ and H^r_{FEM} is finite dimensional space consisting of piecewise polynomials P_r of degree r, over suitable chosen elements e_i in Ω:

$$H^r_{FEM}(\Omega) = \{t_h:\ t_h \in C(\bar{\Omega}),\ t_h|_{e_i} = P_r(e_i), \quad i = 1,\dots,N_{el}$$

$$t_h - t_0 = 0, \quad (x,y) \in \partial_1\Omega\} \subset H_0^1$$

The region Ω is discretized into a set of first order triangular elements. The shape functions P_r are linear due to their simplicity, i.e.:

$$P_1(x,y)|_{e_i} = C_{i0} + C_{i1}\, x + C_{i2}\, y$$

The equilibrium matrix equation have the form of

$$([\lambda] + [\alpha])\,\{t\} = \{p_v\} + \{q\} + \{t_u\} \qquad (3)$$

where $[\lambda]$ - thermal conductivity matrix calculated for all elements,

$[\alpha]$ - film heat-transfer matrix calculated for boundary elements only, derived from the temperature,

$\{t\}$ - nodal temperatures vector,

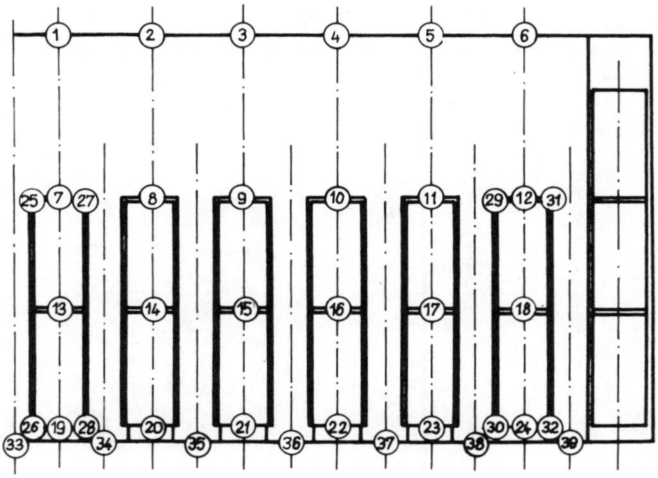

Fig. 1. Linear motor cross section, 1,...,39 termocouples

$\{p_v\}$ - heat sources vector for all elements, derived from the mean element temperature,

$\{q\}$ - heat flux vector for boundary elements only,

$\{t_u\}$ - temperature of ambient air vector for boundary elements only, derived from the temperature.

The system of algebraic equations yielded by the finite element discretization is solved by the Gaussian direct elimination method for linear problems or first step of non-linear calculation.

For non-linear cases the set of equations represented by (3) is solved iteratively by means of the non-stationary succesive overrelaxation (SOR) algorithm.

Its rate of convergence was fast enough in cases when non-linearity of the object under investigation was not so high. The same effect can be performed when the first approximation of the solution is preliminary estimated or calculated by the other methods.

EXAMPLE OF APPLICATION

Figure 1 shows the cross-section of a linear motor used in the example of calculations.

Figures 2 and 3 compare calculated temperatures with those obtained from thermocouples embedded in the slots and located on the external surface of the motor.

It must be added that both the results calculated and measured are more accurate inside the motor than on the boundary.

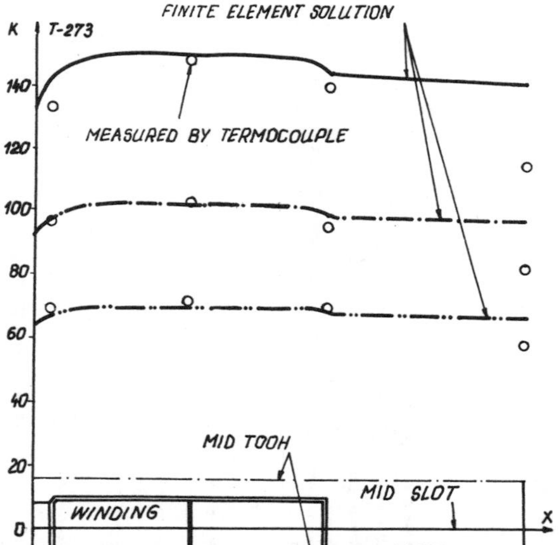

Fig. 2. Comparison of finite element solution with measured temperatures in central slot for different current supply

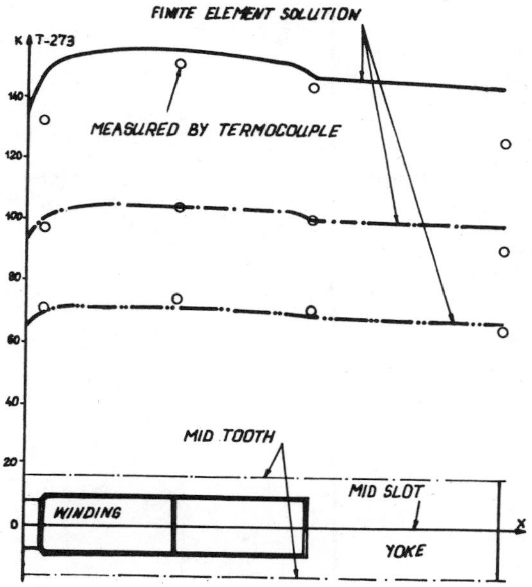

Fig. 3. Comparison of finite element solution with measured temperatures in last slot for different current supply

REFERENCES

1. J. Mukosiej, Equivalent Thermal Network of Totally-enclosed Induction Motors, in: "Proceedings of International Conference on Electrical Machines," Lausanne (1984)
2. A. F. Armor and M. V. K. Chari, Heat Flow in the Stator Core of Large Turbine-Generators by the Method of Three-dimensional Finite Elements, IEEE Trans. on PAS (1976)
3. M. V. K. Chari and A. F. Armor, The Stator Core of a Turbine-Generator by a Hybrid Finite Elements Model, in: "Proceedings of International Conference on Numerical Methods in Electrical and Magnetic Field Problems," S.Margherita Ligure (1976)
4. P. C. Kohnke and J. A. Swanson, Thermo-electric Finite Elements, in: "Proceedings of International Conference on Numerical Methods in Electrical and Magnetic Field Problems," S.Margherita Ligure (1976)
5. T. Yamamura, Y. Saito and H. Nakamura, Calculations on the Temperature Distribution of DC-Machine Armature by Finite Element Method, in: "Proceedings of International Conference on Electrical Machines," Lausanne (1984)
6. P. Ciarlet, "The Finite Element Method for Elliptic Problems," North Holland Publishing Company, Amsterdam - New York - Oxford (1978)
7. G. Krusz, "The Calculation of Temperature Fields by the Method of Finite Elements for the Linear Motor as an Example," Doctor's Thesis, Łódź (1984)

FORCE CALCULATION IN ELECTROMAGNETIC DEVICES

Yvan Lefevre, Michel Lajoie-Mazenc and Bernard Davat

Laboratoire d'Electrotechnique et d'Electronique Industrielle
(U.A. au C.N.R.S. n° 847)
I.N.P.T. - E.N.S.E.E.I.H.T.
2, rue Camichel, Toulouse, France

ABSTRACT

In this paper the authors show how the Maxwell stress tensor can be used, in practice, in order to determine local magnetic forces distribution in an electromagnetic system. An experimental set-up has been designed in view of comparing the theoretical results with the measurements.

INTRODUCTION

In electrical machines and actuators, the magnetic force is an important quantity which should be determined with precision. Generally this force is applied to a moving part of a system. Under this global aspect, it constitutes the main useful quantity, like the torque of an electrical motor. But this force presents also a local aspect, and can then be the cause of vibrations in actuators and electrical machines. In fact the rotor rotation and the variations of currents passing through conductors which are usually placed in the slots produce fluctuations of the magnetic forces applied to the different parts of the electromagnetic structure.

The first step in the study of vibrations of magnetic origin is therefore the knowledge of the distribution of these forces within the machine, in the space and time domains.

In this paper, the authors present an approach which leads to the determination of this distribution. First they recall the different methods which allow the magnetic force calculations in an electromagnetic structure. Then they show how the Maxwell stress tensor, which is usually used for global force calculations[1], can be employed in order to determine local forces applied to a specific part of the device. Finally, they illustrate the possibility of using the Maxwell stress tensor, by setting up an experimental device and comparing the theoretical results with those obtained from the measurements.

MAGNETIC FORCE CALCULATION

In pratice three methods are available for the calculation of magnetic forces[2,3,4] :

- Laplace's law for the calculation of forces applied to a current carrying conductor :

$$F = \int_V J \wedge B \cdot dv$$

- The coenergy derivative with respect to the space coordinates[5] :

$$Fm = \delta/\delta x \int_V \int_0^H B.dH \cdot dv$$

- The integration of Maxwell stress tensor over the surface enclosing that part of the device over which the forces are applied :

$$Fm = \int_S T_m \cdot ds$$

where T_m is a column vector of the Maxwell stress tensor :

$$T = 1/u \cdot \begin{vmatrix} B_1{}^2 - B^2/2 & B_1.B_2 & B_1.B_3 \\ B_1.B_2 & B_2{}^2 - B^2/2 & B_2.B_3 \\ B_1.B_3 & B_2.B_3 & B_3{}^2 - B^2/2 \end{vmatrix}$$

B_i (i=1,2,3) are the components of the induction vector B.

Among these methods, only Laplace's law allows theoretically a local force calculation, when it is applied on conductors. On the other hand the local aspect of the force obtained by the coenergy derivation can be easily underlined since it is derived from a volume integration. It seems therefore interesting to associate to the volume element dV the force dF calculated on it. In contrast, the force calculation by this method needs usually for each force component two successive solutions of the electromagnetic field equations, for two adjoining positions. Finally when the force is calculated by Maxwell stress tensor, it should be noted that even if only one solution of the field equations is sufficient for the determination of the different force components, the integration is carried out over a surface which passes in the air and encloses the considered device part, so that, theoretically, it allows only the calculation of the global force applied to this part.

It is now shown that it would be possible, in practice, to determine a local force by means of Maxwell's tensor.

LOCAL ASPECT OF THE MAXWELL'S TENSOR

Let's consider the periodic structure in figure 1 in which the force applied to the tooth D is to be calculated. It is theoretically impossible to calculate the force applied to this tooth since it cannot be enclosed by a surface passing in the air. However if we solve the electromagnetic field equations for the structure of figure 1 and then for an identical structure in which the tooth D is slightly separated (figure 2) the obtained results are in practice very similar.

In this latter case Maxwell's tensor integration over the surface S is carried out over the surface S_1 and S_2 separately, giving the following results :

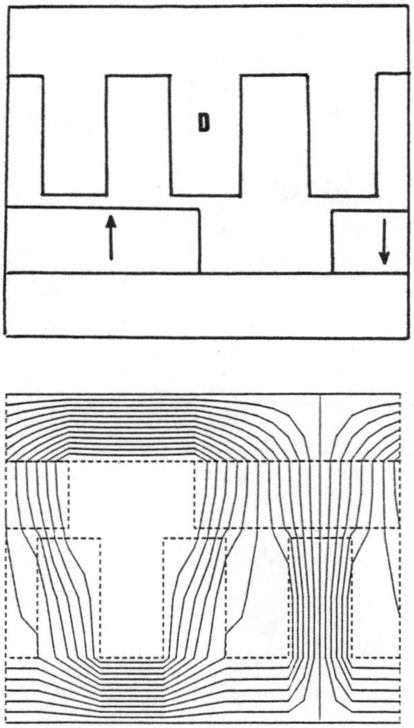

$\Delta A = 0.936\ 10^{-3}$ Wb/m
$A_m = 0.127\ 10^{-1}$ Wb/m

Fig. 1. Considered device.

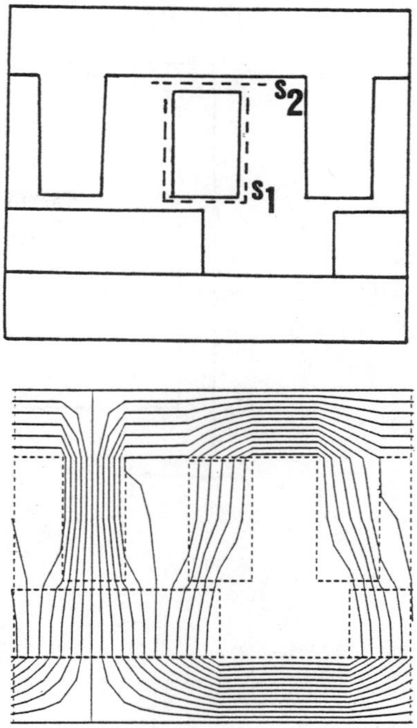

$\Delta A = 0.927\ 10^{-3}$ Wb/m
$A_m = 0.126\ 10^{-1}$ Wb/m

Fig. 2. Sticking out a tooth.

- integration over S_1 : $F_x = -\ 312.$ N,
  $F_y = -\ 540.$ N

- integration over S_2 : $F_x = 0.$ N,
  $F_y = 670.$ N

These results can be compared to those obtained for the original structure in which the tooth D is attached to the device body :

- integration over S_1 : $F_x = -\ 312.$ N,
  $F_y = -\ 540.$ N

This comparison shows that the integration over S_1 yields practically to the same result in the two cases : it can be considered as being the force applied on D by the magnet. The integration over S_2 does not involve any x components which would correspond to a magnet attraction. This force is applied in the y-direction corresponding to the attraction force which tends to stick back the tooth to the body device.

<u>Experimental device</u>

The above example has allowed us to demonstrate that the integration over a surface which encloses partially a device part should lead to the determination of the force applied to it. This result should now be verified on an experimental device. The model used for this purpose is represented in figure 3.

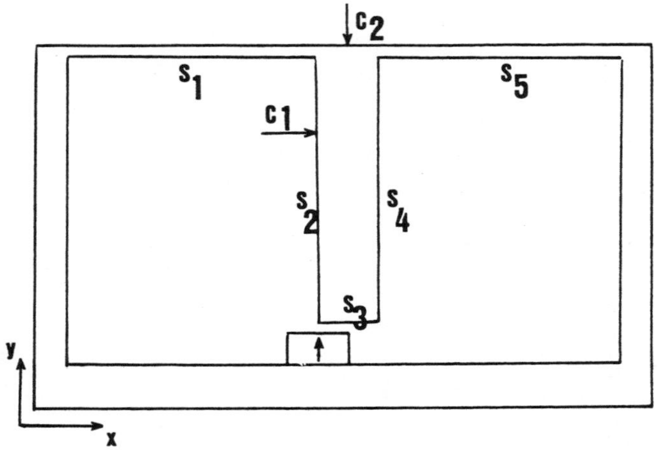

Fig. 3. Experimental device.

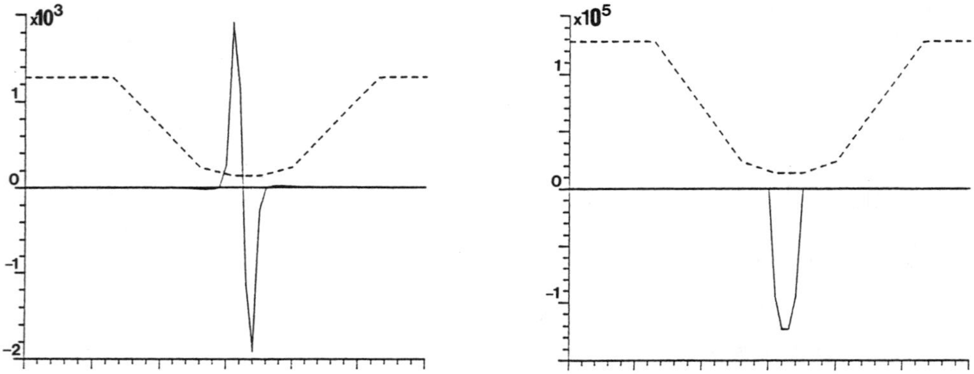

Force density in x-direction　　Force density in y-direction

Fig. 4. Force densities along the tooth (N/m²)

It is composed of a closed magnetic circuit, including a tooth, in front of which a magnet can be displaced. The tooth has been intentionally very long in order to permit simple measurements of components Fx and Fy of the force applied to the tooth by the magnet. The upper part of the magnetic circuit is composed of a thin plate so that it could suffer measurable deformations under the electromagnetic forces.

The force applied to the tooth is obtained numerically by integrating the Maxwell's tensor along the surface S_1, S_2, S_3, S_4 and S_5. Figure 4 illustrates the computed force density distributions over these surfaces for one magnet position. The force densities are decomposed into DF_x and DF_y. The dashed curves correspond to the developed integration surfaces and allows one to locate the point where the force density is applied. It can be observed that the magnetic force is pratically applied only on the surface S_3.

The force has been experimentally measured from the displacements measurement by means of comparators C_1 and C_2, which have been already calibrated by applying known forces to the tooth end along x and y axes.

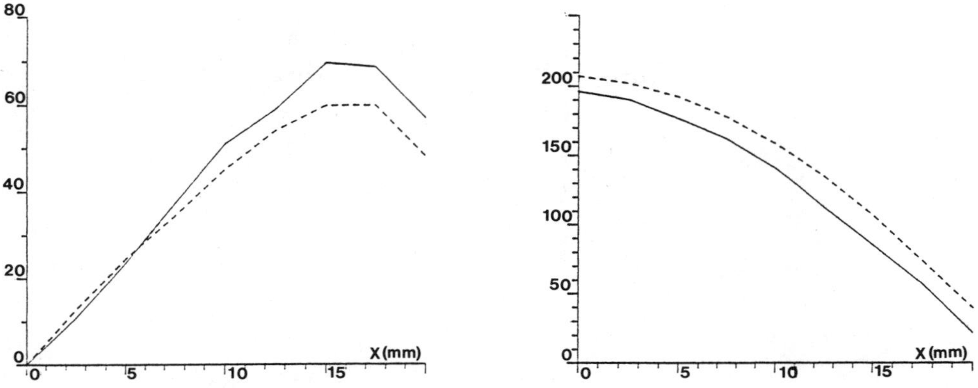

Force in x-direction

Force in y-direction

Fig. 5. Force versus magnet displacement (N).

The results obtained experimentally : the values of forces as a function of the magnet position with respect to its equilibrum position are compared in figure 5 with the results obtained from numerical analysis.

The deviation between the different results is less than 10 % and principally due to the dispersion of the characteristics of the different magnet bars, whereas a mean value has been taken for the numerical calculations.

CONCLUSION

The two examples of electromagnetic structures presented show that the integration of the Maxwell's tensor over a surface enclosing partially a device part allows the calculation of electromagnetic forces applied to this part. The calculation is considered accurate since the stresses are applied especially to the end parts. It would be therefore possible to make use of the Maxwell stress tensor, in spite of its global character, for the evaluation of magnetic forces applied to the teeth of an electrical machine.

REFERENCES

1. C. J. Carpenter, Surface integral methods of calculating forces on magnetized iron parts, I.E.E. Proc. monograph 342 (1959).
2. E. Durand, "Magnétostatique," Masson, Paris (1968).
3. M. Jufer, "Transducteurs électromécaniques," Editions Georgie, Lausanne (1979).
4. H. H. Woodson and J.R. Melcher, Part 2 : Fields, forces and Motion, in : "Electromechanical Dynamics," Wiley, New York (1968).
5. Z. Ren, Contribution à la modélisation des machines électriques par résolution simultanée des équations du champ et des équations du circuit électrique d'alimentation, Thèse de Doctorat I.N.P. Toulouse (1985).

ELECTROMAGNETIC FIELDS AND FORCES IN A LINEAR INDUCTION MOTOR DURING DIRECT CURRENT BRAKING

Krzysztof Pieńkowski

Institute of Electromachine Systems

Technical University of Wrocław, Poland

INTRODUCTION

Linear induction motors (LIM) find great application in many power transmission systems. Besides the motor operation the performance of the LIM during braking is also very important. The characteristics of LIM have been studied in recent years [1,2,3], but considerably less attention has been paid to the direct current braking (DC braking) performance of LIM. This paper presents the results of the analysis of LIM during DC braking with regard to the influence of the longitudinal end-effect. The relationships describing electromagnetic fields in the individual zones of the motor and forces acting on the secondary have been determined. The influence of the longitudinal end-effect on DC braking characteristics has been discussed.

THE MODEL OF THE MOTOR AND SIMPLIFYING ASSUMPTIONS

The model of double-sided LIM is shown in Figure 1 and coordinate axes are chosen as indicated in the figure. To facilitate mathematical analysis, the primary iron cores are considered to extend infinitely in both directions of the x-coordinate. However, only the actual core portion is considered to contribute to the machine performance. It has been assumed that the actual slotted airgap of the motor is replaced using Carter's coefficient by a fictitious unslotted one, the conductivity of the primary core is negligible, and the permeability is infinitely large. The coordinate axes are immovable in relation to the primary iron cores. The secondary moves with a velocity v which is directed along the axis x. Three different zones and regions have been distinguished and denoted by indices I, II, III and 1, 2, 3 respectively as shown in Figure 1. The active part of the length L exists between points x=0 and x=L and has been denoted as zone II. The primary excitation is modeled by sinusoidally distributed infinitely thin current sheets of the length L.

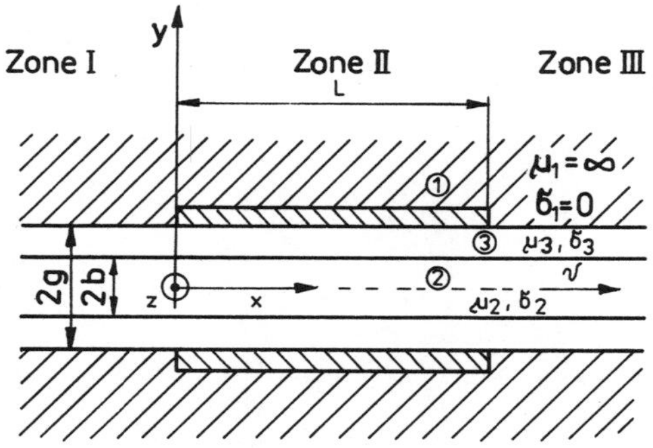

Fig.1. Model of double-sided LIM

During DC braking the motor is fed by a direct current source. It is considered that all variables are independent from time and are only the functions of space. All functions are represented by complex functions of x.

The primary current density in the zone II may be described in the following form:

$$\underline{j}_1 = J_{1m} \exp[-j(kx+\beta_u)] \tag{1}$$

where J_{1m} is the amplitude of the primary current density, $k=\pi/\tau$, τ pole pitch and β_u the angle dependent upon the used scheme of primary winding connection of LIM during DC braking. In the remaining two zones $\underline{j}_1$ is equal to zero.

ELECTROMAGNETIC FIELD EQUATIONS

One-dimensional Analysis

One-dimensional theory is of a great importance because it provides a quick understanding of the most important phenomena and a quite good estimate of machine performances [1,2]. In this theory the skin effect and transverse-edge effect can be taken into account by using appropriate correction factors[1].

The electromagnetic field equation in the airgap of the LIM in the zone II based on 'second-order theory[3]' has the following form[4]:

$$(\partial^2/\partial x^2)\underline{b}-\sigma_e\mu_0 v\,(\partial/\partial x)\underline{b}=-j\mu_0(k/g)\,J_{1m}\,\exp[-j(kx+\beta_u)]. \tag{2}$$

In zones I and III the right-hand side of equation (2) is equal to zero. Here $\underline{b}$ is the magnetic flux density, 2g the airgap length, μ_0 the permeability of the air, and σ_e the equivalent conductivity of the secondary.

After solving the equation (2) and considering adequate boundary conditions we get:

– for the zone I:

$$\underline{b}_I = \underline{B}_s \; [1 - \exp(-rL)] \; \exp(rx) \tag{3a}$$

– for the zone II:

$$\underline{b}_{II} = \underline{b}_{IIc} + \underline{b}_{IIe} = \underline{B}_s \; \exp(-jkx) - \underline{B}_s \; \exp[r(x-L)] \tag{3b}$$

– for the zone III:

$$\underline{b}_{III} = 0 \tag{3c}$$

where: $$\underline{B}_s = -\mu_o \; J_{1m} / g / (r+jk) \quad \exp(-j\beta_u) \tag{4}$$

r – different from zero root of the characteristic equation (2)

$$r = \sigma_e \; \mu_o \; v \; . \tag{5}$$

The exemplary distribution of the magnetic flux density in the airgap determined on the basis of equations (3)-(5) is presented in the Figure 2.

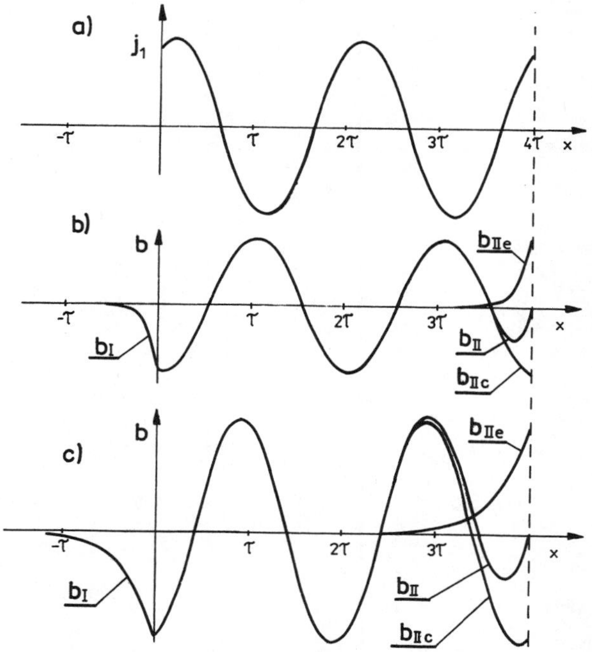

Fig.2 Distribution of field in the airgap of the LIM during DC braking: a – $j_1(x)$; b – $b(x)$ for $v=0.75\,v_s$; c – $b(x)$ for $v=0.25\,v_s$

The LIM supplied with direct current produces in the airgap the fields which are immovable in the space. The field in the zone I represents the longitudinal entry effect. It decays in the negative direction of the x-coordinate. The field in the zone II consists of the components $\underline{b}_{IIc}$ and $\underline{b}_{IIe}$. The first one has the sinusoidal distribution along the motor length and represents the field of infinitely long motor. The second component decays from point x=L in the negative direction of x-coordinate and can be considered as a field brought about by longitudinal exit effect. The coefficient of attenuation r is dependent on the speed of secondary. In the zone III the field does not exist and is equal to zero.

Two-dimensional Analysis

The two-dimensional analysis of the LIM has been conducted in the x-y plane. The equations of the electromagnetic field have been formulated for the vector potential $\underline{A}_i$ in the individual regions of the motor (i=1,2,3). It has been assumed that the vector potential $\underline{A}_i$ has the z-component only.

We obtain the following equations for the electromagnetic field:

-in region 1 and 3 (i=1,3):

$$(\partial^2/\partial x^2)\underline{A}_i + (\partial^2/\partial y^2)\underline{A}_i = 0 \tag{6a}$$

-in region 2 (i=2):

$$(\partial^2/\partial x^2)\underline{A}_i + (\partial^2/\partial y^2)\underline{A}_i = \sigma_2 \mu_2 v (\partial/\partial x)\underline{A}_i . \tag{6b}$$

The equations (6) have been solved by the Fourier transformation. Acting in the same way as in references 2 and 3 one gets from the residue theorem the following expression for the vector potential in the airgap (region 3):

- in the zone II:

$$\underline{A}_3(x,y) = \mu_3 J_{1m} \exp(-j\beta_u) \{G(-k,y)/H(-k) \exp(-jkx)+$$
$$G(\xi_1,y)/(\xi_1+k)/H'(\xi_1) \exp[j\xi_1(x-L)]\} \tag{7}$$

where G,H,H' - functions as presented in reference 2,

$$\xi_1 = -j \mu_3 \sigma_2 vb/g \tag{8}$$

ξ_1 - the root of equation $H(\xi)=0$.

Equation (7) determines at the same time the magnetic flux density in the airgap of the LIM during DC braking. Comparing the solution (7) with the relationships (3) it is possible to state that two-dimensional analysis yields the same qualitative description of the field in the airgap as does one-dimensional analysis.

DRAG FORCES ACTING ON THE SECONDARY

The total drag force acting on the secondary in the x-direction can be considered as the sum of the conventional force F_c of infinitely long motor and the force F_e brought about by the longitudinal end-effect.

From the one-dimensional analysis the force F_c acting on the width unit of the secondary is described by the equation:

$$F_c = -L/v_s/\sigma_e/g \ [\nu G^2/(\nu G^2 + 1)] \ J_{1m}^2 =$$

$$2F_{cm}/(\nu/\nu_m + \nu_m/\nu) \tag{9}$$

where: $\nu = v/v_s$ – relative velocity of the secondary,

$\nu_m = 1/G$,

$G = \mu_o \sigma_e v_s/k$ – factor of motor goodness.

The expression for the force F_e is as follows :

$$F_e = \mu_o/(gk^2) \ [1-\exp(-rL)]*$$

$$[(\nu G)^2 - \tan^2\beta_u]/[(\nu G)^2 + 1] \cos^2\beta_u \ J_{1m}^2 . \tag{10}$$

If the two-dimensional analysis is used , we obtain:

$$F_c = \mu_o \ L/k \ 2\alpha\beta b/[sh^2(g-b)k]*(sh2\alpha b/2\alpha b + \sin 2\beta b/2\beta b)/$$

$$(ch2\alpha b + \cos 2\beta b)/(1+C_\alpha+C_b) \ J_{1m}^2 \tag{11}$$

where factors $\alpha, \beta, C_\alpha, C_b$ are the same as described in reference 5.

The results of calculations of DC braking drag forces of a double-sided LIM for two different ways of primary winding connection are presented in Figure 3.

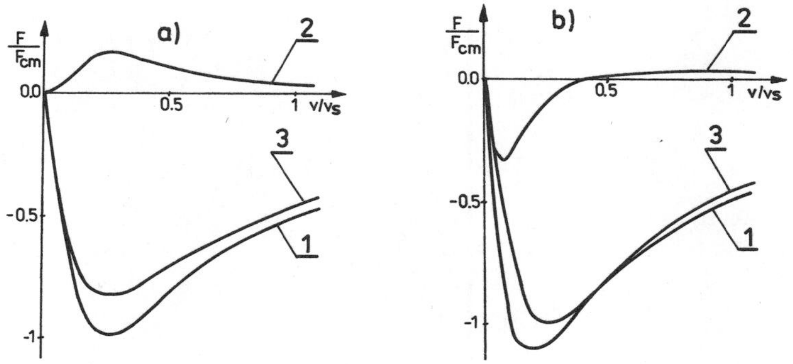

Fig.3.DC braking drag forces versus relative velocity of the secondary: a – $\beta_u=0$; b – $\beta_u=-2\pi/3$. 1 – F_c, 2 – F_e, 3 – F_c+F_e. The data of the LIM: $p=1$, $\tau=0.09$ m, $2g=0.016$ m, $2b=0.005$ m, $\sigma_2=35\ 10^6$ S/m (Al), $v_s=9$ m/s, $J_{1m}=10^5$ A/m

From the analysis and calculations it is clear that the conventional force F_c in the whole range of the velocity of the secondary is the braking force . The force F_e brought about by the longitudinal end-effect can be the braking force which increases the total braking force or it can be the drive force which decreases the total braking force. As the force-speed curve of F_e is dependent on the value of β_u, then the effect of this force is different for particular arrangements of primary winding connection during DC braking.

CONCLUSIONS

From the conducted analysis of the DC braking of the LIM it is clear that the longitudinal end-effect can have significant influence on the flux density distribution along the airgap and the total braking force. The results obtained in this paper compare favourably with those derived by much more complicated methods.

REFERENCES

1. I.Boldea and S.A.Nasar, Quasi 1-dimensional theory of linear induction motors with half-filled primary endslots, Proc. of IEEE, V.122, No.1, (1975).
2. S.Yamamura, Theory of linear induction motors , University Tokyo Press, Tokyo (1978).
3. M.Poloujadoff, The theory of linear induction machinery, Clarendon Press, Oxford (1980).
4. K.Pieńkowski, Dynamic braking performance of double-sided linear induction motors, Modelling, Simulation and Control, A, AMSE, V.13, No.2, (1987).
5. A.Singh, Theory of eddy-current brakes with thick rotating disc, Proc. of IEE, V.124, No.4, (1977).

NUMERICAL ANALYSIS OF ELECTROMAGNETIC AND TEMPERATURE FIELDS IN INDUCTION HEATED FERROMAGNETIC SLABS

A. Stochniol* and V.S. Nemkov**

* Technical University of Kielce, Kielce, Poland
**Lenin Electrical Engineering Institute of Leningrad, Leningrad, USSR

INTRODUCTION

Induction heating is widely used for quick heating of conductive materials. Because temperature distribution of the heated body is governed by the eddy current distribution, it is very important to analyse the relation between the two distributions. For heating of ferromagnetic billets, depending on the dimension of the workpiece, typical field strengths H can range from 50 to 200 kA/m and typical supply frequency f can range from 50 to 3000 Hz. At the resulting deep levels of saturation the waveforms for flux density and especially current density become highly distorted.
The distribution of induced power throughout the heating period is required in order to predict the temperature rise and temperature distribution within the workpiece (load). However, it is not practical to obtain power distribution by solving the nonlinear eddy current problem to otain the distorted flux density and current density waveforms. It has been shown[1] that despite the occurrence of highly distorted flux densities and current density waveforms, a simple time harmonic solution to the eddy current problems provides a very cost effective and reliable estimate of the total and distributed losses.

THE MATHEMATICAL MODEL

The process of induction heating is described by strongly coupled Maxwell and temperature equations. The particular problem being considered (typical in induction heating) is that of a long ferromagnetic conducting slab of rectangular cross-section located in an axially directed magnetic field. It is assumed that the magnetic field strength at the surface of the conductor is known and varies sinusoidally in time with an angular frequency ω (it is found from the solution of the exterior problem). The problem geometry is illustrated in Fig. 1. If the applied field is saturating the surface region of the conductor, the flux density waveform will be non-sinusoidal, in the case of deep saturation, it will approach a square wave. For the application being considered in this paper, the detailed flux density waveform is not required. The solution is posed by defining an effective permeability μ_e at each point of the cross section of the load based upon the material characteristic $\underline{B} = \mu_e(\underline{H})*\underline{H}$, and the rms flux density $\underline{B}$ resulting from sinusoidal

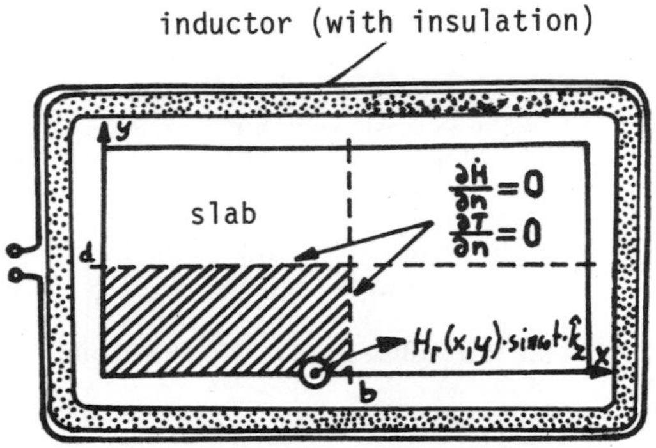

Fig.1. The region under consideration.

H-excitation. By considering rms $\underline{B}$ and $\underline{H}$ magnitudes Maxwell equations for the considered problem can be cast in a harmonic time form (inside the load):

$$\frac{\partial}{\partial x}\left(\rho\frac{\partial \underline{H}}{\partial x}\right) + \frac{\partial}{\partial y}\left(\rho\frac{\partial \underline{H}}{\partial y}\right) = j\omega\mu_e\underline{H} \tag{1}$$

where ρ is the resistivity of the material (dependent on temperature). The heat transfer problem is described by:

$$c\gamma\frac{\partial T}{\partial t} = \frac{\partial}{\partial x}\left(\lambda\frac{\partial T}{\partial x}\right) + \frac{\partial}{\partial y}\left(\lambda\frac{\partial T}{\partial y}\right) + w(x,y) \tag{2}$$

where T is the temperature, γ is the density, $c=c(T)$ is the thermal capacity (the energy of the phase transition at the Curie point is simulated by variation of c), $\lambda=\lambda(T)$ - is the thermal conductivity, $w(x,y)$ - is the thermal source induced by eddy currents. Distribution of induced power is given by:

$$w(x,y) = \rho\left[\frac{\partial \underline{H}}{\partial x}\cdot\frac{\partial \underline{H}^*}{\partial x} + \frac{\partial \underline{H}}{\partial y}\cdot\frac{\partial \underline{H}^*}{\partial y}\right] \tag{3}$$

where $\underline{H}^*$ is the conjugated complex value of $\underline{H}$. For the eddy current problem (1) Dirichlet boundary conditions are given:

$$\underline{H}|_{\Gamma} = \underline{H}_{\Gamma}(x,y) \tag{4}$$

The convective and radiative boundary conditions of the thermal problem must be expressed by:

$$-\lambda\frac{\partial T}{\partial t} = \alpha(T-T_s) + \varepsilon_e\sigma\left[(T+273)^4 - (T_s+273)^4\right] \tag{5}$$

where T_s is the temperature of the inner surface of the inductor (i.e. inductor heat insulation). The calculation method of T_s is given in[2]. The exchange surface coefficient α and the effective emissivity ε_e are also dependent on the temperature. Efficient emissivity is calculated taking into account emissivities of the load and the inductor insulation, as well as the configuration factor [2,3]. At the Curie point ($\approx$ 750°C) steel has a transition phase with associated rapid changes in such characteristics as resistivity and magnetic behaviour. In fact μ_e is the function of $|\underline{H}|$ and T. For steel we use an analytical approximation of μ_e

$$\mu_e = \mu_e(|\underline{H}|,T) = \begin{cases} \mu_o\cdot\left[1+(\hat{\mu}-1)\cdot\left[1-\left(\frac{T}{T_c}\right)^2\right]\right] & , \text{ for } T < T_c \\ \mu_o & , \text{ for } T \geq T_c \end{cases} \tag{6}$$

where

$$\hat{\mu} = \hat{\mu}(|\underline{H}|) = 5\cdot 10^5\cdot|\underline{H}|^{-0.894}$$

This completes the description of the inner problem for induction heating, i.e. mathematical model of distribution of electromagnetic and temperature fields inside the load. The set of equations is time dependent, non-linear (particularly near Curie point) and coupled.
Because of the symmetry of the problem, only a quarter of the load is studied. For this purpose the symmetry boundary conditions at the symmetry lines are introduced (i.e. $\frac{\partial H}{\partial n} = 0$, $\frac{\partial T}{\partial n} = 0$, see Fig. 1).

For the solution of the inner problem (1-6) the finite difference (FD) method with the nonuniform grid is applied.
To solve algebraic equations with the five diagonal matrices having complex coefficients that result from the FD discretization of the eddy current problem (1) (at each time step) the complex version of the approximate LU-type factorization procedure is succesfully used[4]. The distribution of induced power w (3) is computed from the solution obtained by spline interpolation.
To solve algebraic equations that result from FD disretization of the temperature equation (2) at each time step, the approximate LU-type factorization method is applied[5].

THE HYBRID METHOD

A major disadvantage of the proposed model (1-6) of induction heating is the necessity to set the magnetic field strength at the boundary of the load. This value is often unknown but it may be obtained from solution of the exterior problem (i.e. the field problem outside the load).
It is possible to construct the iterative algorithm in which the exterior problem is solved by integral or approximate methods and the interior problem by the finite difference or finite element methods[6]. Boundary conditions for the interior problem are found (or only corrected) from the solution of the exterior problem.
At every step of this iterative process correction of the exterior and interior problems takes place until a test for convergence is reached.
In the considered problem for solution of the exterior problem the approximate total flux method is used[7]. It is a variant of equivalent circuit methods[3,7]. Because the interior problem is solved by computational methods, effective resistance and inductive reactance of the load needed for the total flux method are determined numerically from the solution of the interior problem. The impedance of the load is computed from the complex power absorbed by it. Complex power is related to the boundary values of the tangential electromagnetic field components by the Poynting theorem. Integral parameters of induction heater are also given by this method. The total flux method is very attractive for the design of induction heaters. It is possible to replace this method of solution of the exterior problem easily by one of the integral methods[3,6].

RESULTS

To simplify the analysis, only the inner problem for a known, constant magnetic strength at the surface of the load $\underline{H}$ is considered.

The ferromagnetic steel slab is assumed to have a nonlinear magnetization characteristic given by (6).

A steel slab of cross-section 0.2 * 0.6m is heated under a crest intensity of $\underline{H}_r = 2 \cdot 10^5$ A/m with frequency f=50 Hz. The airgap is 0.02 m (thermal resistance of the heat insulation is 0.015 m^2K/W).

Results of computation are presented in Figs. 2 - 3. Fig. 2 shows the temperature evolution on the most characteristic regions of the load, i.e. on both sides, on the symmetry lines and on the corner bisector. Fig. 3 shows temperature rises at the most characteristic points of the load (1 - the centre of the shorter side, 2 - the corner, 3 - the centre of the longer side, 4 - the centre of the cross-section).

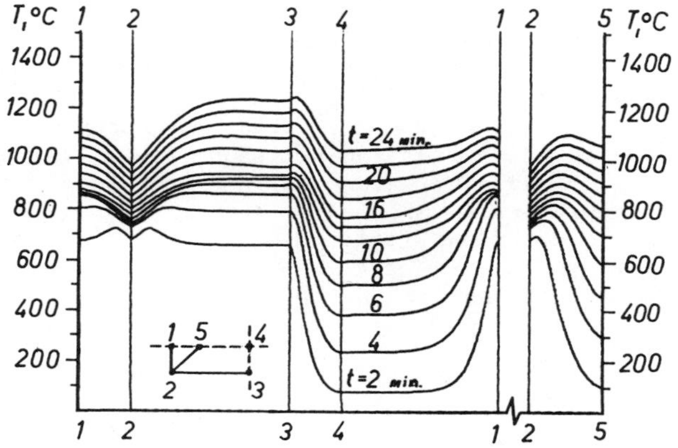

Fig.2. The temperature evolution on the most characteristic load regions.

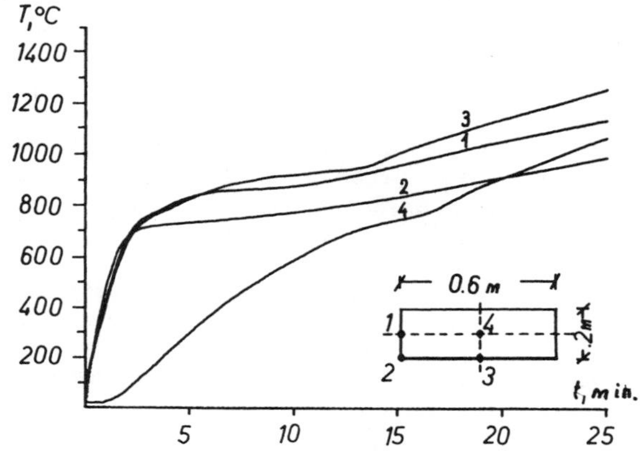

Fig.3. Temperature rises at the most characteristic points of the load.

In the Fig. 4 and Fig. 5 another results of computation are presented. The Fig. 4 shows induced active power evolution on the most characteristic regions of the load (the same as for Fig. 3). Fig. 5 shows variations of active and reactive power induced in the load versus time. There are three distinct phases of the process (intervals I, II, III of the time axis in Fig. 5). In the first one the power rises and falls, rate of temperature is big, particularly near edges. In phase III (linear behaviour) a steady electric state settles, the great skin-depth causes the power fall to a minimum and temperature rises are linear and slow (see Figs. 2, 3). Phase II is characterised by the simultaneous existence

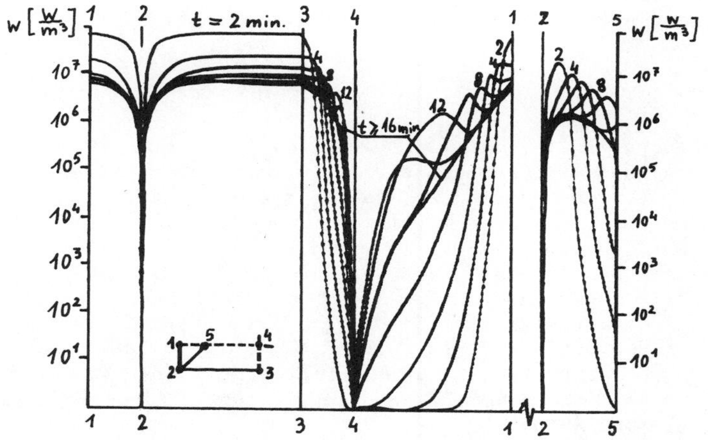

Fig.4. Active power evolution on the most characteristic load regions.

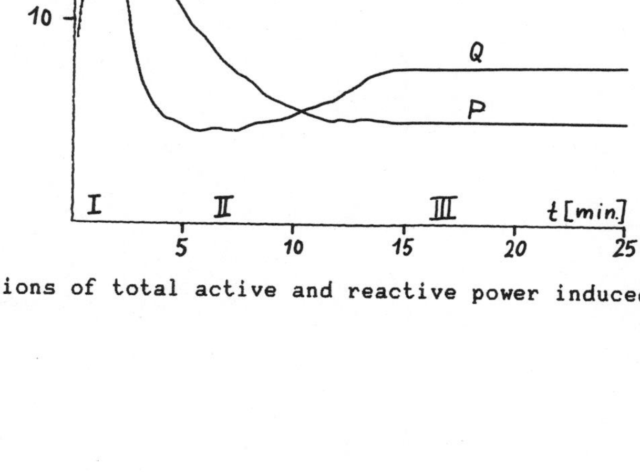

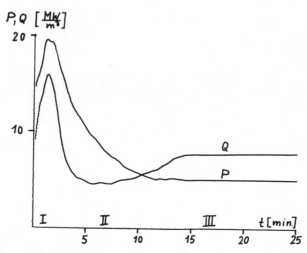

Fig.5. Variations of total active and reactive power induced in the load.

of zones above (edges) and near (centre and maybe corner) the Curie point. For this phase simple relations do not exist. For the whole process great changes in the distribution of induced power occur. The power profiles for different phases are completely different (see Fig.4) and they cause nonuniform distribution of temperature fields.

CONCLUSION

The proposed method of solving the inner problem of induction heating of ferromagnetic slabs fully takes into account nonlinearities of magnetization characteristic and electric and thermic properties. Coupling this method with the exterior solution into a hybrid algorithm gives a powerful tool which may be used in the design of induction heaters and for optimal control of induction systems.

REFERENCES

1. J. D. Lavers, M. R. Ahmed, M. Cao, and S. Kalaichelvan, An evaluation of loss models for nonlinear eddy current problems, IEEE Trans. Magn., Vol. MAG-21, No. 5 (1985), pp. 1850-1852.
2. N. A. Pavlov, "Engineering thermic computation of inductor heaters", Energia, Moscow (1978) (in russian).
3. H. Barber, "Electroheat", Granada, London (1983).
4. A. Stochniol, Efficient methods for the calculation of the two-dimensional electromagnetic field, in: "Proc. Dynamic Processes Simulation - 2nd Conference in Poland", Zakopane (1985), pp.37/1-8.
5. D. J. Evans, Iterative sparse matrix algorithms, in: "Software for Numerical Mathematics", ed. D.J. Evans, Academic Press, London (1974), pp. 49-83.
6. A. Stochniol, V. S. Nemkov, V. B. Demidovitch, The efficient method for the solution of induction heating problems, in: "Proc. Int. Symp. on Electromagn. Fields in Electr. Eng. ISEF'85", Warsaw (1985), pp. 347-350.
7. V. S. Nemkov, Computation of induction heating systems by the equivalent circuit methods, Izvestia VUZ of USRR - Electromechanika, No. 12 (1978), pp. 36-39, (in Russian).

6. VARIOUS APPLICATIONS

Introductory remarks

M. D'Amore

Department of Electric Energy
University of Roma
via Eudossiana 18 - 00184 Roma, Italy

In this chapter papers have been grouped dealing with specific applications of various nature. Some of the papers, however, though describing particular devices, could belong to other chapters, if one considers that generally the results obtained can be extended to a wider class of devices.

This is the case for istance of the paper by P.P. Campostrini and A. Stella. The authors investigate stray capacitances in large magnets for fusion experiments and identify the parameters of the model by means of the frequency response of the coil system. The methodology and even the results could, be applied, for istance, also to high voltage transformers as well.

R.D. Findlay and J.H. Dableh describe an original electromagnetic technique to repositioning, remotely, a set of annular spacers used to maintain the concentricity of two tubes. The technique consists in inserting a coil through the inner tube, to a location adjacent to the spacer, so that the latter is forced to move when a suitable current is supplied. The complex coupled electromagnetic and mechanical problem is studied using a finite element simulation.

In another paper R.D. Findlay with other coauthors tries to solve, analitically, the problem of determining the eddy currents induced in a finite, solid, conducting, rectangular plate, knowing the incident flux, developed in a double Fourier series. The results can be applied to the case of the tank walls in large transformers. Many researchers have attempted the same problem using various other methods which sometimes are to general and so consuming both time and resources. The papers opportunely stresses the necessity of selecting, for each application, the most appropriate method of computation, making effective use of resources and at the same time offering the designer good insight into the problem.

The last three papers deal with computer aided analysis of various devices. Z. Haznadar and S. Berberovic analyse the quasi static current field in earthing systems. Being the field region three-dimensional and having it open and irregular boudaries, the boundary element method appears well suited to model, in particular, the complex earthing systems of large industrial plants.

M.M. Radulescu et al. perform a finite element analysis of the magnetic field in a axisymmetric electromagnet which is a part of proportional hydraulic device.

Finally V.R. Rais et al. use the reluctance network method to analyse the two-dimensional magnetic field in a reverse electromagnetic pump. Different networks with different degree of refinement are investigated and forces of electromagnetic origin are evaluated.

The presence of experimental results confirming the validity of the computed results, and so of the models assumed, is a well appreciated point with is common to all the papers of the chapter and, in particular to the last three ones.

COMPUTER - AIDED MODELLING AND SIMULATION

OF FAST TRANSIENT PHENOMENA IN LARGE COILS

P.P. Campostrini* and A. Stella**

*Istituto Gas Ionizzati (Associazione CNR - EURATOM)

**Dipartimento di Ingegneria Elettrica - Università di Padova
Piazza Salvemini, 13 - 35131 Padova - Italy

ABSTRACT

This paper deals with resonance problems due to stray capacitances in large magnets, such as those used in fusion research machines or big H.V. transformers. The paper presents the method set up to investigate frequency response of the coils of the Poloidal Field System of the RFX fusion experiment.

For each coil an equivalent network is first identified which takes into account, for each turn, self and mutual inductances and stray parameters. The frequency spectrum of such a network is then numerically analysed and the corresponding resonant frequencies are found in order to identify a much simplified model with similar electrical behaviour at the coil terminals, and to allow the overall winding to be analysed at once.

INTRODUCTION

RFX is a device for research into controlled thermonuclear fusion, presently under construction in Padua (Italy) [13]. It has a complex system of coils whose function is to create, sustain and control the magnetic field necessary to generate and to confine the hot plasma ring inside a torus of 2 m and 0.5 m of major and minor radius respectively, with a plasma current of 2 MA.

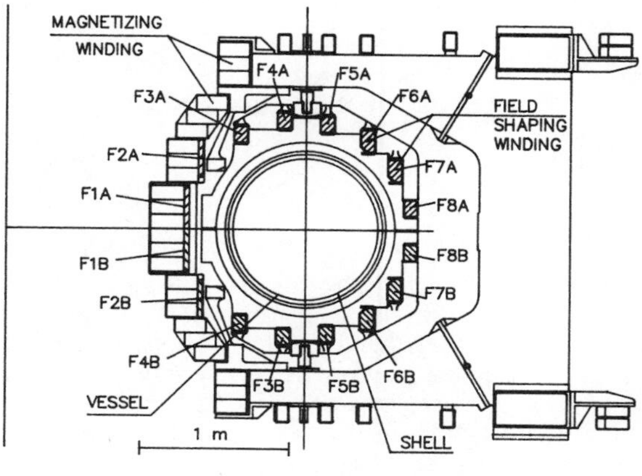

Fig. 1. RFX Poloidal Field System

The Poloidal Field System [10, 4] consists of two sets of coils: the Magnetizing winding (M) and Field Shaping winding (F) as shown in Fig. 1. The first is designed to release the flux swing needed to produce and maintain the toroidal plasma current, the second, consisting of 16 coils closer to the plasma, has to provide the proper magnetic boundary conditions at the shell surface as required for plasma equilibrium. Magnetizing and Field Shaping windings are connected to each other and to the power supplies as shown in Fig. 2. In operation, in order to ionize the gas in the vessel and to cause the plasma current to rise, all the coil terminals are subject to transient high voltages, up to 17.5 kV to ground with a maximum derivative of 4 kV/μs, decaying thereafter with a typical time constant of about 30 ms.

The stray capacitances of the coils cannot be disregarded because overvoltages can be easily produced both in normal and fault operating conditions.

On the other hand a detailed model of the overall coil system would be extremely complex and its computer analysis very time-consuming. For this reason the analysis is approached in steps. In the first instance a detailed model of each individual coil is identified, including self and mutual inductances as well as capacitances to the adjacent turns and to ground. A second step consists of a numerical analysis process which allows a reduction of the network complexity, without substantially changing its response to the main resonant frequency.

Using the same procedure for each coil a reduced model of the complete winding is obtained and, as far as the behaviour at the coil terminals is concerned, the overall system can be analysed.

Finally the analysis of the actual behaviour of any internal point can be performed using the voltage at the coil terminals as an input to the detailed model of the coil concerned.

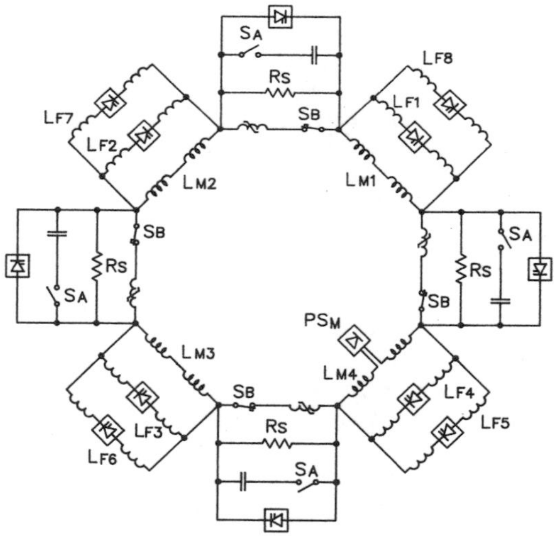

Fig. 2. RFX Poloidal Circuit.

COIL MODELS

For a proper design of the insulation structure of large coils, the transient voltage stress to which any point of the structure will be exposed needs to be known.

However, for a proper transient analysis a coil cannot be simply represented by its self inductance. At high frequencies a coil is essentially a capacitive inductive device, the resistances having little effect. The capacitances to ground and the capacitances between turns and coil sections must be taken into account. In addition, mutual coupling between coil turns cannot be disregarded [1, 7, 11].

Even if the coil is actually a three-dimensional distributed network of capacitances and mutually coupled inductances, models with lumped capacitances and inductances are used, in order to give a simpler description and to fit the pratical limitation of computers. A number of models have been proposed. The simplest and most common approach is to represent the coil as a chain of series connected inductances with shunt capacitances and capacitances to ground [2, 11].

In our case this approach has been followed, since it appears appropriate and convenient because of the availability of large computer network simulators: in the past, the number of sections of the chain used to represent a similar model was limited (usually to ten) [1, 12], due to the difficulties of an analytical representation and solution of a complex circuit.

Fig. 3a shows the cross section of one of the 16 Field Shaping RFX coils, having 5.4 m diameter. The coil is composed of 24 turns in series, wound in four coaxial layers, and is mechanically supported by 24 metal rings evenly spaced along the circumference, which are also the ground reference for the coil. Its lumped equivalent network is shown in Fig. 3b, and takes into account:

- the self inductance of each turn (L_1-L_4);
- the mutual inductances between turns;
- the capacitance between turns (Cs);
- the capacitance between layers (Ca);
- the capacitance to ground (Ct).

Since the circuit behaviour strongly depends on the values of the mutual inductance between turns, particular care was devoted to calculating them to a very high level of accuracy. This has been achieved using an algorithm derived from Garret's formulae [9], based on Gauss weight integration over the cross sectional area of each turn, suitably subdivided into subelements [6]: the number of subelements was chosen in order to provide the required degree of accuracy that is checked by means of a convergence test.

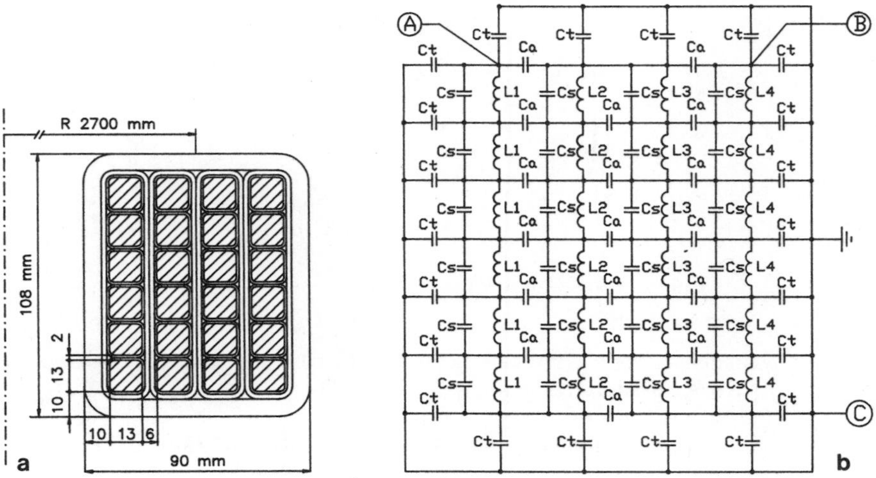

Fig. 3. Cross Section of RFX Field Shaping Coil F8 (a) and its equivalent network (b), where mutual couplings are not indicated for clarity.

Capacitances have been calculated by a FEM field analysis. It should be noticed that, in our particular winding geometry, the simple assumption of uniform electric field between turns leads to capacitances between adjacent turns that differ by only a few percent from the actual values. In addition, the cross capacitances, being at least one order of magnitude lower, have been disregarded in the equivalent network.

The network parameter values are:

L_1 = 20.8 μH, L_2 = 21.0 μH, L_3 = 21.2 μH, L_4 = 21.4 μH

C_a = 1.2 nF, C_s = 4.3 nF, C_t = 40 pF

The network of Fig. 3b is an accurate enough representation of the coil for design purposes, because if a transient analysis is performed no information about turn to turn, layer to layer or turn to ground voltages is lost.

COIL MODEL ANALYSIS

The network of Fig. 3b was then analysed using the SPICE 2 code, by feeding the network through points A (a coil terminal) and C (ground terminal) with a unit frequency-varying AC current. The approach is to some extent similar to the method used to experimentally measure resonance frequencies of real coils [3].

The network can be considered as a passive double-bipole (the ground terminal being common to the two gates of the bipole). The voltage measured between points A and C is the value of the bipole impedance Z_{11}, while the voltage between B (the other coil terminal) and C is the mutual impedance Z_{12}.

Varying the current frequency we obtain the value of Z_{11} and Z_{12} in magnitude and phase, as frequency functions. In Fig. 4, where the magnitudes of the bipole impedances are plotted against frequency, the zeros and poles are easily identified: they correspond to coil resonance frequencies, and are summarized in Tab. 1.

It should be noticed that as frequency increases, phenomena such as skin effects, neglected in the model, become more and more important, causing high frequency oscillations to be damped.

In any case, it is possible to predict very accurately the most important resonance frequencies of every coil. Moreover, if the voltage waveform to which the coil has to be subjected during operation is known, the dielectric stress in every point of insulation can be predicted during design through a transient simulation.

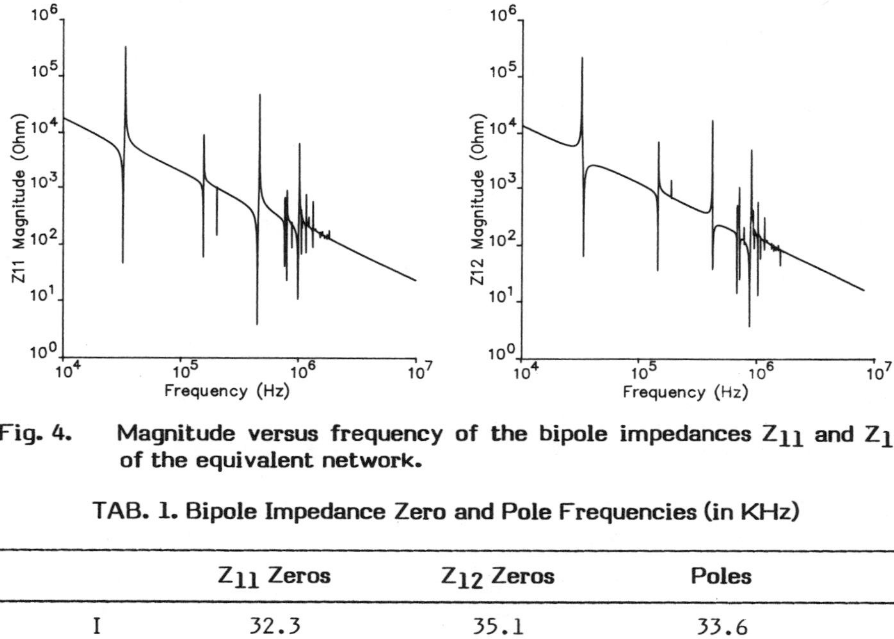

Fig. 4. Magnitude versus frequency of the bipole impedances Z_{11} and Z_{12} of the equivalent network.

TAB. 1. Bipole Impedance Zero and Pole Frequencies (in KHz)

	Z_{11} Zeros	Z_{12} Zeros	Poles
I	32.3	35.1	33.6
II	157.4	157.1	158.7
III	205.1	205.1	205.1
IV	453.0	472.2	470.0
V	809.0	815.9	811.7
VI	1008	999.4	1032

COIL MODEL SIMPLIFICATION

In a very complex system such as the RFX Poloidal Circuit (Fig. 2), in order to determine the voltages appearing in every condition at the coil terminals, it is necessary to perform a computer simulation of the overall circuit. In order to have a reasonably low computer load, the dimension of the model described above must be substantially reduced, while still retaining the voltage behaviour at the coil terminals.

In our case we have considered that the first resonance is by far the most important, mainly because during machine operation no waveform able to significantly excite higher frequency modes is expected to be applied to the coils. Moreover, as a result of the overall circuit simulation performed with the simplified coil model it has been found that for a proper operation the first resonance has also to be damped, inserting linear resistors of an appropriate value between each coil terminal and ground [5]. In this way higher frequencies will be damped even further.

For our purposes the analytical expression of the bipole impedances Z_{11} and Z_{12} can then be written considering the first zero-pole couple only and the behaviour towards zero and infinity.

$$Z_{11}(s) = \frac{K_{11}}{s} \cdot \frac{s^2 + \omega_{01}^2}{s^2 + \omega_p^2} \qquad Z_{12}(s) = \frac{K_{12}}{s} \cdot \frac{s^2 + \omega_{02}^2}{s^2 + \omega_p^2} \tag{1}$$

In the above expressions, written using Laplace transforms, ω_{01} and ω_{02} are the first zero frequencies of Z_{11} and Z_{12} respectively, ω_p is the correspondent pole frequency, which is the same for Z_{11} and Z_{12}, since the bipole is reactive.

The problem is now to find a lumped circuit network presenting an impedance matrix $[Z^*]$ with an impedance expression of the type given in (1).

A simple model consisting of an inductance with shunt capacitance and a capacitance to ground from each terminal (Fig. 6) is suitable for the purpose: in fact the following expressions can be easily found:

$$Z^*_{11}(s) = \frac{K_{11}}{s} \cdot \frac{s^2 + \dfrac{1}{L(C_g + C_s)}}{s^2 + \dfrac{1}{L(C_s + C_g/2)}} \qquad Z^*_{12}(s) = \frac{K_{12}}{s} \cdot \frac{s^2 + \dfrac{1}{L\,C_s}}{s^2 + \dfrac{1}{L(C_s + C_g/2)}} \tag{2}$$

Where:

$$K_{11} = \frac{C_s + C_g}{C_g(2C_s + C_g)} \qquad K_{12} = \frac{C_s}{C_g(2C_s + C_g)} \tag{3}$$

Finally the values of C_s and C_g which make the expressions (1) and (2) identical are easily found.

The parameters calculated for the RFX Field Shaping coil described above are: $\mathbf{C_s}$ = 2.40 nF, $\mathbf{C_g}$ = 430 pF, **L** = 8.562 mH.

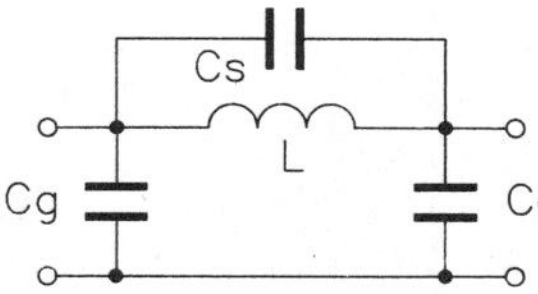

Fig. 5. Reduced coil model.

The values of the coefficients K_{11} and K_{12} can be also calculated from the result of the network analyses performed in the frequency domain (Fig. 5) with the following formulae:

$$K_{11} = \left| Z_{11}(j\,\omega_v) \right| \cdot \omega_v \cdot \frac{\omega_{01}^2 - \omega_v^2}{\omega_p^2 - \omega_v^2} \qquad K_{12} = \left| Z_{12}(j\omega_v) \right| \cdot \omega_v \cdot \frac{\omega_{02}^2 - \omega_v^2}{\omega_p^2 - \omega_v^2} \tag{4}$$

where ω_V is the frequency for which the magnitude of the impedance is evaluated. It should be noted that in general the values of the coefficients calculated in this way vary slightly from those given in (3). Their difference becomes more and more negligible as the higher harmonic frequencies move far away from the first or are small in amplitude. If the complete coil model presented a single resonance frequency they would be identical. For this reason the gap between the coefficents calculated in the two ways represents a good test for evaluating the soundness of the simplified model.

CONCLUSIONS

The main advantage of the method proposed is that the model parameters are identified through the actual resonance frequencies, which can be computed to a very high accuracy.

The basic coil model used to set up the full model of the RFX Poloidal Field System, in spite of being very simple, proved to be suitable for an overall transient analysis. Moreover it should be pointed out that the method is quite general and can still be used to identify the parameters of more complex models to be used if more resonance frequencies have to be taken into account.

REFERENCES

1. Abetti P.A., Maginiss F.J.: "Natural Frequencies of Coils and Windings Determined by Equivalent Circuit". AIEE Trans. on Power Apparatus and System, June 1953.
2. Babare A., et al.: "Resonance Behaviour of High Voltage Transformers". Proc. of CIGRE 1984.
3. Babare A., Ciolli P., Nadali C., Sandrate G.: "Diagnostica nei trasformatori di grande potenza" (in Italian). L'Energia Elettrica, n. 10, 1985.
4. Bellina F., Chitarin G., Guarnieri M., Stella A.: "The RFX Field Shaping Winding Design". Proc. of the 9th International Conference on Magnet Technology MT-9, Zürich (Switzerland), 1985.
5. Campostrini P.P., Chitarin G., Stella A.: "Fault Analysis and Protection Concepts for RFX Poloidal Magnetic Field System". Proc. of 11th Symposium on Fusion Engineering, Austin, Texas (USA), 1985.
6. Chitarin G., Guarnieri M., Stella A.: "Transient Behaviour of Thick-Walled Axisimmetric Windings:a Lumped Parameter Approach", to be published in the January 1988 issue of the IEEE Transaction on Magnetics.
7. Chowduri P., Anderson M.: "Performance of Large Magnets under Transient Voltage". Proc. of the 9th Symposium on Engineering Problems of Fusion Research, Chicago, Illinois (USA), 1981.
8. Chowduri P.: "Transient Voltage Oscillation in Coils". Fusion Technology, vol. 8, no. 1, 1985.
9. Garrett M.W.:" Calculation of Fields, Forces, and Mutual Inductances of Current System by Elliptic Integrals". Journal of Physics, vol. 34, no. 9, 1983.
10. Guarnieri M., Modena C., Schrefler B., Stella A.: "Electromagnetic and Mechanical Design of RFX Magnetizing Winding". Fusion Technology, vol. 8, no. 1, 1985.
11. Mc Nutt W.J., Blalock T.J., Hinton R.A.: "Response of Transformer Winding to System Transient Voltages". IEEE Trans. on Power Apparatus and Systems, vol. PAS-93, no. 2, 1974.
12. Owen E., Schimer D.: "Switching Transient in a Superconducting Coil". Proceeding of the 10th Symposium on Fusion Engineering, Philadelphia, Penn., 1983.
13. Rostagni G. et al.: "The RFX Project: a Design Review". Proc. of the 13th Symposium on Fusion Technology, Varese (Italy), 1984.

REMOTE ELECTROMAGNETIC PROPULSION IN THE PRESENCE OF A METALLIC SHIELD

R.D. Findlay and J.H Dableh

Department of Electrical and Computer Engineering
McMaster University
Hamilton, Ontario, Canada L8S 4L7

ABSTRACT

This paper describes an electromagnetic technique to reposition the spacer springs in CANDU reactors. The need to perform this manoeuvre in non-commissioned reactors was required subsequent to the failure of a pressure tube in a reactor at a Canadian Nuclear Generating Station. A contributing factor in the failure of the tube was the fact that the annular spacers used to maintain the coaxial configuration between the metallic pressure tube and its surrounding metallic calandria tube, had been displaced. Afterwards it was realized that displacement of the spacers had also occurred in the non-commissioned reactors. It will be recognized that the spacers were not accessible for mechanical repositioning, leaving only the possibility of an electromagnetically coupled remote repositioning procedure. This paper describes the analysis and solution of the complex coupled electromechanical problem. The mathematical analysis problem was ultimately solved using a finite element process.

INTRODUCTION

On August 1, 1983, a sudden pressure tube (P/T) rupture was experienced in one fuel channel of unit number two at the Pickering Nuclear Generating Station (NGS) in Ontario, Canada [1]. A contributing factor to the rupture of the tube was the fact that the annular spacers, known as garter springs (G/S), used to maintain the coaxial configuration between the P/T and its surrounding calandria tube (C/T), had been displaced from their design location for a number of years. Displacement of the G/S allowed the P/T and C/T to come into contact which was thought to have caused hydride blistering and helped to accelerate the propagation of hydride layers through the wall of the zirconium-niobium P/T. Subsequent to this finding, it was discovered that about eighty percent of the spacers in non-commissioned reactors had been dislocated during the construction stage and thereafter hot conditioning of the primary heat transport system of the reactor. After loading of the fuel bundles inside the P/T, the G/S are pinched between the P/T and C/T, thus preventing further significant movement [2,3].

Repositioning of the dislocated G/S in five new reactors was deemed necessary before the reactors were brought into service. Since there is no direct mechanical access to the spacers after assembly of the reactor fuel channels, the only available option at that time to rectify the problem would have involved major dismantling of the fuel channels and replacement of the P/T and associated end fittings: a major and costly undertaking To circumvent what seemed an insurmountable difficulty, a novel electromagnetic technique to remotely reposition the spacers from within the P/T was conceived and demonstrated experimentally on very short notice [2,4]. Further optimization and computer simulation of the electromagnetic technique, for potential future use, were completed in 1985 and 1986 [2, 5–10]. The technique to reposition the G/S consisted of inserting a solenoidal-type coil through the P/T, to a location adjacent to the spacer, and passing a time-varying electrical current having appropriate frequency and magnitude to interact with the G/S and cause it to move in the axial direction. The ensuing transient electromagnetic field problem was ultimately analyzed and solved using a finite element process [2, 8–10].

General Description of the Problem and Solution Technique

The core of a CANDU nuclear reactor consists of a large array of fuel channels, 390 or 480 channels depending on reactor size, housed in a large vessel called the calandria. Each fuel channel consists basically of two concentric tubes approximately 6.1 m long: the inner coolant tube known as the P/T, which holds the uranium fuel bundles is 4.2 mm thick and has an inner diameter of 104 mm. The outer tube, known as the calandria tube, is only 1.5 mm thick and its inner diameter is approximately 129 mm. Four G/S spacers are used in each channel to maintain the concentricity of the P/T and C/T and prevent them from coming into contact when the heavy fuel bundles are loaded in the P/T, and subsequently as the P/T grows due to thermal expansion (operating temperature is approximately 300°C) and neutron bombardment. The support structure and end fittings at both ends of the horizontal fuel channels are arranged in a fashion such as to allow the flow of a gas in the annular space between the P/T and C/T but does not provide for any mechanical access to the spacers.

The P/T, C/T and G/S are made of a non-magnetic zirconium alloy which has an electrical conductivity of 1.246×10^{6} S/m (approximately 47 times smaller than the conductivity of copper). The spring itself is open circuited; however the spacer is held in circular form by the tie wire which is made of zirconium-niobium alloy as well and has a thickness of 0.9 mm. Therefore, the main conceptual difficulty was to interact electromagnetically with a thin member, having poor electrical properties, from within a thick tube made of poorly conducting material, and exert a force of sufficient magnitude to overcome the static friction at the seating of the spacer to move it longitudinally in either positive or negative axial direction. In addition to the electromagnetic shielding problem presented by the P/T, there existed several restrictions imposed by the geometry and dimensions of the fuel channels on the design of the electromagnetic coil, transmission line and locations of the equipment delivering the electrical current. Also, the electromagnetic forces exerted on the P/T itself had to be maintained well within specified limits established to ensure the metallurgical integrity of the fuel channel components.

The concept of the electromagnetic solution technique consisted of inserting a solenoidal-type coil through the P/T, to a location such that the first or the last turn of the coil was positioned under the G/S, and passing an oscillatory current pulse through the coil to generate a time-varying electromagnetic field. The main requirements were to keep the frequency of the current low enough to minimize the shielding effects of the P/T, yet high enough to induce a secondary current in the closed path of the thin G/S, and to generate a time-varying magnetic field sufficiently strong to exert the required axial force. This was achieved initially by using current pulses generated by discharging a capacitor bank in the oscillatory mode in conjunction with coils having a very small number of turns (as little as four turns). Figure 1 shows a short section of the P/T and C/T with cutaways in them to show the G/S and the inserted solenoidal coil.

Although the required electrical current could have been supplied by a variety of sources, capacitor bank systems were used to demonstrate the electromagnetic technique because they were readily available at Ontario Hydro Research Laboratory. Figure 2 shows a simplified circuit diagram of a typical capacitor bank system used in this project. The parameters of the capacitor bank system were selected to generate an oscillatory current [5,6] for a load coil having nine turns, an outer diameter of about 92 mm, a length of about 90 mm, a dc resistance of 2.2 mΩ, and an inductance of about 2.5 μH. For a G/S displacement of approximately 10 cm per capacitor discharge, the first current peak was set at 140 kA, the current reversal (defined as the ratio of the second current peak over the first one) at 72% and the frequency at approximately 1000 Hz.

Figure 3 illustrates the structure of a coil prototype having 12 turns before it was encapsulated and restrained to endure the strong electromagnetic forces exerted on it in the axial and radial directions. The two stainless steel bolts are inserted inside the winding and are

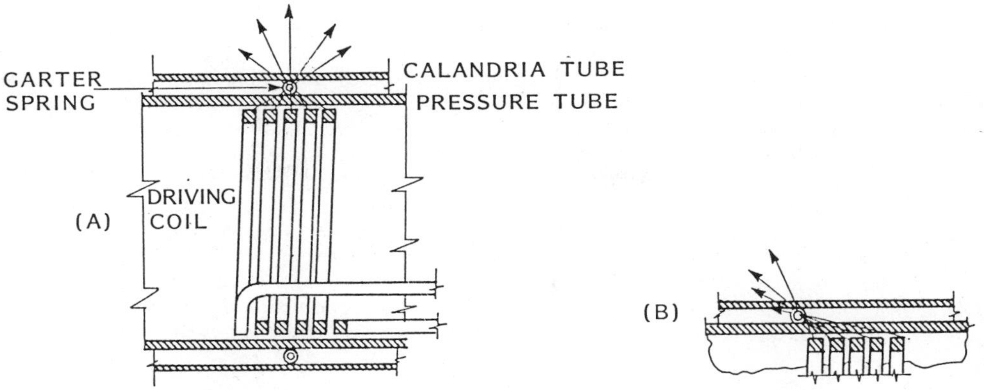

Fig. 1. Photograph of a short fuel channel section with a solenoidal coil inserted inside the P/T.

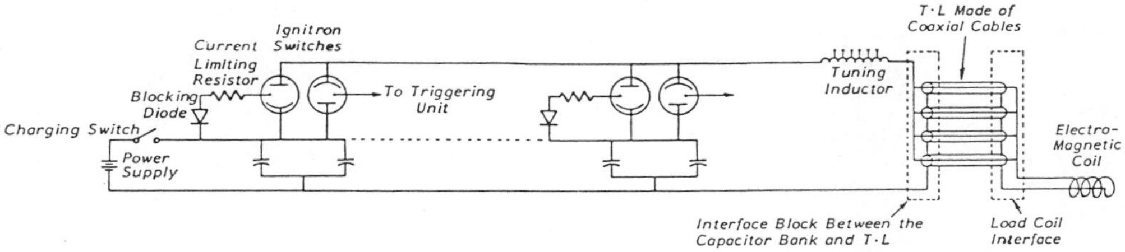

Fig. 2. Simplified circuit diagram of the capacitor bank system

tightened against the end caps of the coil to provide axial pre-stressing for the coil. A coil prototype (pre-stressed in the radial direction by wrapping it with several layers of epoxy-loaded fibreglass bands under tension) is capable of withstanding over five thousand capacitor discharges at the current level of 140 kA. Full details on the design and construction of these various coils are provided in reference 7.

Fig. 3. Photograph of a coil prototype before insulation and pre-stressing.

For any particular coil design, the amount of G/S displacement increased, as the magnitude of the current and the current reversal were increased and the frequency of the current was decreased. Quantitative correlation of all the influential parameters related to the design of the coil, current characteristics and the capacitor bank system is documented in references 2, 3, 7 and 10

Theoretical Analysis and Simulation of the Coupled Electro-mechanical Problem

The theoretical analysis of the electromagnetic repositioning of the G/S consisted mainly of developing a general solution for the electromagnetic field problem involving an impulse or transient current input in a cylindrical geometry. Such a solution enables the determination of various electromagnetic, thermal and mechanical quantities of interest as functions of both space and time. Although the duration of the effective current pulse was only about 2 ms, the repositioning process involved a significant amount of energy that was condensed in a relatively small volume. The temperature of the coil and P/T section surrounding it increased by approximately 40°C for each capacitor discharge. Therefore, in its broadest sense, the repositioning process involved a complex coupled electromagnetic, thermal and mechanical transient problem. Moreover, the process depended heavily on the accuracy of locating the G/S, the initial orientation and accurate positioning of the coil with respect to the location of the G/S.

Eddy-current probes were used to determine the location of the G/S. These probes provided an indication as to where the centre of the G/S was with an accuracy of ± 1.0 cm but could not determine the direction or degree of inclination of the G/S. In most cases the G/S were tilted because they were loose with respect to the P/T: they conformed to the inner diameter of the C/T but were not restricted to or from leaning to an inclined plane. The objectives of the solution technique were to substantiate that axial forces of sufficient magnitude to displace the spacer could be generated in spite of the presence of a thick shielding tube made of the same material as the G/S, and to provide a mechanism to optimize the G/S repositioning process.

The electromagnetic field problem was formulated in terms of the magnetic vector potential, in order to minimize the number of unknown variables involved in the solution process. Certain assumptions related to the geometry of the helical coil were made to reduce the problem from a three dimensional field problem to an axisymmetric vector potential problem. The magnetic field components were calculated from the spatial distribution of the magnetic vector potential, while the electric field component in the circumferential direction was determined from the variation of the magnetic vector potential in time. Other electrical, thermal and mechanical parameters of interest can then be easily determined. The finite element technique was selected to solve for the magnetic vector potential in the space domain. First order triangular elements were chosen for their simplicity, flexibility and suitability for the fuel channel geometry. References 2, 8 and 9 contain full documentation of the solution technique and the major results.

The transient variation of the current input was represented using a time discretization technique which does not place any restriction on the rate of change of the current signal or its shape. The Crank-Nicholson recurrence scheme was used to represent the variation of magnetic vector potential in the time domain. The size of the time step used was 25 μsec. All of the desired variables were calculated for each time iteration and their values were available for each instant of time. Of particular interest for this application was the axial force exerted on the G/S. Hence, the axial, as well as the radial force density were calculated at the middle of each time step for all the metallic elements in the finite element mesh representing the coil and the fuel channel components. This information was quite sufficient to establish the necessary indicators for optimization of the G/S repositioning process. However, in order to enable the comparison of the numerical results to those achieved experimentally, the G/S displacement had to be calculated from the numerical results since it was the most important parameter in the process and the one that could be measured easily with the most accuracy. To ensure a high degree of accuracy in the establishment of the major results and conclusions, a number of experiments were conducted using a short section (60 cm) of P/T specimen and actual G/S without the C/T. This facilitated the accurate positioning of the coil with respect to the G/S and saved a significant amount of experimental time. A simplified theoretical model to correlate the G/S displacement to the computed instantaneous axial force density exerted on the G/S was developed. Figure 4 presents a graph of this force density, as computed by the program for a typical current pulse in a nine-turn coil.

It should be noted that the above force density was calculated assuming the original position of the G/S, i.e. the finite element mesh was not modified in between the time steps to update the new position of G/S during the current impulse. This has the implication of overestimating the instantaneous force on the G/S after it has started to move. To minimize the error that would result from this assumption, the model was used only to investigate cases

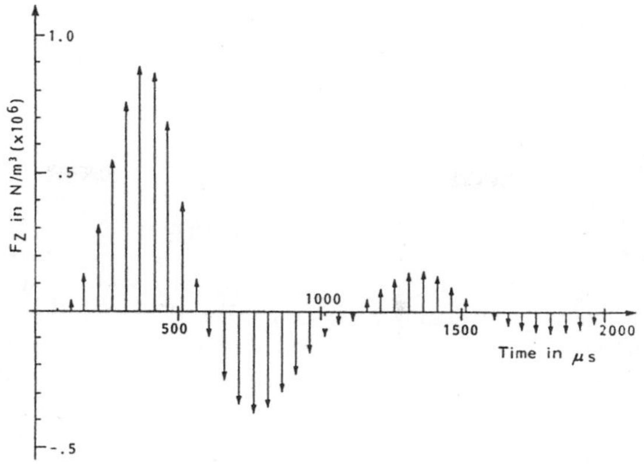

Fig. 4. Instantaneous axial force density exerted on the spacer

having minimal G/S movement (below 5 cm for each capacitor discharge). The impulse energy applied to the G/S via the electromagnetic induction process can be defined as:

$$\text{I.E.} = \int_0^{t_1} F(t)\,dt = \frac{w}{g}(v_1 - v_0) \tag{1}$$

where F(t) is the instantaneous axial force in newtons, t_1 is the duration over which the force is exerted in seconds, w is the weight in newtons, g is the gravitational constant ($g = 9.81$ m/s^2) and v_1 is the velocity in m/s [11]. Noting that the initial velocity v_0 is zero since the G/S is initially at rest and that the integral expression represents the area under the curve F(t) in the time interval 0 to t_1, equation (1) can be used to determine the velocity v_1 of the G/S during its displacement. Once v_1 is known, the energy balance equation which states that the energy supplied to the G/S at its initial position is equal to its energy at the new position plus the losses incurred during its movement, can be used to calculate the G/S displacement in the form

$$\frac{1}{2}\frac{w}{g}v_1^2 = F_r d \tag{2}$$

where F_r is the friction force exerted on the G/S and d is the amount of G/S displacement. The mass of a G/S is 33 g. The friction force encountered by the G/S when it is moved in the axial direction with only the P/T present was measured and found to be 19.6 mN The volume of the G/S is 10.06 mm^3. The impulse energy defined in equation (1) was computed by summing the product of the instantaneous force (calculated at each time step) by the size of the time step (25 μs). The instantaneous force was obtained by multiplying the computed axial force density in the G/S by the volume of the G/S.

The impulse energy for the case presented in this paper, that is the area under the curve joining the tips of the arrows in Fig. 4, was found to be 4.35 mJ. Thus, using equations (1) and (2) yields a G/S displacement of approximately 18 mm. Therefore, the computed G/S displacement is approximately 17% lower than the measured value. This is quite satisfactory considering the simplicity of the mechanical model and the several sources of error involved in various measurements. In addition to the above comparison, there were other important indicators which illustrated that the experimental and simulation findings were in agreement. For example, the influence of the current magnitude on the G/S displacement was found to play a primary role during the experimental program: this finding was confirmed by the numerical results.

The accuracy of the numerical solution was checked by considering several cases for which an analytical solution exists. For example, the case of an infinitely long solenoidal coil with a direct current input (static case) was analyzed. The external forcing function was represented in the computerized solution by a step function. Comparison between the analytical and numerical results for steady-state conditions showed a difference of less than 1% for points that are away from the axis of the solenoid by a distance exceeding one-third of the radius, and a maximum difference of less than 2% for points closer to the axis. A detailed description of the tests performed to check the accuracy of the numerical solution and to ensure that the computer program is free from error is provided in references 2 and 12.

CONCLUSION

The concept of the electromagnetic technique to reposition remotely a thin non-magnetic spacer located outside a thick metallic tube has been demonstrated theoretically and experimentally. Development of this technique alleviated the need to perform large scale fuel channel retubing operations in five non-commissioned CANDU reactors at Ontario Hydro A rigorous and accurate theoretical model, based on finite elements, has been developed to analyze the electromagnetic field problem involved in the G/S repositioning process. The numerical results have confirmed that the axial impulse force exerted on the G/S is sufficient to cause it to move.

ACKNOWLEDGEMENT

The authors wish to express their great appreciation to Ontario Hydro for the opportunity to work on this problem. There are many individuals in that organization whose contributions to this work have been significant.

REFERENCES

1. D. Mosey, "Tube Failure at Pickering Nuclear Generating Station", Energy Newsletter, Vol. 5, No. 2, June 1984, ISSN 0711-3366, pp. 56-65, published by the McMaster Institute for Energy Studies.
2. J.H. Dableh, "A Novel Electromagnetic Technique for Remote Repositioning of Coolant Tube Spacers in CANDU Nuclear Reactors", Ph.D. Thesis, McMaster University, Hamilton, Ontario, Canada, May 1986.
3. Joseph H. Dableh, "Novel Electromagnetic Technique for Repositioning of Coolant Tube Spacers in CANDU Nuclear Reactors", Review of Scientific Instruments, Vol. 57, No. 6, June 1986.
4. Joseph H. Dableh, "Methods of Repositioning Annular Spacers in Calandria Structures and Apparatus Thereof", United States Patent Number 4,613,477, Sept. 23, 1986.
5. J.H. Dableh, R.D. Findlay, N.T. Nicholson, N.T. Olson, "Experience with High Pulse Power Systems for Remote Repositioning of Concentric Coolant Tube Spacers", Jordan International Electrical and Electronic Conference, April 28-May 1, 1985, Amman, Jordan.
6. J.H. Dableh, R.D. Findlay, I.L. Colquhoun, and M.E. Treumner, "Cable for High Pulse Power Application", IEEE Transactions on Power Apparatus and Systems, Vol. PAS-194, No. 8, August 14, 1985.
7. J.H Dableh, R.D. Findlay and G.S. Klempner, "Design and Development of Compact, Durable Electromagnetic Pulse Power Coils for Repositioning of Coolant Tube Spacers in CANDU Nuclear Reactors", IEEE Transactions on Energy Conversion, Vol. EC-1, Number 4, December 1986.
8. E. Tarasiewicz, R.D. Findlay, and J.H. Dableh, "Finite Element Approach to the Solution of Axisymmetric Vector Field Problems", Paper No. GB-11, Intermag '86 Conference, Phoenix, AZ, April 14-17, 1986.
9. E. Tarasiewicz, R.D. Findlay and J.H. Dableh, "Accurate Computation of Axisymmetric Vector Potential Fields with the Finite Element Method", paper No. 86 SM 406-3, 1986 IEEE Summer Power Meeting, July 20-25, 1986, Mexico City.
10. R.D. Findlay and J.H. Dableh, "Summary of Theoretical and Experimental Development of a Novel Electromagnetic Technique for Remote Repositioning of Coolant Tube Spacers in CANDU Nuclear Reactors", paper accepted for presentation at the IEEE 1987 Summer Power Meeting, July 12-17, in San Francisco, and for publication in the IEEE Transactions on Energy Conservation.
11. W.T. Thomson, "Theory of Vibrations with Applications", Prentice-Hall Inc., Englewood Cliffs, N.J., 1981.
12. E. Tarasiewicz, R.D. Findlay and J.H. Dableh, "The Variational Treatment of the Diffusion Equation for Vector Field Problems", IEEE Transations on Magnetics, Vol. MAG-23, No. 4, July 1987.

APPLICATION OF THE HARMONIC ANALYSIS TECHNIQUE TO DETERMINING EDDY CURRENTS IN CONDUCTING PLATES

R.D. Findlay, B. Szabados, I. ElNahas, and M.S. ElSobki

Department of Electrical and Computer Engineering
McMaster University
Hamilton, Ontario
Canada L8S 4L7

M. Poloujadoff

E.N.S.I. Paris
France

ABSTRACT

A novel analytical approach to predict eddy currents in a conducting plate is proposed. Based upon the knowledge of the incident flux to the rectangular plate, a curve fit to a double Fourier series is obtained. This closed formula representation is used to solve the diffusion equation using the physical boundary conditions of the plate. Limitations and applicability of the method are then discussed.

INTRODUCTION

Losses in conducting plates has been a major concern for power equipment designers since the turn of the century. With the advent of computers and new approaches to analysis and evaluation of such losses, the problem has become somewhat more tractable. However, the computational effort to establish eddy current losses in some applications is still too extensive to be practicable. The method requires only the values of the incident flux density on the surface of the plate: these values may be obtained by any flux solving program, or, if required, by experimental means. This method assumes that the incident flux distribution is known for a grid on the surface of a rectangular plate. The equations leading to the flux density distribution within the plate are derived, then the eddy currents in the plate are found. The paper describes how the incident flux may be analytically represented, using a least mean error fit to a double Fourier series, to produce the values of eddy currents and further losses within the plate. Some discussion on the expansion of this method to piece-wise linear modeling is given, based upon experimental results which were in excellent agreement with the predicted values. This paper discusses an alternative approach, one which makes stray loss evaluation an economically feasible computation.

FLUX IN A RECTANGULAR PLATE

Basic Equation for Eddy Currents and Boundary Conditions

The equation used to determine the induction of eddy currents in conductors is a limiting form of Maxwell's equation. The full set of Maxwell's equations leads to a general second order equation which is extremely difficult to solve. Because the displacement current within conductors may be neglected with respect to the conductive current, the equations take on the characteristics of a diffusion equation. The basic set of equations consists of the quasisteady form of Maxwell's equations:

$$\nabla \times \mathbf{E} = -\frac{\partial \mathbf{B}}{\partial t}, \quad \nabla \times \mathbf{H} = \mathbf{J}, \quad \nabla \cdot \mathbf{E} = 0, \quad \nabla \cdot \mathbf{B} = 0 \qquad (1\text{-}4)$$

To this we add Ohm's law for a linear, isotropic, stationary conductor where σ is the electric conductivity of the material (5), and then constitutive relation (6) between the flux density **B** and magnetic intensity **H**

$$\mathbf{J} = \sigma \mathbf{E} , \qquad \mathbf{B} = \mu \mathbf{H} \tag{5-6}$$

Using Equations (5) and (6) to eliminate the current density **J** and the electric field intensity **E** in equation (1-4), one can easily show that

$$\nabla^2 B = \mu \sigma \frac{\partial \mathbf{B}}{\partial t} \qquad \text{where } \nabla^2 = \frac{\partial^2}{\partial x^2} + \frac{\partial^2}{\partial y^2} + \frac{\partial^2}{\partial z^2} \tag{7-8}$$

Let us consider a finite solid rectangular plate of length L, width ℓ, and depth a, with linear characteristics for μ_r and σ. The flux density is considered to be established by an outside source such that it impinges normally on the plate on the upper face (see figure 1). At each point [x,y] on the upper face the incident flux is assumed normal and has a known value. Since we have chosen the centre of the surface of the plate as the origin for our coordinates, we also assume a symmetrical incident flux pattern.

$$B_{oi}(-x,y,a/2) = -B_{zi}(x,y,a/2) , \qquad B_{oi}(x,-y,a/2) = -B_{zi}(x,y,a/2) \tag{9}$$

Now the method of separation of variables together with the following boundary conditions are used to obtain a solution for equation (7). These boundary conditions are:

1. Since the incident flux density, actually denotes the flux density in the air at the bottom plate surface (z=0), **B** has only a z component which satisfies the conditions in equation (9).

$$B_i = B_z(x,y) \tag{10}$$

2. Since no flux crosses the top surface of the plate (z=a), $B_z = 0$.
3. J_y is zero for $y = \pm \ell/2$ and J_x is zero for $x = \pm L/2$

Development in Fourier Series: We may seek to develop an expression for the flux as a double Fourier series [10]. Hence we are effectively seeking a solution in the form of an harmonic function. We express the steady state normal component of the flux density in the plate as:

$$B_z = \sum_{\substack{m=1,3,..\\ n=1,3,..}} A_{mn} \sin\left[\frac{2m\pi x}{L}\right] \cos\left[\frac{n\pi y}{\ell}\right] \left[\frac{e^{\alpha_{mn} z} - e^{\alpha_{mn}(2a-z)}}{\left(1 - e^{2a\alpha_{mn}}\right)}\right] \tag{11}$$

where A_{mn} is a coefficient obtained from the known incident flux pattern, and

$$\alpha_{mn}^2 - \left(\frac{2m\pi}{L}\right)^2 - \left(\frac{n\pi}{\ell}\right)^2 - j\omega\mu_0\mu_r\sigma = 0$$

The sinusoidal functions account for the lateral boundaries while the exponential functions account for the field penetration. Since $(1 - e^{2a\alpha_{mn}})$ is a constant we may use the final formulae:

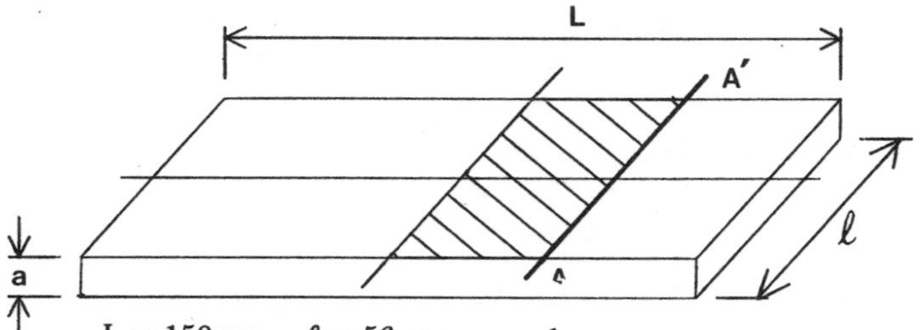

$L = 150$ cm , $\ell = 56$ cm , $a = 1$ cm

Fig. 1. Simplified Rectangular Plate

$$B_z = \sum_{\substack{m=1,3,..\\ n=1,3,..}} C_{mnz} \sin\left[\frac{2m\pi x}{L}\right]\cos\left[\frac{n\pi y}{\ell}\right]\left[e^{\alpha_{mn} z} - e^{\alpha_{mn}(2a-z)}\right]$$

$$B_y = \sum_{\substack{m=1,3,..\\ n=1,3,..}} C_{mny} \sin\left[\frac{2m\pi x}{L}\right]\sin\left[\frac{n\pi y}{\ell}\right]\left[e^{\alpha_{mn} z} + e^{\alpha_{mn}(2a-z)}\right] \tag{12}$$

$$B_x = \sum_{\substack{m=1,3,..\\ n=1,3,..}} C_{mnx} \cos\left[\frac{2m\pi x}{L}\right]\cos\left[\frac{n\pi y}{\ell}\right]\left[e^{\alpha_{mn} z} + e^{\alpha_{mn}(2a-z)}\right]$$

Since $\nabla \cdot \mathbf{B} = 0$ and $J_z = 0$ on the boundaries, the upper and lower faces at $z = \{0,a\}$, we can write:

$$C_{mnx} = \frac{\alpha_{mn} C_{mnz}}{\pi} \frac{2mL\ell^2}{n^2 L^2 + 4m^2 \ell^2}, \quad C_{mny} = \frac{\alpha_{mn} C_{mnz}}{\pi} \frac{nL\ell^2}{n^2 L^2 + 4m^2 \ell^2} \tag{13}$$

Hence A_{mn} can be obtained from the incident flux on the upper surface, leading to a closed form solution for the flux density within the plate $[B_x, B_y, B_z]$ using (12) with (13).

DEVELOPMENT OF THE INCIDENT FLUX

Here we assume that the incident flux has been obtained at discrete points on the surface of the upper face of the plate. We rely on a data matrix B(x,y), either coming from measured values in an experimental setup, or a flux simulation such as reported in [2]. The incident flux may be represented by the double Fourier expansion.

$$B_z = \sum_{\substack{m=1,3,..\\ n=1,3,..}} A_{mn} \sin\left[\frac{2m\pi x}{L}\right]\cos\left[\frac{n\pi y}{\ell}\right] \tag{14}$$

The coefficients A_{mn} were computed according to the following least error fit at all points of the grid.

$$(\varepsilon)^2 = \sum_{i,j} \left\{ B_z(i,j) - \sum_{m,n} A_{mn} \sin\left[\frac{2m\pi x(i)}{L}\right]\cos\left[\frac{n\pi y(j)}{\ell}\right]\right\}^2 \tag{15}$$

If equation (15) is differentiated with respect to the coefficients A_{mn} the error can be established as zero, resulting in a set of algebraic equations with arguments A_{mn}:

$$\begin{bmatrix} S_{m_i n_j} & S_{M_i N_j} & \cdots\cdots \\ \cdots & \cdots & \cdots\cdots \end{bmatrix} \begin{bmatrix} A_{mn} \\ \\ \end{bmatrix} = \begin{bmatrix} \sum_{i,j} B_{z(i,j)} \cdot S_{m_i n_j} \\ \cdots\cdots \end{bmatrix}$$

where

$$S_{m_i n_j} = \sum_{i,j} \sin\left[\frac{2m\pi x_i}{L}\right]\cos\left[\frac{n\pi y_j}{\ell}\right], \quad S_{M_i N_j} = \sum_{i,j} \sin\left[\frac{2M\pi x_i}{L}\right]\cos\left[\frac{N\pi y_j}{\ell}\right],$$

$i = 1$ .. number of points on the X axis on plate section and $j = 1$ .. number of points on the Y axis on plate section. The solution of the coefficients, obtained using standard numerical methods,leads to the evaluation of the flux density at any depth in the plate.

Eddy Current Losses: $\mathbf{J}(x,y,z)$ is readily found from

$$J_x = \frac{1}{\mu_0\mu_r}\left[\frac{\partial B_z}{\partial y} - \frac{\partial B_y}{\partial z}\right], \quad J_y = \frac{1}{\mu_0\mu_r}\left[\frac{\partial B_x}{\partial z} - \frac{\partial B_z}{\partial x}\right], \quad J_z = 0 \tag{16}$$

after the flux density has been determined (equation 12), equation 16 becomes

$$J_x = \sum_{\substack{m=1,3,..\\ n=1,3,..}} J_{xmn} \sin\left[\frac{2m\pi x}{L}\right] \sin\left[\frac{n\pi y}{\ell}\right]\left[e^{\alpha_{mn} z} - e^{\alpha_{mn}(2a-z)}\right]$$

$$J_y = \sum_{\substack{m=1,3,..\\ n=1,3,..}} J_{ymn} \cos\left[\frac{2m\pi x}{L}\right] \cos\left[\frac{n\pi y}{\ell}\right]\left[e^{\alpha_{mn} z} - e^{\alpha_{mn}(2a-z)}\right] \tag{17}$$

where

$$J_{xmn} = \frac{-1}{\mu_0\mu_r}\left[\alpha_{mn} C_{mny} + \frac{n\pi}{\ell} A_{mn}\right] \text{ and } J_{ymn} = \frac{-1}{\mu_0\mu_r}\left[\alpha_{mn} C_{mnx} - \frac{n\pi}{\ell} A_{mn}\right]$$

The eddy current losses in the volume of the plate are given by:

$$P = \frac{1}{2\sigma} \sum_{\substack{m=1,3,..\\ n=1,3,..}} \int_0^a \int_{-\frac{\ell}{2}}^{\frac{\ell}{2}} \int_{-\frac{L}{2}}^{\frac{L}{2}} (|J_x|^2 + |J_y|^2)\, dxdydz \tag{18}$$

where

$$|J_x|^2 = J_x \times J_x^* \quad \text{and} \quad |J_y|^2 = J_y \times J_y^* .$$

The data constituting figure 2 is used as input to the program. The program is used to smooth the data fitting the experimental values to a closed form representation of the incident flux density as given by equation (11). The surface plot shown in figure 3 illustrates the results of the data conditioning. Due to the limitations of the personal computer used for the purpose, and the size of the matrix required, fewer points were computed in this case than experimental points taken. Finally, using equations (12), the vector flux density inside the plate on the surface is computed. Then the vector current density induced is found as in equations (17). Figures 4 and 5 show the comparison of the measured surface current densities on the plate, and the results for the above calculations.

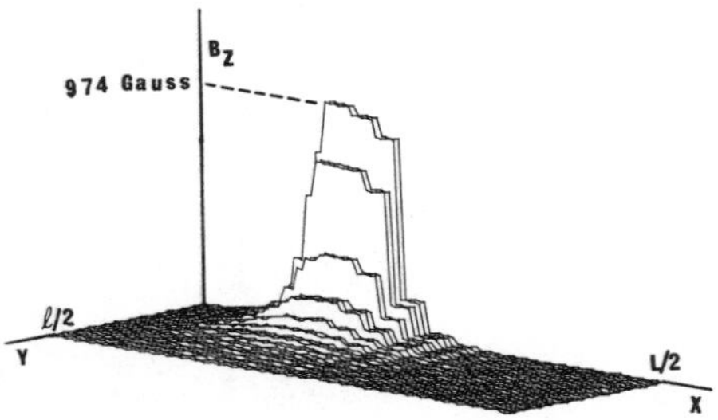

Fig. 2. Experimental Values of the Incident Flux Density on the plate

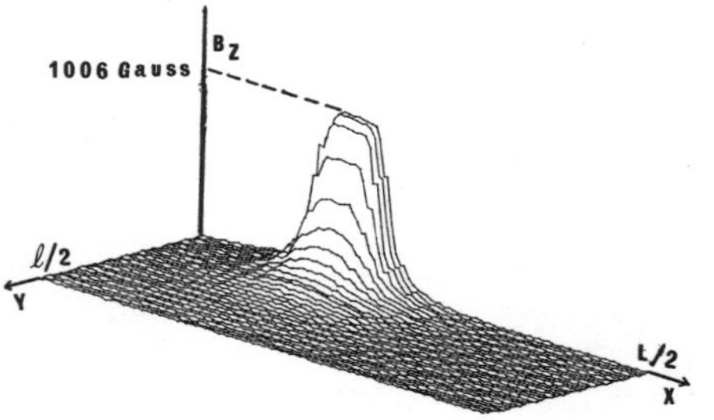

Fig. 3. Incident Flux Density Resulting from the Fitted Data

DISCUSSION

It is understood that this is the first step of the analytical formulation of a problem many researchers have attempted using various other procedures. There still remains to be investigated those practical cases involving nonlinear properties such as permeability and resistivity. As a first approach we describe the possibility of application of piece-wise linear analysis. In the application sought, there is heavy concentration of flux (usually at the ends of the field-producing coils [2]). Hence a refined grid is required to obtain sufficient accuracy for calculations in these regions. As well, flux lines in low density regions can be represented with a coarse grid. If one attempts to use the condensed grid size dictated by the higher flux concentration area over the entire region, the size of the matrices required for solution would become unmanageable.

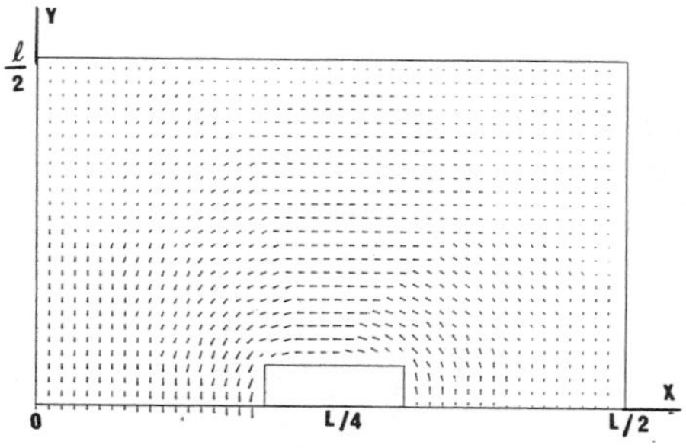

Fig. 4. Experimental Surface Current Density on the plate (Scale = 20 A/mm^2)

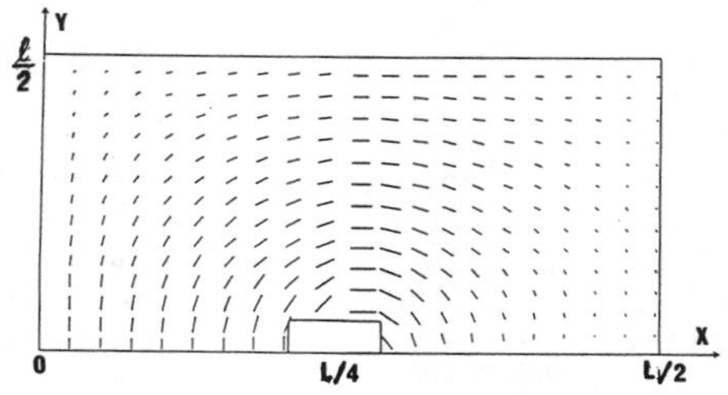

Fig. 5. Predicted Surface Current Densities (Scale = 5 A/mm^2)

A close examination of the coefficients known, show that they depend upon the physical dimensions L, and ℓ. Since the medium is considered linear, superposition may be applied. Figure 1 shows that the plate may be divided into a high concentration grid and a low concentration grid. Values on each surface are found independently using equation 17, with J1(x,y,z) and J2(x,y,z) respectively for each surface. Since the boundary conditions at the separation area AA' must be the same, we can equate the values of J1(x,y,z) and J2(x,y,z) on those boundaries, and hence obtain the proportionality factor between the coefficients A_{mn1} and A_{mn2}. Hence a simple rescaling of the current densities is all that is needed to merge the two areas of the plate. From this computational trick, we may derive invaluable advantages:

1. The size of the matrices may be considerably reduced using plate fragmentation (which reduces storage space and computing time by a square law), with the rescaling needed for the data merge as being only a proportional law.
2. There is no restriction in assuming a piece-wise linear model for the resistivity and for the permeability of the material used between fragmentations provided the boundary conditions between the fragmented plate match.
3. The data required for the eddy current calculation consists of the incident flux pattern obtained from a data file, typically produced by a flux calculating program [2].

ADVANTAGES OF THE METHOD: Transformer design engineers increasingly use finite element programs to solve for the flux and eddy current losses. Because of the inherent properties of the method, a global solution is appropriate. Any variation sought by the designer requires a fresh run of the entire process. This consumes much time and many resources. Furthermore, the size of the mesh is limited in modern programs to 2000 in axi-symmetry two dimensional problems, which require only approximate geometries to be considered. Despite the fact that data is required only on the tank walls, the programs need to result in solutions over the entire space. In a recent paper [3] we have proposed a simpler version leading to a true three dimensional calculation based upon magnetic flux concepts. This will give the flux only where needed, hence results in a much faster, more efficient use of resources. Having obtained the data, the calculation of the losses in the tank wall is a separate problem, tackled with ease now. Furthermore, a designer may wish to concentrate efforts only on certain areas of the plate, (probably in the high flux density areas). Using superposition as described earlier, only partial answers need by calculated. This method, although devised for tank walls, is obviously not limited to that narrow application. Other structural parts in the path of the transformer leakage fluxes can be analyzed in exactly the same manner. For example, the mechanical clamping end frame structures which hold the magnetic core together are covered in this method. Currently we are limiting ourselves to mild steel structures only, and exclude the conducting eddy current screens.

CONCLUSIONS

This paper has presented a novel approach to the solution for eddy current losses in tank walls. An analytical solution of the problem has been derived with good corroboration with an initial experiment. Suggestions are given on how to implement this method in the practical case of large transformer design. This paper presents the first step in a process which will be extended to studies of coefficient sensitivity, as well as the definitions of magnetic penetration, and the problems created by local saturation in actual transformer tank walls. The target of this research is to provide the design engineer with a much more powerful analytical method than has been available to calculate losses in tank walls.

ACKNOWLEDGEMENTS: This project is undertaken with the aid of a Natural Sciences and Engineering Research Council of Canada grant. The valuable contributions of Mr. S. Spencer, research engineer at McMaster, and Mr. P. Birke, Mrs. S. Lie and Mr. W. Lam from Westinghouse Canada are acknowledged.

REFERENCES

1. Agarwal P.D., "Eddy Current Losses in Solid and Laminated Iron", AIEE Trans., May 1959, pp. 169-181.
2. El-Nahas I., Szabados B., et.al., "Three Dimensional Flux Calculations on a Three Phase Transformer", IEEE PES 85-SM-378-5 July 1985.
3. El-Nahas I., Szabados B., et.al., "A Three Dimensional Electromagnetic Field Analysis Technique Utilizing the Magnetic Charge Concept", submitted publication to the IEEE.
4. Forsythe G.E. and Wasow W.R., "Finite-Difference Methods for Partial Differential Equations", J. Wiley and Sons, 1960.
5. Gibbs W.J., "Theory and Design of Eddy Current Slip Couplings", the Beama Journal, London, Eng., vol. 53, Apr.-May-June 1946, pp. 123-172-219.
6. Hayt W. , "Engineering Electromagnetics" McGraw- Hill, 1967.
7. Malti M.G., Ramakusar R., "Three Dimensional Theory of the Eddy Current Coupling", AIEE trans., Oct. 1963, pp. 793-800.
8. Poritsky H., Jerrard R.P., "Eddy Current Losses in a Semi-Infinite Solid due to Nearby Alternating Currents", AIEE trans., May 1954, pp. 97-106.
9. Rosenberg E., "Eddy Currents in Iron Masses", Electrician, London, Eng., Aug. 1923.
10. Roth E., "Etude Analytique du Champ de Fuites des Transformateurs et des Efforts Méchaniques Exercés sur les Enroulements", RGE 23, 773 (1928).
11. Valkovic Z., "Calculation of the Losses in Three-phase Transformer Tanks", IEE proc., vol. 127, no. 1, Jan. 1980, pp. 20-25
12. Vogel F.J. and Adolphson E.J., "A Stray Loss Problem in Transformer Tanks", AIEE trans., August 1954, pp. 760-764.

NUMERICAL FIELD CALCULATION OF EARTHING SYSTEMS

Zijad Haznadar and Sead Berberovic

Electrotechnical Faculty
University of Zagreb
Yugoslavia

ABSTRACT

This article describes results of the long-standing research on computer aided design, analysis and calculation of earthing systems. Numerical calculations of the quasistatic current field have been carried out by the boundary element method (current simulation method). Analyses of the complex earthing systems of the large industrial plants have been based on these calculations. The results obtained by measuring models and real objects have proved of satisfactory accuracy for the numerical procedure.

INTRODUCTION

The solution of the quasistatic earthing current field is a complex problem because of the three-dimensional earthing geometry, which results in a three-dimensional field distribution, as well as due to the open and irregular boundaries (boundary earth-air or boundary between earth layers with different conductivity). Experience based on numerous calculations shows that it is the most suitable to substitute the complex non-homogeneous earth structure by one-or two-layer earth structure with horizontal flat boundaries. In this case the boundary value problem can be solved in two ways: either applying the method of images or the equivalent boundary sources method.

In the first method the potential distribution of the earthing system is described by the Fredholm integral equation of the first kind:

$$\phi(\bar{r}) = \frac{1}{\kappa_1} \int_S J(\bar{r}') \, G(\bar{r},\bar{r}') \, dS' \qquad (1)$$

where: $J(\bar{r}')$ is the current density, S is the earthing surface and κ_1 conductivity.

Green's function in (1) is determined by:

$$G(\bar{r},\bar{r}') = \frac{1}{4\pi}\frac{1}{|\bar{r}-\bar{r}'|} \qquad (2)$$

By applying the method of images for the solution of boundary value problem, the contribution of image sources must be added to the Green's function. Corresponding to Green's function for homogeneous earth model is:

$$G(\bar{r},\bar{r}') = \frac{1}{4\pi}\left[\frac{1}{|\bar{r}-\bar{r}'|} + \frac{1}{|\bar{r}-\bar{r}_1'|}\right] \qquad (3)$$

where $\bar{r}_1'$ is the radius vector of the image source position.

At the second approach, the equivalent boundary sources are introduced to calculate the potential distribution, according to the equation (1), where S is the sum of all surfaces (surface of all earthings and the surface of all boundaries).

The surface density of current sources on the boundary between two layers "i" and "k" is given by the following Fredholm integral equation of the second kind:

$$J_{ik}(\bar{r}) = 2k_{ik} \int_S J(\bar{r}') \frac{\partial G(\bar{r},\bar{r}')}{\partial n_k} dS' \qquad (4)$$

$$k_{ik} = \frac{\kappa_i - \kappa_k}{\kappa_i + \kappa_k}$$

where n_k is normal on the boundary.

If there is a passive (not-connected) electrode of surface S_p in the earthing field it gets an unknown "floating potential" described by the integral equation (1). By introducing a continuity equation:

$$\int_{S_p} J(\bar{r}')\, dS_p' = 0 \qquad (5)$$

one gets the complete equation system.

In the numerical procedure of solving the integral equations, the easiest way is to replace the source distribution on the boundary elements by constant amount - so called "simplex" procedure. Such a procedure asks for a very fine division on the boundary elements with a consequence of a very large number of unknown matters in corresponding system of the linear algebraic equations. We obtain a considerably better approximation of the unknown current distribution on the boundary element by the cubic spline functions, Fig. 1. The cubic spline function on the boundary element, defined in 5 points, approximates the unknown current distribution on the "j" boundary element as follows:

$$I_j^{(i)} = \sum_{k=1}^{7} b_{jk}^{(i)}(s)\, I_k \qquad (6)$$

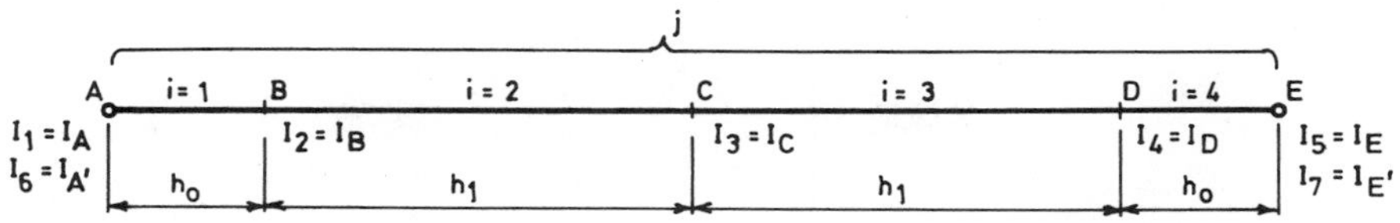

Fig. 1. The boundary element

where: j - boundary element index;
i - index of a boundary element part; i = 1,...,4;
I_k - current and current derivation in node points of the boundary element parts (Fig.1); k=1,...,7;

$b_{jk}^{(i)}(s)=b_{jk}^{(i)}(1,s,s^2,s^3)$ - function that describes the cubic spline.

Function that describes the cubic spline can be, in case of the boundary element division according to Fig. 1., expressed by means of third-order polinomial of the independent variable "s" from the definition formulas for a cubic interpolation spline function.

In most cases the earthing dimensions are such, that one can assume the whole earthing to be of constant potential. Potential along the very long earthing parts changes because of conductance and voltage drop. In order to take into account the variable earthing potential it is advisable to observe the earthing parts as a no-load line and modelled mathematically by the chain of quadripoles with distributed parameters G, R and L. Conductance G is numerically calculated by supposing that the earthing potential is constant. After calculating the potential distribution along the earthing parts, we solved the integral equation (1).

ANALYSIS OF COMPLEX EARTHING SYSTEMS

Mutual earthing influence is analyzed on the example of two large industrial plant earthings, Fig. 2. The calculation results show that, with regard to the earthing system resistance and potential distribution, it is best to connect both earthings galvanically. Potential differences between separated earthings and potential gradients close to the earthings are especially large when the smaller earthing (A) is on the potential of 100% and the bigger earthing (B) is passive, Fig. 2.

The influence of metal parts on the earthing and potential distribution, as well as the carrying in of low potential and carrying out of high potential are analyzed on the example of the earthing system formed by the object earthings, metal fence in the ground and underground pipeline, Fig. 3.

In the case when the fence and pipeline are disconnected from the earthing system, there is a possibility that by means of a pipeline a low potential from the area out of the plant circle is carried into an area of plant circle with relatively high ground potential, Fig. 3.

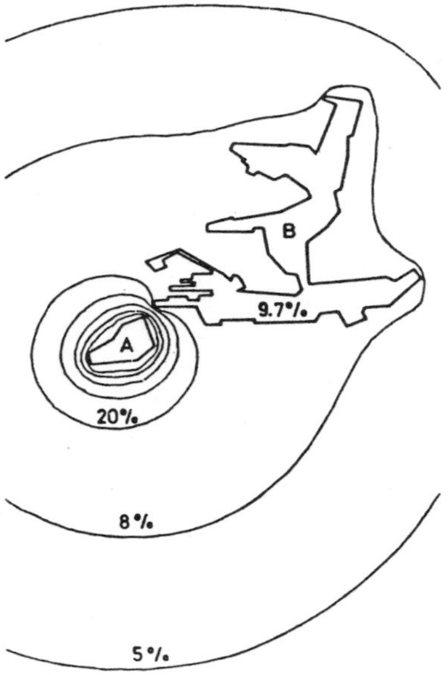

Fig. 2. Mutual earthing influence

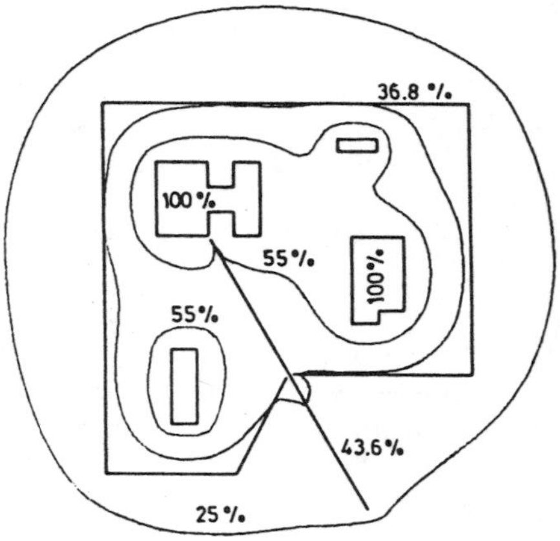

Fig. 3. Carrying in of a low potential

In the case when the fence is disconnected to the earthing system and the pipeline is connected, there is a possibility that by means of pipeline a high potential is carried out in areas with low ground potential, Fig. 4.

The earthing length influence is analyzed in the example of the earthing system of a gas plant, Fig. 6. Because of long earthing of the casing tubes connected in the earthing system, it is necessary to determine potential distribution on them. Fig. 5. shows the potential distribution along the earthing of one casing tube, and Fig. 6. shows the potential distribution around the gas plant.

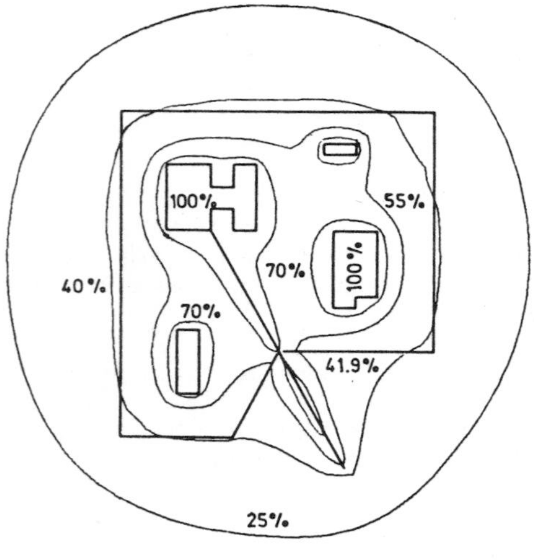

Fig. 4. Carrying out of a high potential

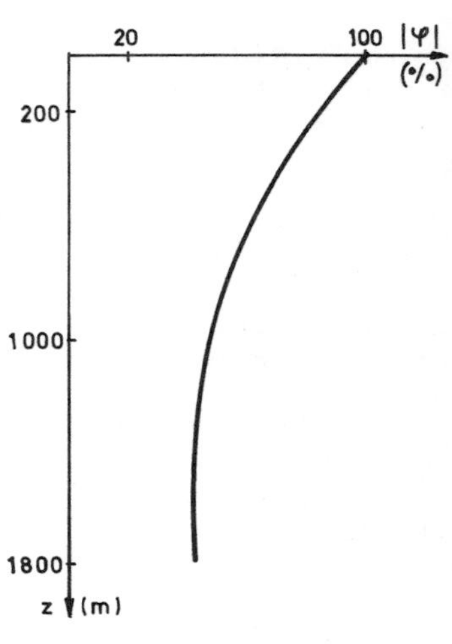

Fig. 5. Potential along the casing tube

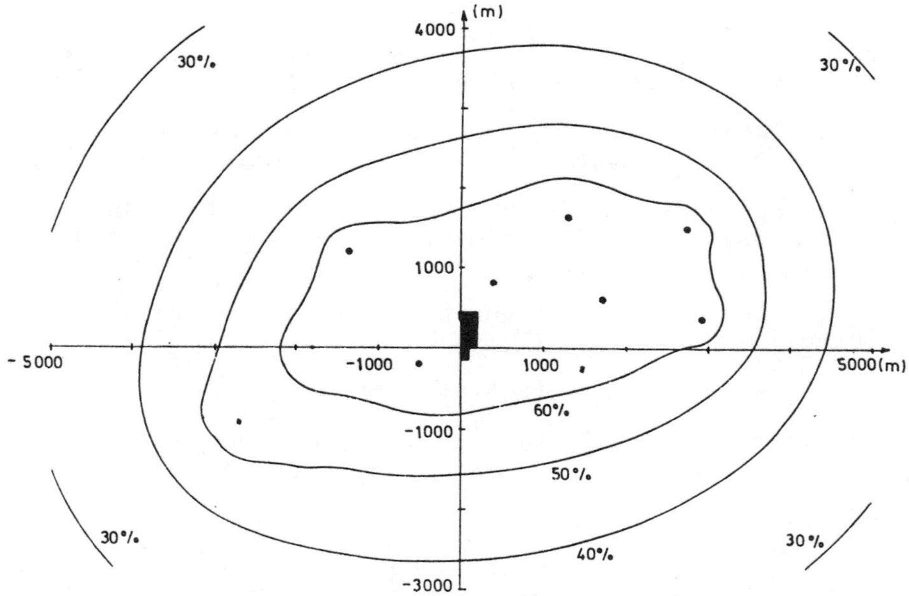

Fig. 6. Potential distribution around the gas plant

The floating passive electrode potential is analyzed in the example of a railway placed close to the earthing, Fig. 7. The calculation results show that the passive electrode may acquire a high touch voltage which becomes greater with the increasing of the conductivity ratio, Fig. 8.

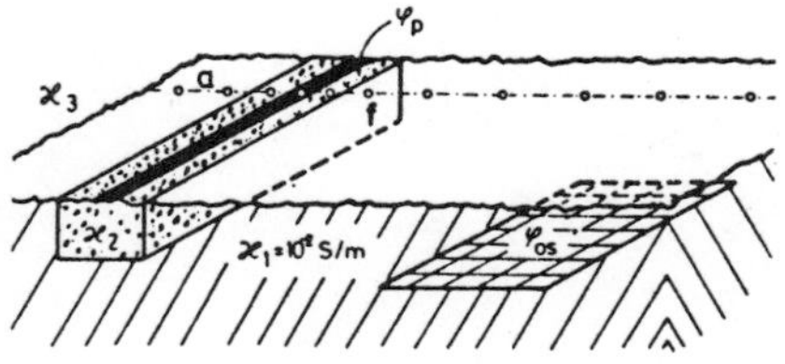

Fig. 7. Railway in vicinity of the earthing

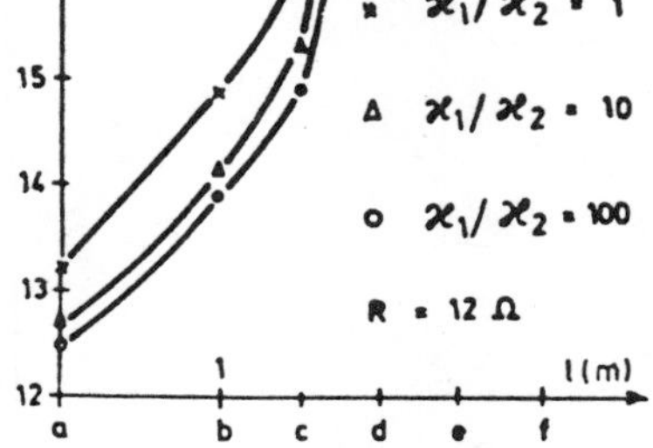

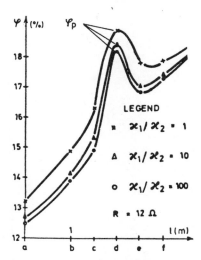

Fig. 8. Potential distribution along the a - f direction

CONCLUSIONS

The interactive CAD system with its theoretical base described in this paper enables:

1. Design, analysis and calculation of the complex earthing systems;

2. Optimization of the earthing systems from the standpoint of minimum material and chioce of most suitable earthing geometry;

3. Simple input of minimum quantity of entering data and its control;

4. Designer-computer interactive relation;

5. Graphic illustration of the calculating results.

REFERENCES

1. S. Berberovic and J.Sindler, Cubic spline and third-order polynomial approximation of source-field distribution in integral equation problems, VII International symposium CAD/CAM, Zagreb (1985).
2. Z. Haznadar,N. Zanic and S. Berberovic, Passive electrode's high voltage in the environment of the earthing system, V Symposium on Numerical Methods in Technique, Zagreb (1983).
3. Z. Haznadar and S. Berberovic, Computer Aided Design of Earthing Systems, Symposium "Numerical Field Calculations in Electrical Engineering", Graz, (1985).
4. Z. Haznadar and S. Berberovic, The Advanced Numerical Procedure for Analysis and Design of Earthing Systems, Elektrotehnika, Zagreb, No. 3, (1986).

COMPUTER-AIDED DESIGN TECHNIQUE

FOR PROPORTIONAL ELECTROMAGNETS

Mircea M. Rădulescu, Vasile Iancu, Ioan-Adrian Viorel, and Károly Biró

Faculty of Electrical Engineering
Polytechnic Institute of Cluj-Napoca
Cluj-Napoca, Romania

INTRODUCTION

A remarkable class of electromechanical converters used in proportional hydraulic devices, at the interface between the electronic control part and the hydraulic drive part, is represented by proportional electromagnets. These are, in fact, plunger-type d.c. electromagnets with a very studied air-gap geometry and intentionally saturated ferromagnetic zones. Such a particular magnetic structure allows a static characteristic adjustment, so that the developed electromagnetic force (i) remains constant with plunger positional changes (within the work domain of the axial air gap), (ii) is proportional to the excitation current, and (iii) has a considerably increased magnitude in a limited volume.

The aim of this paper is to prove that the finite element magnetic field analysis is a very convenient computer-aided design technique for proportional electromagnets, avoiding the difficulties encountered in experimental investigations and the cost of prototypes.

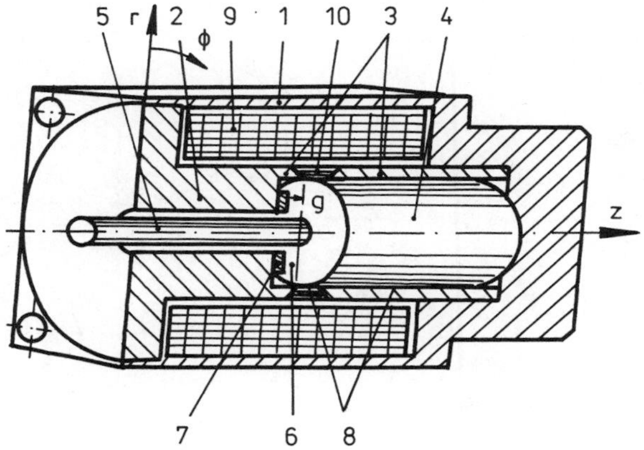

Fig. 1. Axial section of a proportional electromagnet.

FINITE ELEMENT FIELD ANALYSIS AND STATIC CHARACTERISTIC COMPUTATION

Figure 1 shows the axial section of a typical proportional electromagnet. The axisymmetric magnetic circuit contains the external yoke 1, the fixed armature (stopper) 2, the guidance sleeve 3, and the moving armature (plunger) 4, all made of ferromagnetic materials. The plunger actuates a nonmagnetic rod 5 inside a coaxial cylindrical channel. The axial (principal) air gap 6 between frontal surfaces of both armatures and the nonmagnetic washer 7 delimit the useful race g of the plunger. Another radial (technological) air gap 8 of very small thickness g' exists between the moving armature and its guidance sleeve. The magnetic flux results from the current I of the solenoidal exciting coil 9 adjacent to the iron magnetic circuit.

For a suitable adjustment of the proportional electromagnet static characteristic F(g) (electromagnetic force-versus-axial air-gap length), one may intervene in the detailed geometry and the saturation level of the magnetic circuit. Such a twofold intervention is made by means of the brazen coniform ring 10 mounted on the guidance sleeve (Fig. 1).

The finite element field analysis may be used to accurately predict the performance of any particular design of the proportional electromagnet and hence leads to an optimized design technique. With this in view, the nonlinear axisymmetric problem of the proportional electromagnet field analysis is formulated in variational terms as an energy-related functional[1]

$$\mathcal{F}(U) = 2\pi \int_R \left[\left(\int_0^B r\,\nu(b)\,b\,db\right) - J_\phi U\right] dr\,dz, \tag{1}$$

the extremization of which yields the required field solution. In the axisymmetric representation given in (1), R denotes the field domain in the z-r plane - of the electromagnet axial section - with homogeneous Dirichlet conditions on its boundary S (Fig. 2); U is a modified magnetic vector potential function defined as the product of the radius r - of a cylindrical coordinate system (r,ϕ,z) - by the azimuthal component of the magnetic vector potential; J_ϕ means the ϕ-component of the excitation current density and ν is the reluctivity, i.e. a single-valued function of the flux density B in ferromagnetic zones of R. In order to numerically extremize the functional (1) by the finite element method (FEM), the field domain of the z-r plane is subdivided into first-order triangular elements (Fig. 2)

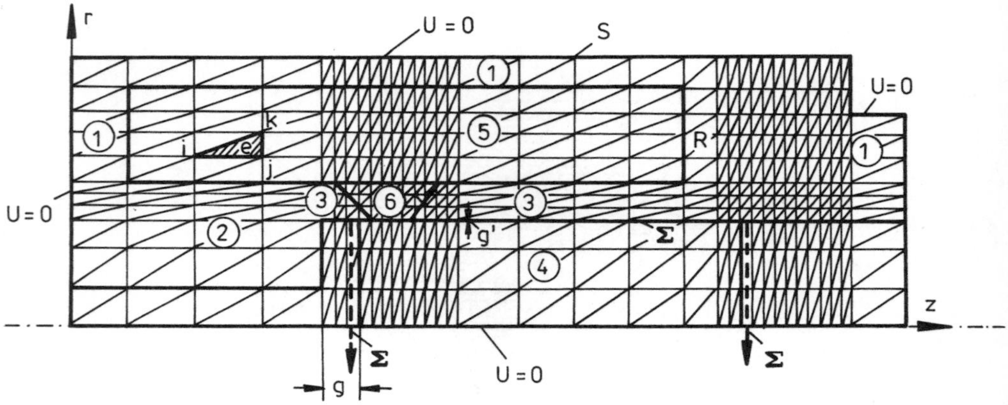

Fig. 2. First-order triangular discretization of the magnetic field domain in the axial half-section of the studied proportional electromagnet.

and the modified magnetic potential function within each element is defined in terms of the standard shape functions[1] and the nodal potential values :

$$U^e(r,z) = \sum_i N_i^e(z,r)U_i^e = \sum_i (n_i^e + p_i^e z + q_i^e r)U_i^e, \qquad (2)$$

where the index of summation ranges over the vertices i,j,k of the prototypal triangular element e. Figure 2 illustrates an automatically drawn triangulation of the field domain into 694 nodes and 1278 elements. The axis of symmetry (z-axis) is chosen as a boundary of the discretized domain with the potential function U along it assumed to be zero. Thus, the singularity problem is effectively removed without any special procedure for elements near the z-axis. A finer mesh is used for the axial, respectively radial, air-gap discretization. In this way, the useful race of the plunger is incremented by the mesh in twelve succesive steps (of 0.5 mm magnitude), allowing the computation of the proportional electromagnet static characteristic F(g).

According to the approximation (2), the FEM extremization of functional (1), subject to homogeneous Dirichlet boundary conditions, leads to the global nonlinear system of equations[1]

$$U_i \sum_e K_{ii}^e + \sum_e K_{ij}^e U_j^e + \sum_e K_{ik}^e U_k^e + \sum_e G_i^e = 0,\ i = 1,\ 694, \qquad (3)$$

where the summation is taken over the triangles adjoining vertex i, and

$$K_{iu}^e = 2\pi \Delta^e \nu^e (p_i^e p_u^e + q_i^e q_u^e)/r^e, \quad u = i,j,k \qquad (4)$$

$$G_i = -2\pi \Delta^e J_\phi^e/3, \qquad (5)$$

with Δ^e and r^e denoting the area, respectively the barycentre radial coordinate, of the triangle e. The iterative solution of (3) involves an inner cycle for nodal potential values and an outer cycle for reluctivities. Cubic splines are used to model the reluctivity characteristic of iron[1].

Based on the previous FEM field analysis performed for each of the twelve succesive positions of the plunger, the corresponding values of the developed electromagnetic force F are evaluated by means of the Maxwell stress tensor method[2]:

$$\bar{F} = \int_\Sigma [\ \nu_0(\bar{B}\cdot\bar{n}_\Sigma)\bar{B} - \frac{1}{2}\nu_0 B^2 \bar{n}_\Sigma] \, d\Sigma, \qquad (6)$$

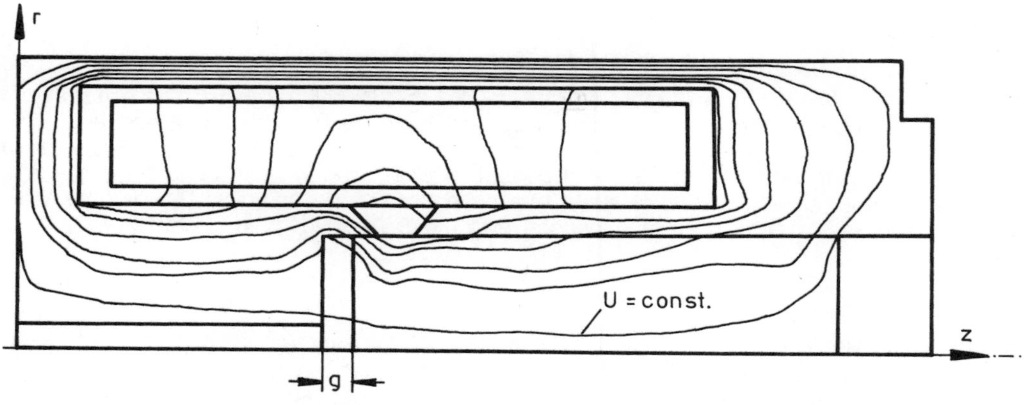

Fig. 3. Magnetic flux distribution in an axial half-section of the studied proportional electromagnet (for a certain position of the plunger).

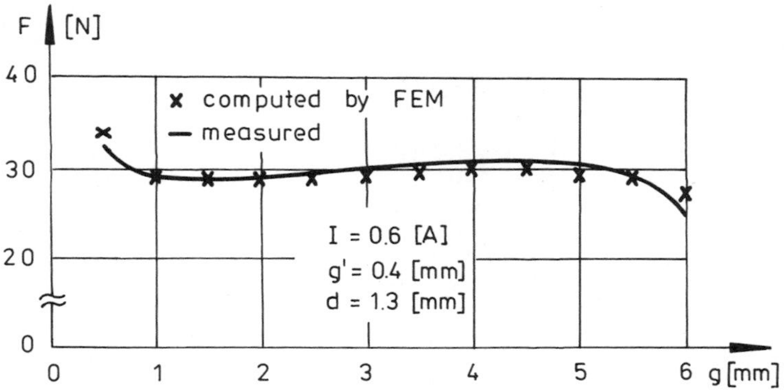

Fig. 4. Comparison between FEM-computed and measured static characteristics of the studied proportional electromagnet.

where the integration path Σ , having the unit outward vector normal $\bar{n}_{\Sigma}$, surrounds the plunger and passes through the side middles of the air-gap mesh triangles (Fig. 2); ν_0 is the free space reluctivity.

RESULTS

The calculated flux distribution (lines of U=const.) in the studied proportional electromagnet, for one of the intermediate positions of the plunger, is given in Figure 3. By performing a dozen of such FEM field analyses for all the positions of the plunger in its useful race and calculating the corresponding values of the developed force, one obtains the static characteristic F(g) of the proportional electromagnet.

For design purposes variations were introduced for (i) the main geometric parameters of the magnetic circuit, like g' and the sleeve length d from the stopper to the brazen ring, (ii) the excitation current value, and (iii) the material properties. Thus, FEM simulations have lead to several computed static characteristics, offering a very useful tool for the designer in identifying the best magnetic structure of the proportional electromagnet. Such a one was found to yield the static characteristic depicted in Figure 4, which also exhibits a good agreement with the values of the force measured on a corresponding prototype.

CONCLUSION

The problems involved in the computer-aided design of sophisticated plunger-type proportional electromagnets have been considered and a suitable technique based on FEM axisymmetric field analyses developed. Satisfactory agreement between predicted and measured static characteristics of the studied proportional electromagnet has been obtained, proving the accuracy and the efficiency of the proposed computer-aided design technique.

REFERENCES

1. G. Mîndru and M. M. Rădulescu, "Numerical analysis of the electromagnetic field"(in Romanian), Dacia Publishing House, Cluj-Napoca (1986)
2. Y.-S. Hong, Berechnung von Proportionalmagneten mit dem Verfahren der finiten Elemente, Ölhydraulik und Pneumatik, 28, 9, 552-558 (1984).

RELUCTANCE NETWORK ANALYSIS OF COUPLED FIELDS

IN A REVERSIBLE ELECTROMAGNETIC MOTOR

Victor R. Rais[x], Janusz Turowski[xx], and Marek Turowski[xxx]

[x] Novosibirsk Institute of Electrical Engineering, USSR

[xx] Institute of Electrical Machines and Transformers
Technical University of Lodz, Poland

[xxx] Institute of Electronics, Techn.University of Lodz, Poland

INTRODUCTION

A large number of reversible electromagnetic, thyristor controlled, linear motors (Fig.1) have been designed and produced by the Novosibirsk Institute of Electrical Engineering. These machines cover a wide range of frequencies (from 0.4 to 50 Hz) and forces (from a few to several thousand newtons) and have been applied in the chemical and mining industries and used in household equipment and professional tools[1,2].

The control circuit (Fig.2) provides for the reversible operation of the motor and the proper frequency for a given shaft speed.

A general reluctance network program[3,4] and a model of the linear motor were developed at the Technical University of Lodz for running on an IBM PC/AT. This program and model were used to investigate the electromagnetic performance of the motor and to optimize its structure.

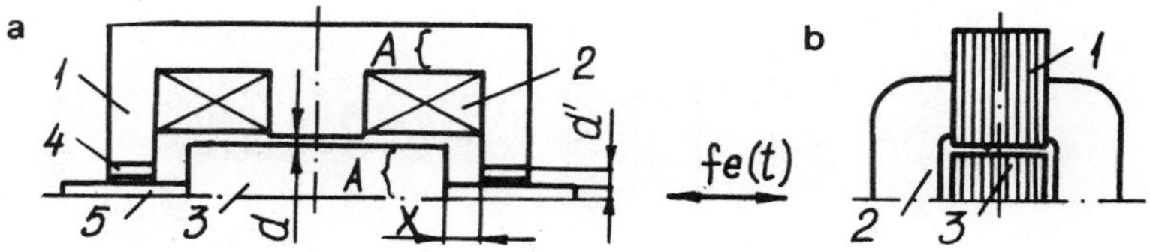

Fig.1. Reversible linear motor[1]: 1 - laminated iron stator core, 2 - excitation coils, 3 - laminated iron plunger, 4 - nonmagnetic sleeves, 5 - solid iron shaft, A - cross-section area.

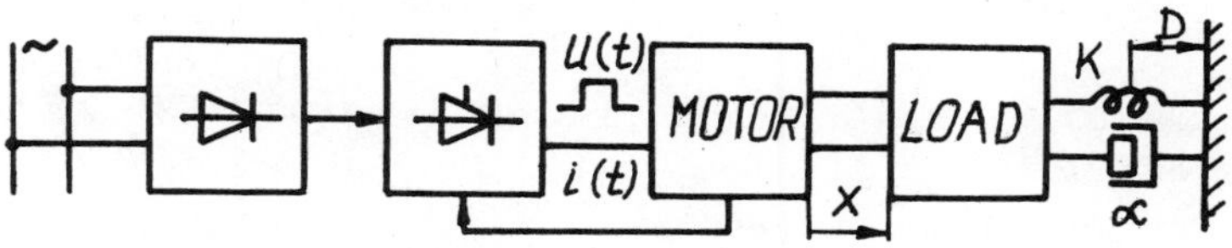

Fig.2. Linear motor and control circuit.

ELECTROMECHANICAL COUPLED FIELD AND ITS EQUATIONS

The dynamics of the motor shown in Fig.1 can be described by applying Euler-Lagrange equations for lumped parameter systems. The two resulting principle equations of motion are[5]:

$$u = Ri + \frac{d\psi}{dt} = Ri + \frac{\partial L}{\partial x}\frac{dx}{dt} + \frac{\partial L}{\partial i}\frac{di}{dt} + L\frac{di}{dt} \tag{1}$$

$$M\frac{d^2x}{dt^2} + \alpha\frac{dx}{dt} + K\,(x - D) = f_e + f(t) \tag{2}$$

where:

$$L(x,i) = \frac{N^2 A\ \mu(x,i)}{d + d' + x} = \frac{\psi(x,i)}{i(x)}\ , \tag{3}$$

u(t) is the input voltage, R - the coil resistance, M - the mass of the plunger and load, i - the current, x - the shaft displacement, α - the viscous mechanical friction coefficient, K - the elasticity (spring) constant, D - the length of the spring when its force is zero, f_e - the electromagnetic force, t - the time, N - the number of turns of one coil, ψ - the flux linkage and f(t) - an externally applied constraining force.

If i(t) is the independent variable, the electromagnetic force f_e applied to a mass M can be evaluated by the method of arbitrary displacement and magnetic coenergy:

$$W'_m = \int_0^i \psi'(i',x)\,di' \tag{4}$$

and

$$f_e = \left[\frac{\partial W'_m(i,x)}{\partial x}\right]_{i=const} \tag{5}$$

Assuming that the magnetic circuit of the motor is not saturated, $\psi = Li$ and

$$W'_m = W_m = \frac{1}{2} L\, i^2 \tag{6}$$

and equations (1) and (2) can be simplified as:

$$u(t) = Ri + L\frac{di}{dt} + i\,\frac{\partial L}{\partial x}\frac{dx}{dt} \tag{7}$$

$$f(t) = M\frac{d^2x}{dt^2} + \alpha\frac{dx}{dt} + K(x - D) - \frac{1}{2}\,i^2\frac{dL}{dx} \tag{8}$$

The response i(t) and x(t) resulting from the application of driving functions u(t) and f(t) are the desired quantities in these equations. As equations (1,2) and (7,8) are nonlinear, some simplifications and parameter analyses were made[5]. A more detailed analysis of the electromagnetic field was necessary, however, to include the nonlinearity of the permeability of the steel.

This was done using a reluctance network method (RNM) and the corresponding program, MSR.

The motor is assumed to be symmetric about both axes so only a quarter cross-section (Fig.1a) needs to be considered. A dense model of 400 nodes (Fig.3a) was used initially to check the convergence of the RNM. Then a simpler model of 95 nodes (Fig.3b) was used for further analysis. Both models were used to obtain solutions for an input current $i = 2$ A and gap dimensions: $x = 0, 1, 2, \ldots, 10$ mm.

The results (Fig.4) indicate that the simpler model is sufficiently accurate, although it only requires about 20% (3 min.) of the run time of the denser model (18 min.).

The nonlinear permeability of the steel is included by assuming first that the permeability is constant at $\mu_r = 1000$ and then a new value of μ_r is found iteratively from the calculated value of B. It was found that only a few iterations were needed to obtain sufficient accuracy.

Elementary reluctances for the equivalent networks (Fig.3) were evaluated as follows.

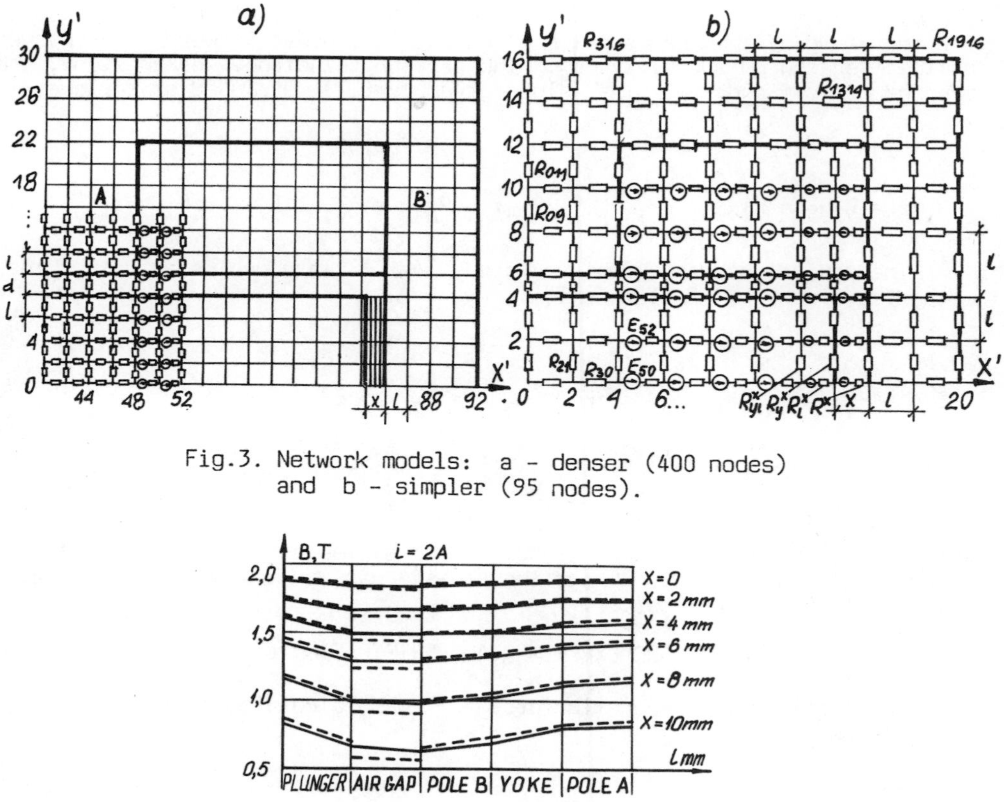

Fig.3. Network models: a - denser (400 nodes) and b - simpler (95 nodes).

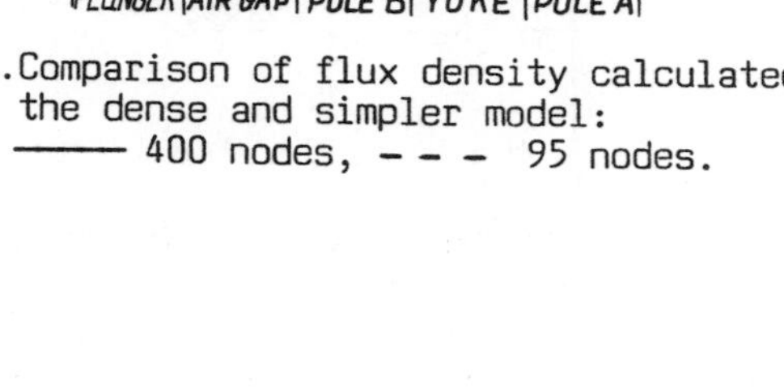

Fig.4. Comparison of flux density calculated with the dense and simpler model:
——— 400 nodes, - - - 95 nodes.

For the laminated core and plunger

$$R_{core} = \frac{1}{\mu_r \, \mu_0 \, A_1} \tag{9}$$

For the air and winding area

$$R_{air} = \frac{1}{\mu_0 \, A_1} \tag{10}$$

For the gap due to shaft displacement

$$R^x = \frac{x}{\mu_r \, \mu_0 \, A_1} \quad \text{and} \quad R^x_y = \frac{1}{\mu_0 \, A_1} \, \frac{2l}{l-x} \, ,$$
$$R^x_l = \frac{l-x}{\mu_r \, \mu_0 \, A_1} \quad \text{and} \quad R^x_{yl} = \frac{1}{\mu_r \, \mu_0 \, A_1} \, \frac{2l}{2l-x} \, , \tag{11}$$

where A_1 is the cross-section of one element.

In equation (11) the reluctances have been described with the indices x and y corresponding to the discrete coordinate points of the model. For example, for the simpler model with element dimensions of 10x10 mm, except for the air gap, the core and the plunger reluctances (9) were only dependent on the iterated permeability μ_r. The air gap reluctance was constant, independent of x, i and μ_r. The values given by (11) were used as follows:

$$R_{150} = R_{152} = \ldots = R_{1510} = R^x, \quad R_{130} = R_{132} = R_{134} = R^x_l \, , \tag{12}$$
$$R_{136} = R_{138} = R_{1310} = R^x_l \, \mu_r, \quad R_{141} = R_{143} = R^x_y, \quad R_{121} = R_{123} = R^x_{yl} \, .$$

The equivalent magnetomotive forces (MMF) were evaluated using the method developed by J. Turowski[3]:

$$\begin{aligned} &E_{50} = E_{52} = E_{54} = E_{56} = E_{70} = E_{72} = \ldots = E_{114} = E_{116} = 3 \cdot E \, , \\ &E_{58} = E_{78} = E_{98} = E_{118} = 2 \cdot E \, , \quad E_{510} = E_{710} = E_{910} = E_{1110} = E = \frac{iN}{15}, \\ &E_{130} = E_{132} = E_{134} = E_{136} = \frac{l-x}{l} \cdot 3E, \quad E_{138} = \frac{l-x}{l} \cdot 2E, \quad E_{1310} = \frac{l-x}{l} \cdot E, \\ &E_{150} = E_{152} = E_{154} = E_{156} = \frac{x}{l} \cdot 3E, \quad E_{158} = \frac{x}{l} \cdot 2E, \quad E_{1510} = \frac{x}{l} \cdot E \, . \end{aligned} \tag{13}$$

While the flux in each branch is calculated by the program MSR, only the flux values in the active gap x and the passive gap d are necessary to calculate the force on the shaft.

ELECTROMAGNETIC FORCES AND EXPERIMENTAL VERIFICATION

The static electromagnetic force F_e was calculated from Maxwell´s formulas as

$$F_e = 2 \, \frac{\phi_1^2}{2 \, \mu_0 \, A} - K_{fr} \, 2 \, \frac{\phi_2^2}{2 \, \mu_0 \, A}, \tag{14}$$

where

$\phi_1 = \phi_{130} + \phi_{132} + \phi_{134}$ - the magnetic flux in the active air gap x,

$\phi_2 = \phi_{07} + \phi_{27} + \phi_{47}$ - the magnetic flux in the passive air gap d

and K_{fr} - the coefficient of friction.

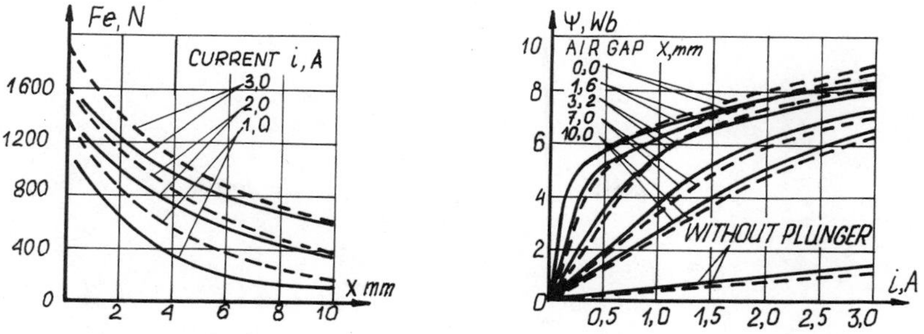

Fig.5. Experimental verification of the force and flux linkages: ——— measured, - - - calculated.

The computed values showed satisfactory agrrement with the experimentally measured values (Fig.5). The larger calculated forces for small gaps is due to the simplifications assumed for the model, that is neglecting holes in the stator for the shaft and bearings, eddy currents in the shaft, and leakage flux in the third dimension z.

The calculated flux linkages $\psi(x,i)$ also show good agreement with experimental values (Fig.5).

Our studies demonstrate that the proposed model (Fig.3) can be used to compute the dynamic and static performances of electromagnetic reversible linear motors. This model will be used to optimize the dimensions and winding parameters of the motor structure for given materials from the viewpoint of energy efficiency.

CONCLUSION

The forces F_e and flux linkage $\psi(x,i)$ necessary for the design and dynamic analysis of reversible linear electromagnetic motors can be calculated quickly and easily using a simple reluctance network model with a small number of nodes.

REFERENCES

1. V. R. Rais, A. I. Smelyagin, V.N.Obukhov and Y. P. Misyuk, Pump for the Artificial Blood Circulation Apparatus. (In Russian). International Conference - Trends in Human Biomechanics Research and Applications in Medicine and Surgery. Riga, USSR (1986).
2. N. P. Ryashentsev and A. T. Malov, "Electromagnetic Hammers", (In Russian). Nauka. Siberian Branch, Novosibirsk, USSR (1979).
3. J. Turowski, Reluctance Network of Leakage Field in Transformers, (In Polish). Rozprawy Elektrot. 4 (1984).
4. J. Turowski and M. Turowski, The Network Approach to the Solution of Stray Field Problems in Large Transformers. Rozprawy Elektrot. 2 (1985).
5. D. C. White and H. H. Woodson, "Electromechanical Energy Conver - sion", J.Wiley and Sons, New York (1959), pp. 93 - 100.

7. SYNTHESIS

Introductory remarks

K. Pawluk

Department of Fundamental Research
Instytut Elektrotechniki
Warsaw, Poland

It is widely known that, in electrodynamics, analysis was developed in preference. Therefore, when the geometric and material structure of a device is known and when the field sources in this device are given, then we may accomplish the analysis of field in it. Synthetizing means searching for the structure producing the desired field. Synthesis is performed in order to design some device or to discover the features of the existing device unless they are not yet known.

The reports treating synthesis problems in ISEF'87 have been brought together. Let us express some general remarks pertaining to the use of synthesis technique in the engineering design. The most part of designers does not appreciate the possibilities offered by the inverse treatment of electrodynamic problems. Really, the electromagnetic synthesis problems, for the most part, lead to nonunique solutions. This nonuniqueness is a consequence of the fact that the synthesis problems are, generally, ill-posed in Hadamard understanding.

The design problems are, generally, of the synthesis type. In traditional design, synthesis was accomplished by an appropriate forming succesive device configurations on the ground of field analysis results. This procedure is continued until the device configuration is satisfactory from an engineering and an economical stand point. Being based on long tradition and developed achievements and making use of the auxiliary technical data (diagrams, experimental coefficients) such a procedure has resulted in numerous excellent electric devices. But such an approach is not sufficient when designing new, modern devices for which the auxiliary data base does not exist as yet. Then the synthesis methods seem to be very useful and they are better tools as compared to the methods based on the examination of expensive physical models.

An interesting discussion occurred during the symposium. I would like to quote here the problems put by Professor Hammond:

"Professor Pawluk and Dr. Rudnicki are to be congratulated on defining and explaining several kinds of synthesis problems. Clearly synthesis is the central feature of engineering design and if general methods could be developed for synthesis this would be a considerable advantage. My question to the authors is whether they believe that in principle such methods can be constructed.

I have a number of difficulties in my mind. The first concerns the general lack of symmetry between induction and deduction, which had been elucidated by Popper[1]. The second is that the practice of engineering design for the last 100 years, has always started with a collection of known solutions and achieved synthesis by superposition. I have in mind the use of series of independent terms as for example Fourier series. Thirdly, I am troubled by the luck of uniqueness of the author's methods. This means that they cannot define any invariant system parameters and this suggests that the very powerful variational techniques of Lagrangian mechanics are not available to them. How will the authors test the convertude of a large set of particles without prior knowledge of the system? Could it be only those problems can be solved by synthesis, which have solutions obtainable by analysis?

Lastly I wonder how the authors would tackle the problem of physical realisability? I have in mind the design of an antenna to give a particular distant field pattern."

The authors' reply to the above problems:

We term "synthesis" such problems which consist in investigating field sources and/or field region structure when the field distribution in a part of the region is given as a desired or measured field quantity. So, we consider the synthesis problems as deduction like those concerning the analysis. We think that induction were in the case if we searched for the unknown field equation form using the information pertaining to the given field distribution. The synthesis equations are then the same field equations (e.g. Laplace, Poisson, Helmholtz equations) like in analysis, but they have got to be solved in opposite way. The synthesis problems can be tested, of course, by analysis using the synthetized sources and/or structure as the input data.

The nonuniqueness of synthesis methods is the consequence of the fact that the field distribution regarded as input data does not pertain, largely, to the whole synthesis region Ω but to a checked subregion Ω_o only. The smaller is this subregion the closer will be a set of solutions and vice versa. We would like to point out that the feature of the synthesis problems (of the field sources) is the unfullness of input information. If, for instance, we search for the Dirichlet boundary conditions when the field in the whole region ($\Omega_o = \Omega$) is given, either a unique one or no solution will be obtained.

The physical realisability of the synthesis problems could be clearly illustrated for the case of two linkage circuits where both physically realisable and unrealisable results may be obtained. The physical unrealisability of field synthesis may occur when we obtain e.g. $\mu < 0$ or $\rho < 0$.

We did not study ourselves the synthesis problems pertaining to hyperbolic equations. The identification of the local antenna damage made by use of the field data measured at short distance has been examined by Mittra[2].

REFERENCES

1. K. R. Popper, The Logic of Scientific Discovery, Hutchinson Publ. Group Ltd, London (1974)
2. R. Mittra, Computer Technics for Electromagnetics, Pergamon Press, Oxford (1973)

THE STATE OF ART IN THE SYNTHESIS OF ELECTROMAGNETIC FIELDS

Krystyn Pawluk and Marek Rudnicki

Department of Fundamental Research
Institute of Electrical Engineering
Warsaw, Poland

INTRODUCTION

One observes that many inverse problems of technical electrodynamics are recently studied. The basic ordination of the synthesis problems might be regarded from an engineering, a physical and a mathematical point of view. There is a wide diversity in formulating the synthesis problems; some of them have been discussed in this paper and a basic classification of engineering synthesis problems has been proposed. A short review of mathematical techniques has been also presented.

INVERSE PROBLEMS IN ELECTROMAGNETISM

We use here the notion "field synthesis" as opposite to "field analysis" which is commonly considered, indeed, as the mathematically modelled physical problem that consists in solving the field equation being, in general, a partial differential equation of second order with respect to some field quantity, scalar or vector. In analysis this solution is accomplished in the enviromentally and geometrically, specified region which might be finite, bounded infinite or unbouned infinite. Field sources in the analysis are known and given in the form of boundary conditions or/and inner excitations.

The synthesis of the field can be formulated in many ways and it should be precisely defined what kind of problem is to be investigated. The field distribution or some integral of it is given a priori in the synthesis problems.

The following quantities synthetizing the field may be searched for:

(a) boundary conditions (i.e. external sources),
(b) inner excitations,
(c) geometric shape of a region including the localization of inner sources,
(d) enviromental parameters.

When searching for any specific sources or region features, the simple synthesis problem is investigated while the occurrence of more unknown quantities of different types leads to the compound synthesis problem.

The simple synthesis problem of the electromagnetic field consists, therefore, in searching for unknown excitations which should be distributed in a given manner or, vice versa, the excitating subregions have got to be designed. These excitations determine the physical phenomenon characterized by the field quantity distribution or by its given integral and are to be measured or desired. Recently solved engineering inverse problems, e.g. Adamiak[1], Palka[2] or Sikora[3] belong to the simple synthesis.

Let us note that fields in inverse problems are rather synthetized not in the whole region under consideration, but in some specific part of it named synthesis checked subregion, frequently less dimensional than the primary region. When the field source is geometrically fixed inside the investigated region we will speak of the synthesis source subregion. Both subregions may be separate, common or partially common.

BASIC FORMULATIONS

Let us present typical inverse problems formulations pertaining, for simplicity, to a stationary field with the scalar potential in the finite region Ω with boundary Γ.

Searching for Boundary Conditions

Consider the region Ω of $\mu = \mu_0$ plunged into a superpermeable environment presented in Fig. 1. The Laplace's equation for magnetic scalar potential ψ

$$\nabla^2 \psi = 0 \tag{1}$$

is valid in Ω and let Ω_0 be the synthesis checked subregion in which we desire to synthetize the field function $\psi = \psi(x,y,z)$. We suppose Ψ to have constant zero value on Γ and try to do the synthesis by determining a distribution of magnetic dipoles on the boundary. Then the simple synthesis problem is as follows: what should be the distribution on Γ of the potential normal derivative $\partial\psi/\partial n = H_n(x,y,z)$ in order to assure the potential in Ω_0 as close as possible to the desired function $\Psi(x,y,z)$?

The problem is governed by the linear Fredholm's integral equation of the first kind

$$\int_\Gamma K(P,Q)\, H_n(Q)\, dQ = \psi(P) \tag{2}$$

where: $Q(\xi,\eta,\zeta) \in \Gamma$ and $P(x,y,z) \in \Omega_0$ are the source and observation points respectively; $dQ = d\xi\, d\eta\, d\zeta$ - differential volume element; $K(P,Q) = 1/4\pi r(P,Q)$ - the kernel being the Green's function, i.e. the free space fundamental solution, here.

From the synthesis problem that was formulated in such a way we should point out that:

(a) function $\psi(P)$ for $P \in \Omega - \Omega_0$ is of no interest,
(b) solution $H_n(P)$ of (2) is, generally, nonunique,
(c) physical sense of a problem will be maintained for some classes of functions $\Psi(P)$ only, e.g. for harmonic in Ω_0 ones.

Variant 1. In the subregion Ω_0 the x-component of the gradient of ψ will be synthethized. Then integral equation (2) will be transformed into another one by performing the operation of gradient to both sides of it

$$\int_\Gamma K_1 (P,Q)\, H_n(Q)\, dQ = H_x (P) \tag{3}$$

where $K_1 (P,Q) = \partial K (P,Q)/\partial x$ - the new kernel.

Variant 2. Let Ω_0 be the y,z - plane for $x=x_1$ and let some quantity of flux type, namely the integral $\Phi = \mu_0 \int H_x (P) dydz$ over Ω_0, be given there. Integrating both sides of (3) yields the degenerate integral equation

$$\int_\Gamma L(P_1,Q)\, H_n(Q)\, dQ = \Phi \tag{4}$$

where $L(P_1,Q) = \mu_0 \int K(x_1,y,z,\xi,\eta,\zeta)$ may be considered as a degenerated kernel. Equation (4) has infinite many solutions. Only those with some constraints imposed on the searched boundary function $H_n(Q)$ are of practical importance.

Searching for Inner Sources

Let us now consider the electrostatic field in a region Ω, comporting a checked subregion Ω_0 and a source subregion Ω_I, presented in Fig. 2. Let both electric potential ϕ on Γ and its normal derivative E_n be known. We could formulate the following synthesis problem: what should be the distribution of the volume electric charge density $\rho(x,y,z)$ in Ω_I in order to ensure the potential $\phi = \phi(x,y,z)$ in Ω_0 ?

The following Fredholm's equation pertains to the problem under consideration

$$\int_{\Omega_I} G(P,Q)\, \rho(Q)\, dQ = \phi(P) + F(P) \tag{5}$$

where: $Q(\xi,\theta,\zeta) \in \Omega_I$ and $P(x,y,z) \in \Omega_0$ are the source and observation points respectively; $G(P,Q)$ - Green's function of free space and $F(P) = \int[G(P,R)\, E_n(R) + (\partial G/\partial n)\, \phi(R)]\, dR$ over Γ, where $R(\xi,\theta,\zeta)$ are the boundary source points.

We notice that both problems that have been discussed above belong to the group of synthesis problems that we term the group of rigid structure. We are going to qualify two following problems to the group of supple structure.

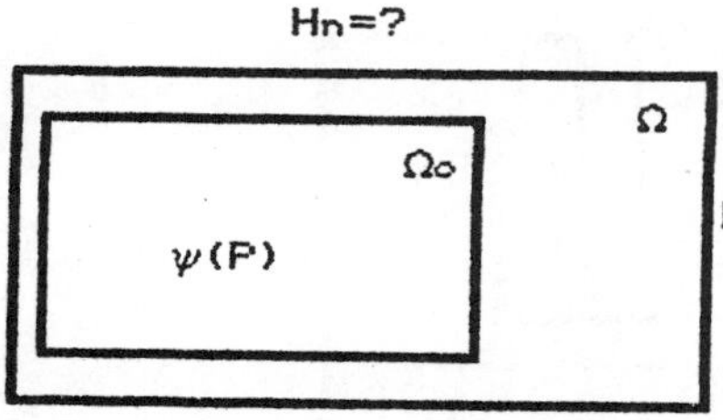

Fig. 1 Magnetic field synthetized in Ω_0 by H_n on Γ

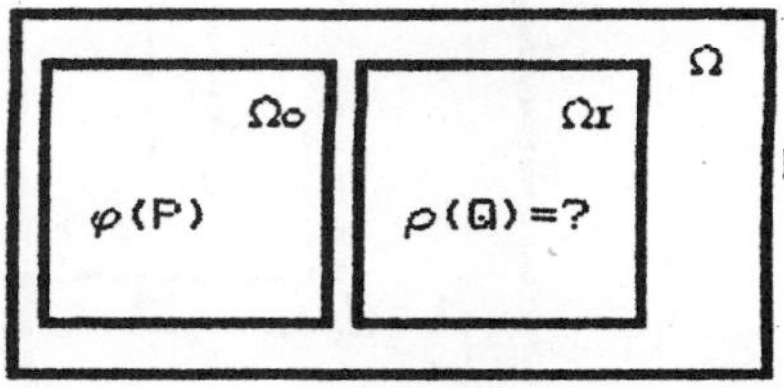

Fig. 2 Electric field synthetized in Ω_0 by ρ in Ω_I

Forming a Boundary

Let us try, to simplify, the x,y-region Ω shown in Fig. 3 in which the Laplace's equation holds. The magnetic field $\psi = \psi(x,y,z)$ in Ω_0 should be synthetized by an appropriate forming of a part of the boundary. Let $\Gamma = \Gamma_1 \cup \Gamma_2$ and $\Gamma_1 \cap \Gamma_2 = 0$; the part Γ_1 spreading from the point $B_1(x_1,y_1)$ to the point $B_2(x_2,y_2)$ is assumed to be the known equation $y=\gamma_1(x)$; Γ_2 from B_2 to B_1 is to be restored. In other words the shape $y=\gamma_2(x)$ placed between Γ_2' and Γ_2'' is searched for. Suppose that $\psi = 0$ on the whole Γ, next $H_{n1}(x,y) = H_{n1}[x,\gamma_1(x)] = H_{n1}(x)$ on Γ_1 and H_{n2} = const on Γ_2. The following nonlinear Fredholm's integral equation of the first kind is valid

$$H_{n2} \int_{x2}^{x1} G[x,y,\xi,\gamma_2(\xi)]\, d\xi = F(x,y) - \psi(x,y) \tag{6}$$

In this equation the unknown function $\gamma_2(\xi)$ is the argument of the kernel, being the Green's function of free space for $\eta = \gamma_2(\xi)$. On the right side

$$F(x,y) = \int_{x1}^{x2} \frac{\partial G[x,y,\xi,\gamma_1(\xi)]}{\partial n} H_{n1}(\xi)\, d\xi \tag{7}$$

In the problem which is put here the whole boundary functions, namely ψ and H_n are assumed. This is the correct form in the synthesis, although it would have excessive boundary conditions in analysis. If the problem is modified in such a way that only ψ on Γ is given, we shall introduce the special Green's function into (6) being equal to zero on the boundary. It is rather an important task to search for such a function and that is why in many engineering synthesis problems the integral equation formulation could be deprived of usefulness for mathematical solutions.

Determination of Enviromental Parameters

We wish to synthetize the electric field $\phi = \phi(x,y,z)$ in a domain $\Omega_0 = \Omega$ restoring an electric permeability function $\varepsilon(x,y,z)$. For simplicity we assume that $\rho = 0$, i.e. no free electric charge is present in Ω. This problem has been investigated by Rawa[4] using the differential equation of the first order with respect to the function $\varepsilon(x,y,z)$

$$\text{grad}\,\phi \cdot \text{grad}(\ln \varepsilon) = -\nabla^2 \phi \tag{8}$$

where both the potential gradient and potential Laplacian are assumed to be known functions.

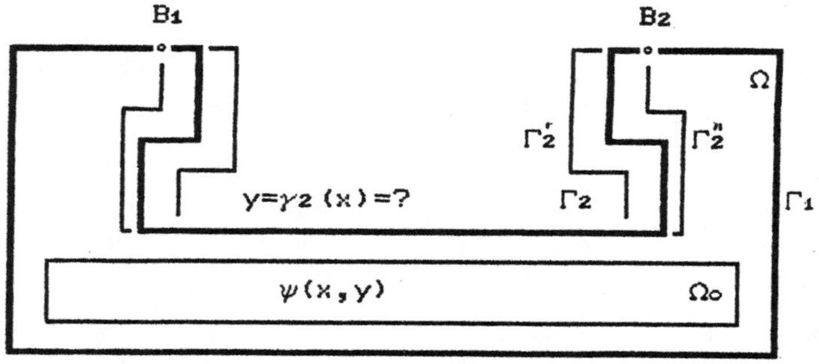

Fig. 3 Field synthesis by forming a pole shape

In general, the physical classification of the synthesis problems coincides with that of analysis problems. Thus, there are also synthesis problems concerning electrostatic, magnetostatic and current flow fields as well as electromagnetic fields (including the monoharmonic case of low frequency and also electromagnetic waves), transient fields and so on.

ENGINEERING PROBLEMS OF SYNTHESIS

From a technical standpoint a synthesis problem may be one of the following:

The design synthesis problems. The design of an element, which is a very important one, of an electric device (machine, apparatus, isolator, antenna) under assumption that its desired features can be formulated in dependence on the field distribution in some part of this device or in dependence on some integral field characteristics; we term such problems the design synthesis problems.

The optimization synthesis problems. When synthetizing the field in the sense of some integral metric (not necessarily desiring the field distribution) one can formulate the optimization synthesis problems.

The identification synthesis problems. Structural or material identification of an existing device accomplished on the ground of field measurements in some part of this device is very important on account of manufactured accuracy verification and also for checking the device after a long exploitation or after failure. Such problems are defined as the identification synthesis problems.

The secondary sources synthesis problems. Aspects of the synthesis may also occur in some analysis problems examined by the method of secondary sources[5]. This method is based on the principle that the continuity of the vector field quantities on the internal and external boundaries is equivalent to the distribution of potentials (of single or double layer) on these boundaries. We suggest to term this kind of analysis the secondary sources synthesis problem.

It is worth while to point out here that a calculation of magnetizing current when designing electrical machines and transformers is, essentially, a synthesis problem: searching for the localized field source for given magnetic flux. It is, however, a very simplified synthesis.

A design problem of the coil which excites the field of the desired distribution along the coil axis may be effectively solved by the use of the Fredholm's equation[1]. It concerns a choice of the coil sections carrying different currents as well as the coil geometric shape.

A good example concerning an application of synthesis methods to an electric machine design is the pole shoe which has to be formed in order to obtain a desired magnetic field distribution on the armature, for instance the sinusoidal one.

We are of the opinion that some problems of magnetic or electromagnetic screening, for instance in the end zone of turbogenerator may be effectively studied like synthesis problems.

We could point out, too, the dipole and quadrupole coils of the iron-bound air-core magnets whose design may be performed using the synthesis technique.

MATHEMATICAL SOLUTION METHODS

Field synthesis problems frequently lead to nonunique solutions. It is due to different reasons. Synthesis problems are, generally, ill-posed in the Hadamard sense. In practical problems of technical importance, we have, by and large, too small input information to obtain the unique solution. One could distinguish three fundamental groups of solution methods for the electromagnetic field synthesis problems.

Firstly, one can single out the methods based on the integral equation approach[1,6,9]. The synthesis problems from this group are formulated in terms of Fredholm or Volterra's integral equations of the first or second kind, linear or not; they are numerically solved in a discretized form.

In the second group the variational approach is employed. Such well-known numerical techniques like FEM (formulated reversely) or BEM may be quoted[2,7,3].

The third group is a special one in the sense that it contains optimization techniques that include some a priori bounds on the solution which has to be optimal in accordance with any chosen criteria[8].

REFERENCES

1. K. Adamiak, Method of the Magnetic Field Synthesis on the Axis of Cylinder Solenoid, Appl.Phys. 16:417 (1978)
2. R. Palka, Synthesis of Magnetic Field due to the Direct Current, Etz Arch. 9:299 (1985)
3. J. Sikora & others, Singular Value Decomposition in Identification of Magnetic Field Modelled with Finite Elements, Sc.Elec. 3:17 (1986)
4. H. Rawa, Direct Synthesis Method of Electric and Magnetic Parameters in Heterogeneous Environment, Pr.Nauk.Pol.War. 83:3 (1986) (in Polish)
5. O. V. Tozoni, "Secondary Sources Method in Electrical Engineering", Energhya, Moscow (1975) (in Russian)
6. W. Groetsch, "The Theory of Tichonov Regularization for Fredholm Integral Equations of the First Kind", Pitman, London (1984)
7. P. Colli-Franzone & others, Finite Element Approximation of Regularized Solutions of the Inverse Potential Problem of Electrocardiography and Applications to Experimental Data, Calcolo. 1:91 (1985)
8. A. Gottvald, Comparative Analysis of Optimization Methods for Magnetostatics, in Proc. of the COMPUMAG'87, Graz
9. M. Rudnicki, The Choice of Regularization Parameter in Reverse Problems of Electromagnetic Field, in Proc. of the ISEF'85, Warsaw

COMPARATIVE ANALYSIS OF NUMERICAL METHODS FOR SHAPE DESIGNING

Jan Sikora, Maciej Stodolski, and Stanisław Wincenciak

Warsaw University of Technology

Electrical Engineering Department, Warsaw, Poland

SUMMARY

Based on the Finite Element Method (FEM) two nonlinear algorithms for the synthesis of the geometry of the investigated region have been presented. One of the algorithms used is the Variable Metric Method (VMM) and the other is based on the Sensitivity Method (SM). The simple numerical example using this methods has been presented.

INTRODUCTION

Until recently, the finite element method (FEM) has been used almost excusively for the analysis of computer generated models[1]. With design optimization, the computer itself can now be used to take an active part in the designing of a model.

This computerized method of integrating design and analysis means that engineers can arrive at better designs and better products in less time.

Design optimization[8] is a computer technique for generating a series of designs which are checked for feasibility and which improve as the series progresses until a "best" design is obtained. This process uses the concepts of:

1. Design Variables (DV). DVs are in our case geometric parameters such as node coordinates. The designer specifies limits as a minimum and maximum value (side constraints) for each DV.
2. State Variables (SV). Electric potential, magnetic vector potential are typical state variables for Electromagnetic Field Problems.
3. Objective Function (OBJ). The OBJ is a value which is to be minimized in the design process. The form of the OBJ depends on the applied method of minimization.

The problems of optimal structural design in Electromagnetic Field Theory belongs to inverse problems and can be divided into two parts.

One part of the problems leads to linear and the other part to nonlinear optimization problems. There is only one of the several different ways of partition. The following belong to linear optimization problems:

1. synthesis or identification of the source function
2. synthesis or identification of the boundary conditions,

and to nonlinear problems belongs:

1. synthesis or identification of physical properties of the investigated region
2. synthesis or identification of the boundary shape.

The first part seems to be easier to solve and has been well presented in literature, although some problems[5,6] exist that till now have not been very well solved.

The second part is much more difficult, but constantly increasing the interest in this topics, can be observed in[2,7,8].

This paper is concerned on the numerical methods to optimal shape design.

STATEMENT OF PROBLEMS

Consider Dirichlet's problem in two dimensional regions. We want to find the boundary shape in order to obtain a certain assumed condition, for example potential or flux density distribution in the region.

Let us assume now that for solving this problem finite element approximation is applied. The main purpose of this paper is to compare two different approaches to the synthesis of the boundary line in the investigated region. The so called classical approach has been discussed first. The nodes coordinates belonging to the boundary lines have been chosen as a design variables. A design optimization program utilizes parametric input and output.

The input parameters are the DVs representing the changing parts of the model. The output parameters are the SVs retrieved from the results of the solution of the model. The optimization routine acts as a controller for the FE analysis program. It determines new values of design variables based upon analysis results (state variables) and the minimization of the OBJ. The user decides which items are design and which state variables, including limits for each, and the objective functions. The objective function in our problem have been constructed as follows

$$f_1(x^k) = \sum_{i=1}^{m} (\varphi_i^k - \varphi_{oi}^k)^2 \qquad (1)$$

or

$$f_1(x^k) = \sum_{i=1}^{p} (\nabla\varphi^k - \vec{E}_i^k)^2 \qquad (2)$$

where m - number of nodes for which the potential (SV) was assumed.

p - number of elements with assumed values of the vector $\vec{E}$.

x^k - vector of moving nodes (DVs) for k-th iteration step.

In order to minimize the objective function (1) or (2) the variable metric method (VMM) was applied. This algorithm is well known so the more detailed description is omitted here. In each step of the VMM (for gradient or even for OBJ) the matrix of the state had to be generated and state equations had to be solved. This method is very simple because without any rearrangements, the FEM existing library could be used. This is a very great advantage. But for industrial problems the state equations dimension may be very large and even for VMM the number of necessary solutions of state equation might exceed computation possibility.

That is, why sensitivity approach introduced by Zienkiewicz and Campell in 1973[1] may seem to be very attractive as an alternate solution. The main advantage of the sensitivity method (SM) is a significant reduction of the dimension of the optimization problem. Thanks to this feature this method was often adopted in practics[7]. The basis of the SM is the generalized Newton algorithm. The corrections added in each step of the algorithm are obtained as the solution of Least Squares Problem (LSP) with the aid of Singular Value Decomposition (SVD)[5,6].

To complete the presentation of this approach, the algorithm of solving the inverse problem is summarized as follows:

step 1. Assume coordinate vector $x^{k=0}$.

step 2. Generate and solve state equation

$$[A^k]\{\varphi^k\} = \{b^k\} \tag{3}$$

step 3. Determine the deviation between state variables and their assumed values

$$\{\Delta\varphi^k\} = \{\varphi^k\} - \{\varphi_0^k\}$$

step 4. Determine sensitivities from Taylor expansion

$$\varphi_i^k - \varphi_{i0}^k = \sum_{j=1}^{n} \frac{\partial\varphi_i^k}{\partial x_j^k}(x_j^k - x_{j0}^k),$$

where n - number of DVs (j=1,...,n)
m - number of nodes with assumed SVs (i=1,...,m).

In matrix form it is given as

$$\begin{bmatrix} \dfrac{\partial\varphi_1^k}{\partial x_1^k} & \cdots & \dfrac{\partial\varphi_1^k}{\partial x_n^k} \\ \vdots & & \vdots \\ \dfrac{\partial\varphi_m^k}{\partial x_1^k} & \cdots & \dfrac{\partial\varphi_m^k}{\partial x_n^k} \end{bmatrix} \{\Delta x^k\} = \begin{Bmatrix} \Delta\varphi_1^k \\ \vdots \\ \Delta\varphi_m^k \end{Bmatrix} \tag{4}$$

where rectangular (m x n) coefficient matrix is the sensitivity matrix.

step 5. Use a quadratic programming (LSP) to find the step size $\{\Delta x^k\}$.

step 6. Update the DVs $\{x^{k+1}\} = \{x^k\} + \gamma\{\Delta x^k\}$

step 7. Check the convergence. If the convergence is not achieved, return to step 2.

The most important problem in this algorithm is the computation of the sensitivity matrix.

SENSITIVITY MATRIX

Finite Element discretization of the Dirichlet problem leads to the following set of algebraic equations

$$[A]\{\varphi\} = \{b\} \tag{5}$$

In this case both matrix A and vector b are dependent on the design variables. Differentiating equation (5) we obtain

$$[A]\frac{\partial\{\varphi\}}{\partial x_j} + \frac{\partial[A]}{\partial x_j}\{\varphi\} = \frac{\partial\{b\}}{\partial x_j} \tag{6}$$

and the sensitivities are described by

$$[A]\frac{\partial\{\varphi\}}{\partial x_j} = \frac{\partial\{b\}}{\partial x_j} - \frac{\partial[A]}{\partial x_j}\{\varphi\} \tag{7}$$

In order to obtain the sensitivity matrix, the equation (7) has to be solved for several right hand sides ($j=1,\dots,n$).

The sensitivity matrix was obtained in a very economical way[6], because the matrix A was formulated and decomposed only once.

The incremental right hand side in eq. (7) was determined by the numerical perturbation method, where perturbation coefficient was usually cut down from 1.E-3 to 1.E-6.

NUMERICAL EXAMPLE

Let us consider a potential distribution in two dimensional region as shown in fig.1.

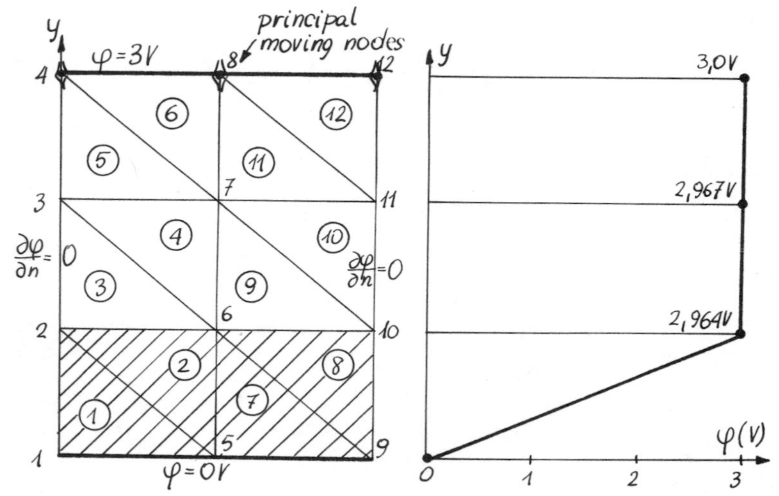

Fig.1. Finite element mesh in the test region.

The inverse test problem is based on searching for "y" coordinates of the upper boundary (on this boundary line the potential is equal to 3V) if in the fixed nodes 3, 7 and 11 potentials are already known. The analysis of the problem shown in fig.1. results in the values of SVs in these nodes, and next we change them substracting 1% of their values.

The test problem apparently is simple (only three design parameters), but material properties were specially selected to obtain a not very well conditioned problem. It means that small changes of assumed DVs cause a great change of the upper boundary. Next difficulties occured in this test are caused by the nonconvexity of the objective function. It is easy to prove that suggested discretization gives us at last two local minima. The probe of different discretization only change the place of those minima as it is shown in fig.2. The main question is, how to avoid local minimum No 2, which, from the physical point of view, leads to the wrong results of the designing process.

The answer is:

1. Very strict side constraints could be imposed on each of the DVs, but even for a simple test it might represent some difficulties.
2. Change the objective function in the manner forcing its convex form. Usually it is the additional expression for energy, length of the design shape and others.

The two presented algorithms were used for solving this test problem and both of them have given positive results for different starting points. Algorithm based on the VMM for the unprofitable starting point achieved convergence after 7 iterations and 40 callings of the objective function.

The second algorithm based on the SM for the same starting point, achieved convergence after 3 iterations. Results are shown in fig.3.

CONCLUSION AND FINAL REMARKS

Two different approaches to optimization of the shape region approximated by FEM or BEM have been presented in this paper.

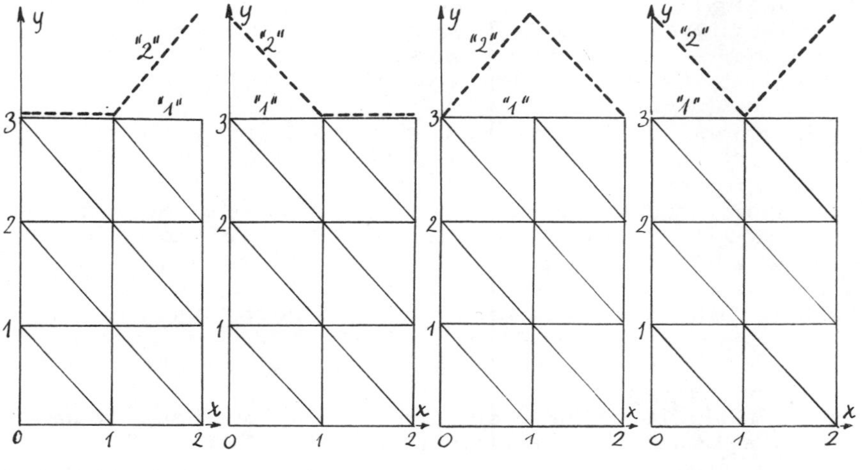

Fig.2. Different local minima caused by different discretization.

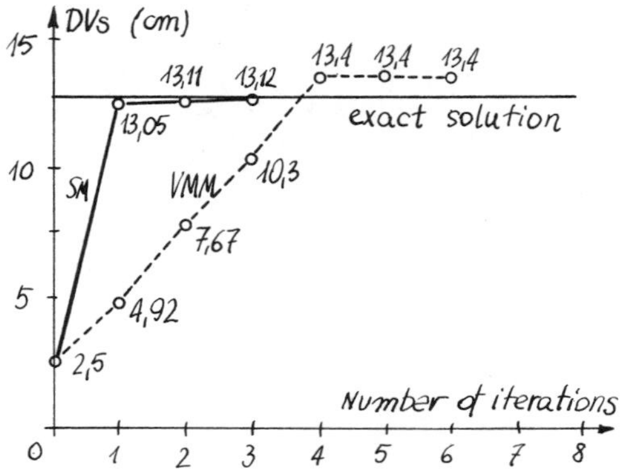

Fig.3. Design variables.

Both algorithms gave numerically stable and very precise results.

The main advantage of SM is linking the vector of assumed values (SVs) with the design variables ("y" coordinates of boundary nodes). Due to this link the problem size is significantly reduced.

To enhance the efficiency of the VM algorithm, the key issue is to reduce the computational effort for reanalysis (40 callings of OBJ). One of the special features of the shape optimization problems is that only a part of the region is changed during the iteration process. Most of the domain occupied by the structure remains the same. Keeping it in mind, we can subdivide the whole structure into some substructures (super-elements) with fixed domain and with varying boundaries. In practice, the varying boundary is usually a small part of the whole boundary, so a sub-structuring technique tremendously reduces the computational effort for assembling and solving the state equations.

Thus both methods become an indispensible tool for the numerical method in shape optimization problems.

REFERENCES

1. R.H.Gallagher, O.C. Zienkiewicz, "Optimum Structural Design, Theory and Applications," John Wiley and Sons, London (1973).
2. A. Gottvald, Comparative analysis of optimization methods for magnetostatics, in: Proceedings of the COMPUMAG ´87 conf. Graz (1987).
3. A. Marrocco, O. Pironneau, Optimum design of a magnet with Lagrangian Finite Elements, in: Proceedings of the COMPUMAG ´78 conf. Grenoble (1978).
4. P. Rabinowitz, Numerical methods for nonlinear algebraic equations, Gordon and Breach Science Publishers, London (1970).
5. J. Sikora, Sensitivity analysis vs. unit distribution method: the application to inverse problems. in: Proceedings of VIII Int. Symp. "COMPUTER at the University", Cavtat/Dubrovnik (1986).
6. J. Sikora, New approach to identification of boundary conditions. in: Proceedings of the Int. Symp. on Electromagnetic Theory-URSI ´86, Budapest (1986).
7. J. Win Hoú, Techniques and applications of shape optimum design, in: Computers and Structures, 1-3:467 (1985).
8. S. Wincenciak, J. Sikora, A. Michalski, Optimal shape design of an electr. coil, in: IX Int.Symp."COMPUTER at the Univ."Cavtat (1987)

SYNTHESIS OF A TURBO-GENERATOR NONLINEAR PARAMETRIC MODEL FOR THE ANALYSIS PROBLEMS OF POWER INDUSTRY SYSTEM REGIMES

V.E.Tonkal, Yu.G.Blavdzevitch, and N.V.Raptsun

Ukrainian Academy of Sciences, Kiev, USSR

INTRODUCTION

At the level of the national economy development achieved, the problem of improving the methods and means to plan and to control the electric power systems (EPS) is of particular concern. The solution of this problem involves the intensive use of mathematical modelling methods and modern computer facilities. In this case, to obtain the results of high quality and reliability it is necessary to have the authentic mathematical models of the EPS elements.

The turbo-generator (TG) is one of the basic elements in the EPS. The electromagnetic and electromechanical processes running within the turbo-generators govern greatly the power system operational regime nature. Because of the complexity of these processes, when modelling the EPS regimes one has to resort to the simplified idealized mathematical description of TG. In particular, at present in calculation and analysis of both steady states and transient conditions of the EPS, the relation between the turbo-generator model parameters and its saturation is not taken into account. At the same time, in his well known work [1], A.A.Gorev by a specific example has shown that under static stability condition, at the cost of saturation of a nonsalient pole machine it is possible to transmit power 30% higher than in the case of a nonsaturated machine.

The works [2,3] have described the TG electromagnetic field numerical models taking into account the core geometry, inhomogenety and saturation. However, these model complexity and computation awkwardness make it difficult to implement them as an element of the EPS regime analysis software. It is possible to improve the efficiency of computer-aided implementation for such problems by the use of another kind of model obtained on the base of numerical models.

The procedure of the mathematical model form transformation may be attributed to the identification problem class solution which is provided by the special mathematical methods and software tools. The organization and execution of the experiments with a real object model is the basic principle of one of the modelling kinds, i.e. simulation. In case when the information necessary for identifying the model of a desired (convenient, simplified) form is obtained as a result of a specially organized computer-aided calculation experiment with the mathematical model describing the object properties with sufficient accuracy, the simulation may be mentioned. This work deals with the problems of applying the simulation identification as a method of the TG parametric model synthesis taking into account the saturation and oriented for the use in the desing problems and the EPS regime analysis.

METHOD AND ALGORITHM

The main variety of the EPS models used in the EPS regime analysis are the equivalent circuits which consist of the equivalent circuits of its individual elements, i.e. generators, transmission lines, transformers, etc.

The TG equialent circuits used for both steady and transient regimes follow from the classical linear theory of synchronous electrical machines which is based on a number of the serious simplifying assumptions. Application of this theory to the high utilization synchronous machines, first of all, to the TG's results in considerable errors. The possibilities of improving the precision of the calculation procedures is related first of all to the more profound analysis of the magnetic field distribution in the ferromagnetic medium of electric machines.

The development of computer facilities allows us to use the effective numerical methods to study the magnetic fields which allow us to avoid introducing most of the assumptions specific to the classical theory of electrical machines. In connection with this, in a number of works, a question is raised to create the non-linear theory of electric machines on the base of these methods. In particular, the authors of the above works suggest dropping one of the basic concepts in the linear theory of synchronous machines about the inductive impedances because of the fact that these impedances are not constants but dependent on the machine's magnetic circuit saturation.

However, it must be noted that these parameters have been in practice for a long time and are widely used in the techniques by the organizations engaged in design work, manufacture and operation of the TG's. For a large-scale software support developed and mastered at present in the electric machine engineering and electric power industry, the inductive parameters are the necessary input or output data.

Therefore, it seems to be reasonable to combine the main merits of the linear theory of synchronous machines namely simplicity and clearness of the equations with the advantages in numerical analysis of a magnetic field, i.e. the possibility of detailed allowance for geometry and structure of an object under study, non-linearity and anisotropy of structural materials, actual distribution of the field source in space and time.

When implementing such an approach, naturally there may be some doubts about the ligitimacy of applying the set of linear algebraic equations to a mathematical model of a device with non-linear electromagnetic couplings. Actually, in this case, some difficulties arise in unequivocal determination of model parameters (coefficient). In particular, it is impossible to determine them unequivocally by physical experiments [4]. This can be demonstrated by the elementary example, i.e. by the double-wound transformer with the core subjected to saturation. The transformer winding magnetic-flux linkage may be defined as

$$\psi_1(i_1, i_2) = L_{11}(i_1,i_2)\cdot i_1 + L_{12}(i_1,i_2)\cdot i_2 ;$$
$$\psi_2(i_1, i_2) = L_{21}(i_1,i_2)\cdot i_1 + L_{22}(i_1,i_2)\cdot i_2 , \qquad (1)$$

where ψ_1, ψ_2 are the winding magnetic-flux linkages; i_1, i_2 are the winding currents; L_{11}, L_{12}, L_{21}, L_{22} are the winding self and mutual static inductances.

If one measures the magnetic-flux linkages ψ_1' and ψ_2' induced by the given currents i_1' and i_2' and then substitutes the values of the currents and magnetic-flux linkages into the expression (1), one will obtain the set of linear equations of the second order with the four unknowns (in the general case for n windings of n equations with n^2 unknowns), which has no unique solution.

At the same time, the problem of unequivocal determination of the static parameters can be solved on the basis of numerical simulation of an electromagnetic field. In the above example of double-wound transformer, the algorithm of determining the static inductances will include the following procedures.

1. The numerical calculation of the TG magnetic field distribution with allowance for the steel saturation when the currents i_1' and i_2' are given.
2. The numerical calculation of the magnetic fields induced separately by the currents i_1' and i_2' at the fixed magnetic state of the steel core defined at the previous stage. The magnetic-flux linkage calculation $\psi_1(i_1',0)$, $\psi_2(i_1',0)$, $\psi_1(0,i_2')$, $\psi_2(0,i_2')$.
3. The determination of static inductances from the following expressions.

$$L_{11}(i_1', i_2') = \frac{\psi_1(i_1',0)}{i_1'}; \quad L_{12}(i_1', i_2') = \frac{\psi_1(0,i_2')}{i_2'};$$
$$L_{21}(i_1', i_2') = \frac{\psi_2(i_1',0)}{i_1'}; \quad L_{22}(i_1', i_2') = \frac{\psi_2(0,i_2')}{i_2'}. \qquad (2)$$

The special feature and merit of the algorithm consist in the possibility of fixing the core magnetic state followed by separating the individual current contribution to establish the resultant field, this cannot be achieved by physical experiments.

The above algorithm of determining the inductance parameters is also applicable to the electric machines including the turbo-generators.

When carring out the practical calculations, the resistance quantities of the armature winding end leakage inductive impedance are usually ignored because of their negligible magnitude. In this case, the armature winding phase voltage is expressed only through the inductive impedances corresponding to the magnetic fluxes which are closed within the TG active zone.

$$\dot{U} = -jx_{af}(\dot{I}_f, \dot{I})\cdot\dot{I}_f - jx_s(\dot{I}_f, \dot{I})\cdot\dot{I} , \qquad (3)$$

where U is the phase voltage across the armature winding terminals; I_f is the field current reduced to the armature winding; I is the armature current; x_{af} is the armature and field winding mutual inductive impedance; x_s is the armature winding synchronous inductive impedance.

The fig.1 shows the enlarged flowchart of the TG nonlinear parametric model simulation identification algorithm for the steady operational regimes.

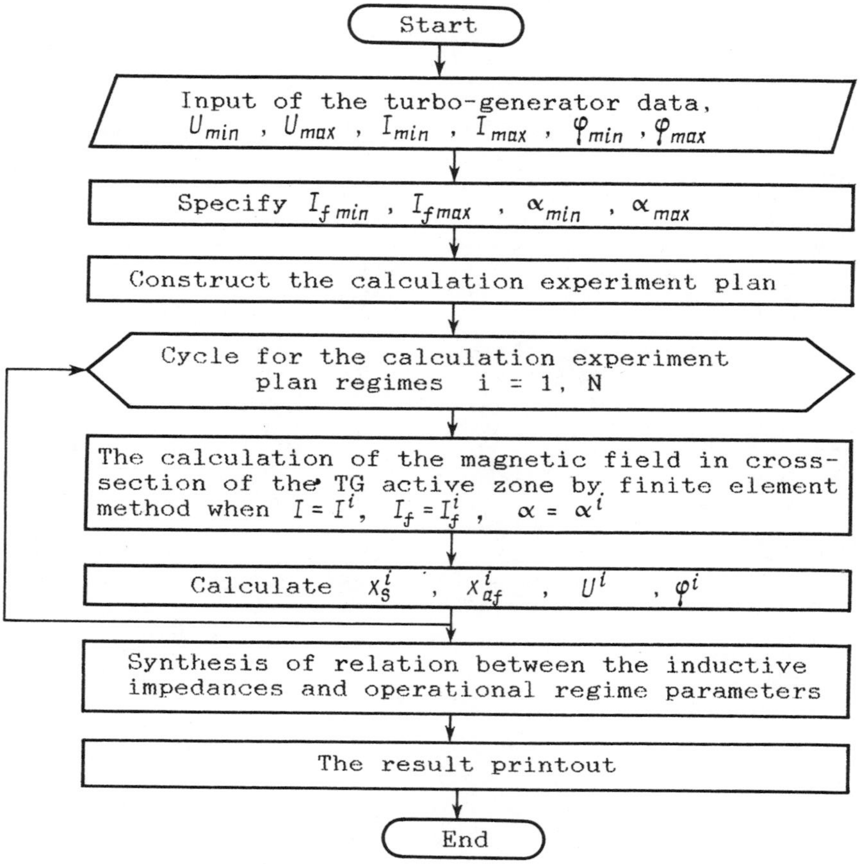

Fig.1 Algorithm of simulation identification

The right part of the complex equation (3) contains three scalar independent variables, i.e. the current moduluses $I_f = |\dot{I}_f|$, $I = |\dot{I}|$ and angle α contained by them.

It is evident that in practice, the attention has been directed to the inductive impedance values corresponding only to the combinations of I_f, I and α which agree with the actual permissable regimes of the TG operation rather than the arbitrary combinations I_f, I and α. As a rule, the operational regimes are given by so called external regime parameters (U, I, and φ), i.e. by the quantities related to the armature windings. Let's mark the boundaries of the permissable regime range through U_{min} and U_{max}, I_{min} and I_{max}, φ_{min} and φ_{max}. The boundaries of the field current (I_{fmin} and I_{fmax}) and the angle α (α_{min} and α_{max}) corresponding to permissable operational regime boundaries may be approximately defined by the vector diagram of the nonsalient pole synchronous electric machine emf.

To carry out the calculation experiment within the variation range for each of the numerical model input variables I_f, I, α some number of fixed values or levels of a variables (n_f, n_a, n_α respectively) are chosen. The application of the experemental design theory methods makes it possible to reasonably organize the calculation experiment and thus to decrease the amount of required calculation considerably. In this case, the number of regimes for which the experiments are carried out is $N \ll \ll n_f \cdot n_a \cdot n_\alpha$.

The inductive impedances X_{af} and X_S are defined by the numerical calculation results of a magnetic field distribution in the cross-section of the TG active zone. In addition to the inductive impedances, the TG numerical model output is the voltage U and power angle φ. This makes it possible to determine the relation between the inductive impedances and external regime parameters U, I, φ.

The synthesis of the relations between the inductive impedances and the regime parameters is carried out by the regression analysis methods.

DIGITAL SIMULATION RESULTS

The algorithm shown in the fig.1 has been implemented in the form of a program set with the algorithmic language FORTRAN.

The table 1 contains the parameters of some typical regimes from the plan of a computation experiment performed on the 300 MW TG. It should be noted that an essential dependence of synchronous inductive impedance X_S on the regime parameters exists.

Table 1. Parameters of some regimes for the 300 MW turbogenerator

N	U/U_N, p.u	I/I_N, p.u	φ, rad	I_f/I_{fN}, p.u	α, rad	X_S, p.u
1	1.001	0.050	0.031	0.350	1.700	1.984
2	0.901	0.050	0.015	0.300	1.700	2.075
3	1.100	0.050	0.564	0.450	2.200	1.777
4	1.009	0.500	-0.505	0.350	2.100	2.009
5	1.098	0.500	-0.459	0.400	2.000	1.874
6	0.928	0.500	-0.251	0.400	2.300	2.066
7	1.040	0.500	0.560	0.650	2.600	1.825
8	1.099	0.500	0.983	0.800	2.800	1.602
9	1.066	0.500	1.497	0.800	3.100	1.660
10	0.902	0.800	-0.263	0.550	2.600	2.103
11	1.076	0.800	0.580	0.900	2.700	1.643
12	1.093	0.800	0.843	1.000	2.800	1.536
13	1.084	0.900	0.787	1.050	2.800	1.539
14	0.988	1.000	0.564	0.950	2.800	1.822
15	0.927	1.050	0.796	1.000	2.900	1.876

As the calculation results have shown, within the range of the permissable regimes, because of saturation, X_S varies from 1.440 to 2.135. For nominal regime X_S = 1.787. The nonsaturated value X_S (the steel permeability is assumed to be an infinitely large quantity) is 2.210.

According to "Handbook of the electric power system design", the 300 MW synchronous inductive impedance magnitude is 2.195 which corresponds to the unsaturated value of the inductive impedance. Therefore, failure to take into account the TG magnetic system saturation in a solution of the EPS regime design and analysis problems results in an inaccurate predetermination of the equivalent circuit parameters. In particular, X_S for the 300 MW TG excesses by 3 - 52 %.

The polinomial relation between the synchronous inductive impedance and armature winding voltage, current and power angle obtained by processing the calculation experiment results is as follows.

$$x_s = 7,767\,U - 4,596\,U^2 - 0,587\,U^2 I\varphi + 0,445\,I\varphi - \\ - 0,223\,UI^2\varphi + 0,077\,U^2 I^2 \varphi^2 - 1,203 . \qquad (4)$$

The relative error in approximation of the calculation experiment results by expression (4) lies within the limits of 2.5 %.

The fig.2 shows some plot relationship constucted on the base of the expression (4).

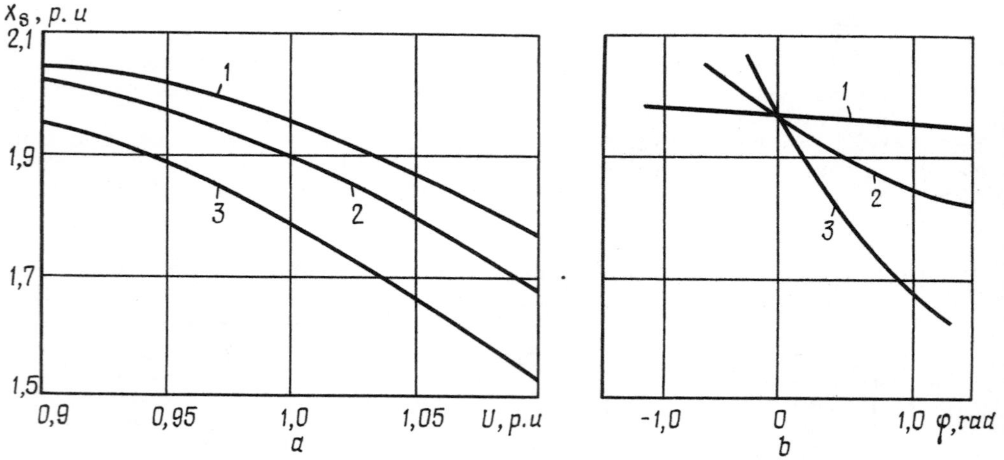

Fig.2. Relations between x_s and the TG regime parameters:
a) $x_s = f(U)$, φ = 0.555; 1. I = 0.1; 2. I = 0.5; 3. I = 1.0;
b) $x_s = f(\varphi)$, U = 1.0; 1. I = 0.1; 2. I = 0.5, 3. I = 1.0

CONCLUSION

The above algorithm of the TG simulation identification and program set developed on its base make it possible to synthesize the TG parametric models with high accuracy taking into account the non-linearity of its regime characteristics (magnetic system saturation). The application of such models in the solution of the EPS regime analysis problems will contribute to improving the calculation accuracy and thus to improving the system operation reliability and efficiency.

REFERENCES

1. Горев А.А. Переходные процессы синхронной машины.-Л.: Наука,1985.
2. Fuchs E., Erdelyi E. Non-linear theory of turboalternators. -IEEE Trans. V.PAS-92, 1973.
3. Chary M., Silvester P. Analysis of turboalternator magnetic field by finite elements.-IEEE Trans., V.PAS-90, 1971, N 2.
4. Фильц Р.В. Математические основы теории электромеханических преобразователей.-К.:Наук.думка, 1979.

INDEX